AF880500

Additive Manufacturing Technology

This series systematically summarizes the technology developments in the additive manufacturing field in China in recent years, introducing the technical development status in terms of additive manufacturing processes, materials, technologies, and applications.

This series is one of the national key publishing projects in China, and has been listed in the national key book projects of China's "13th Five-Year Plan", supported by the National Publishing Fund.

Di Wang · Yongqiang Yang · Yang Liu ·
Yuchao Bai · Chaolin Tan

Laser Powder Bed Fusion of Additive Manufacturing Technology

国防工業出版社
National Defense Industry Press

Springer

Di Wang
School of Mechanical and Automotive Engineering
South China University of Technology
Guangzhou, Guangdong, China

Yongqiang Yang
School of Mechanical and Automotive Engineering
South China University of Technology
Guangzhou, Guangdong, China

Yang Liu
Faculty of Mechanical Engineering and Mechanics
Ningbo University
Ningbo City, Zhejiang, China

Yuchao Bai
School of Mechanical Engineering and Automation
Harbin Institute of Technology
Shenzhen, Guangdong, China

Chaolin Tan
Singapore Institute of Manufacturing Technology (SIMTech)
Agency for Science, Technology and Research (A*STAR)
Singapore, Singapore

ISSN 2731-6114 ISSN 2731-6122 (electronic)
Additive Manufacturing Technology
ISBN 978-981-99-5512-1 ISBN 978-981-99-5513-8 (eBook)
https://doi.org/10.1007/978-981-99-5513-8

Jointly published with National Defense Industry Press
The print edition is not for sale in China (Mainland). Customers from China (Mainland) please order the print book from: National Defense Industry Press.

© National Defense Industry Press 2024

This work is subject to copyright. All rights are solely and exclusively licensed by the Publisher, whether the whole or part of the material is concerned, specifically the rights of reprinting, reuse of illustrations, recitation, broadcasting, reproduction on microfilms or in any other physical way, and transmission or information storage and retrieval, electronic adaptation, computer software, or by similar or dissimilar methodology now known or hereafter developed.
The use of general descriptive names, registered names, trademarks, service marks, etc. in this publication does not imply, even in the absence of a specific statement, that such names are exempt from the relevant protective laws and regulations and therefore free for general use.
The publishers, the authors, and the editors are safe to assume that the advice and information in this book are believed to be true and accurate at the date of publication. Neither the publishers nor the authors or the editors give a warranty, expressed or implied, with respect to the material contained herein or for any errors or omissions that may have been made. The publishers remain neutral with regard to jurisdictional claims in published maps and institutional affiliations.

This Springer imprint is published by the registered company Springer Nature Singapore Pte Ltd.
The registered company address is: 152 Beach Road, #21-01/04 Gateway East, Singapore 189721, Singapore

Contents

Chapter 1
Introductions

1.1 Historical Development Process of Laser Powder Bed Fusion Technology

Powder bed fusion (PBF) is an additive manufacturing (AM) technology in which a heat source is used to selectively melt/sinter powder bed. Typical PBF technology [1] includes selective laser sintering (SLS), selective laser melting (SLM), electron beam melting (EBM), etc. Among them, SLM uses a laser as the heat source to selectively melt the powder bed to fabricate metal parts, which is also known as laser powder bed fusion (LPBF) technology.

LPBF technology is developed based on the SLS technology. Their difference is that the powder material in the former is directly heated by the laser to above its melting point. Therefore, low melting point alloy or non-metallic powders are not required as a bonding material for sintering, and one kind of metal powder is needed on the powder bed. The SLS technology based on plastic powders was invented by C. R. Dechard from the University of Texas at Austin in 1986, and he first proposed a utility patent (No. WO1992010343 A1) in 1989 [2]. The LPBF technology then was developed at the Fraunhofer Institute for Laser Technology (ILT) in Germany in 1995. The corresponding patent (No. DE19649865) was applied in Germany in 1997 and was granted in the following year [3]. The first LPBF system was developed in 1999 by Fockele&Schwarze (F&S) and Fraunhofer Institute, Germany, based on stainless steel powder [4].

The main countries that study LPBF technology are Germany, the United States, Britain, Japan, France, etc. [5]. Following the launch of the first LPBF machine, several companies introduced related equipment, such as the first commercial SLM machine MCP Realizer 250 was released in 2004 by F&S together with the MCP company (now MTT), which was later upgraded to SLM Realizer 250 [6]. In addition, Germany's EOS company, SLM Solutions company, and Laser Concept company, British RENISHAW company also released similar products [7]. In terms of scientific research institutions, many universities and research institutes, such as

© National Defense Industry Press 2024

D. Wang et al., *Laser Powder Bed Fusion of Additive Manufacturing Technology*, Additive Manufacturing Technology, https://doi.org/10.1007/978-981-99-5513-8_1

Langhoff Institute, RWTH Aachen University in Germany, the British Welding Institute, the University of Liverpool, the University of Leeds, the University of Leuven in Belgium, Osaka University in Japan and other famous organizations have carried out related research works on the materials, processes and properties [8, 9].

In China, there are mainly Huazhong University of Science and Technology, South China University of Technology, Nanjing University of Aeronautics and Astronautics, Northwest University of Technology and other universities and research institutes that researched the LPBF technology. Each has its own advantages and features. South China University of Technology developed its SLM equipment DiMetal-240 in 2003, commercialized equipment DiMetal-280 in 2007, and precision equipment DiMetal-100 in 2012 [10]. Its research focuses on the application in medical field.

The optical system is the core part of the LPBF equipment, and the laser is the key component of the optical system. The technical progress of laser has promoted the technical level of LPBF equipment. Fiber laser has excellent beam quality, high photoelectric conversion efficiency, almost maintenance-free, and other significant advantages. After 2005, the application of fiber lasers in LPBF equipment gradually was improved and has become the mainstream choice. With the development of fiber lasers, the power of fiber lasers tends to increase gradually. From the initial 50 W to the current mainstream 200–500 W, fiber lasers equipped in some large equipment even exceed 1000 W. With the increase of power, faster scanning speed can be obtained to improve manufacturing efficiency.

With the technological progress of each subsystem of the LPBF equipment, metal powder materials available for the LPBF process are also being developed simultaneously, which changes from the earliest water atomization and aerosolization of stainless steel powder and titanium alloy powder to dozens of commercial metal powders, greatly broadening the applications of LPBF technology [11]. At present, the LPBF technology has broken through the processing of copper alloy with extremely high laser reflectivity, tungsten alloy with high melting point metal, brittle intermetallic compounds, and other materials [12, 13].

With the progress of materials and technology, the commercial application of LPBF technology has become possible. In the past ten years, this technology has been successfully applied in aerospace, biomedicine (especially dentistry), mold, and other fields.

1.2 Principle of Laser Powder Bed Fusion Technology

The schematic diagram of the common LPBF equipment is shown in Fig. 1.1, which mainly includes a sealed manufacturing cavity (powder supply cylinder, part forming cylinder and powder recovery tank), laser and optical systems, a powder laying system and a protective gas circulation system, etc. The main process flow of this technology is as follows:

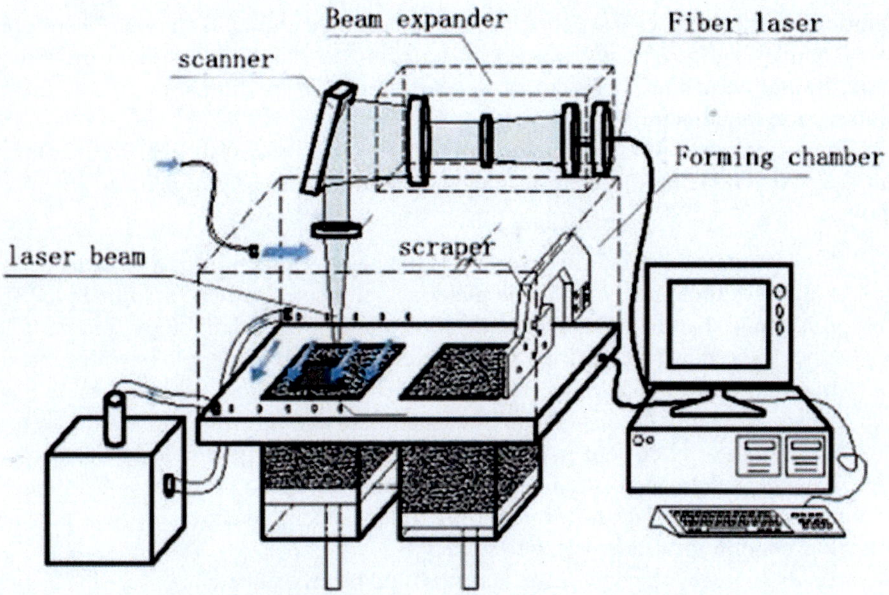

Fig. 1.1 Principle of LPBF equipment

(1) Use computer modeling software or computed tomography (CT), 3D scanner, and other scanning equipment to obtain digital 3D models.
(2) Use common AM software to analyze the 3D model, including repairing, adding support, and other processing.
(3) The computer stratification software is used to slice the designed digital 3D model along the Z axis to layers, and then a large number of single-layer data is obtained.
(4) After conducting the laser scanning path planning of the two-dimensional layer data, the computer-controlled laser manufacturing system will scan the metal powder material on the powder bed according to the set scanning path.
(5) A high-temperature metal molten pool will form in a very short time under laser irradiation, and metallurgical bonding with the next layer will also form. Then the molten pool condenses quickly to form a layer of the metal entity.
(6) After completing the melting and forming of the layer, the forming cylinder drops a layer thickness, and the powder laying system lays a new layer of powder on the layer of metal entity. The same operation will be performed on the next layer of data, and the metal entity will be manufactured finally.

Usually, to avoid failure caused by metal reaction with gases such as oxygen at high temperature melting state, the manufacturing process will be conducted in a closed manufacturing cavity, where an inert protective gas is used to prevent oxidation. Because the hierarchical scanning data is directly obtained from the digital 3D model, as long as the hierarchical accuracy is high enough, the dimension of the fabricated solid part will be very close to the designed digital 3D model. Therefore, a 3D model can be directly formed into a complex shape part without the structural limitation of the 3D model as well as the tool shape limitation of traditional subtraction manufacturing.

There are dozens of factors influencing the forming process of LPBF technology. In the past ten years, the main technical difficulties can be summarized into five points:

(1) Powder material: Material selection is the key to the application of LPBF technology. Although any moldable material can theoretically be applied in LPBF, it is found that the requirements on the composition, morphology, and particle size of the powder are strict. It was found that alloy materials (stainless steel, titanium alloy, nickel alloy, etc.) are easier to form than pure metal materials, which is mainly because the alloy elements in the material increase the wettability of the molten pool and the oxidation resistance. It is known that the oxygen content in the composition has a great impact on the LPBF process. The spherical powder is easier to form than irregular powder because spherical powder has good fluidity and is easy to spread.
(2) Laser source with good beam quality: Good beam quality can ensure the acquisition of finely focused spots, which is of great significance to improve manufacturing accuracy. Due to the finely focused spots, the manufacturing process can adopt a laser with 50–250 W to realize the fabrication manufacturing of the mainstream metal powder materials, while limiting the thermally affected area of the scanning process, and inhibiting the thermal deformation during the manufacturing.
(3) Precision powder laying device: During LPBF manufacturing, it is necessary to ensure the complete metallurgical bonding between the current layer and the previous layer, as well as between the adjacent melting tracks in the same layer. However, some defects such as spattering and spheroidization will occur in the laser-melting process, and some spattering particles will be mixed in the melt pool, which will deteriorate the surface quality of the parts. During the powder-laying process, the diameter of the spattering particles is generally greater than the thickness of the powder-laying layer, resulting in a collision between the powder-laying device and the formed surface. Therefore, unlike the SLS technology, a specially designed powder-laying device is needed for the LPBF technology, such as a flexible powder-laying system, special structure scraper, etc. And, the requirements of the LPBF process for the quality of the powder laying are flat, compact, consistent layer thickness, and as thin as possible after the powder paving.
(4) Gas protection and circulation system: Metal materials are easy to react with the oxygen in the air at high temperatures. Oxide has a very negative impact on manufacturing quality, resulting in a reduction of the wettability of materials and hindering the metallurgical bonding quality between layers and melting tracks. Therefore, the LPBF manufacturing process must be performed in a protective gas atmosphere, which can be argon or less-cost nitrogen, depending on the

manufacturing used material. The generation of spatters and metal plume dust during the manufacturing process will pollute the optical lens and powder bed in the cavity. So, it is necessary to use the circulation system to purge the protective gas above the powder bed and filter the gas in the cavity to keep the gas clean.

(5) Appropriate manufacturing process: It is generally considered that the main process parameters include laser power (P) and laser scan speed (v_s), laser hatch space (h), layer thickness (t) and laser scanning strategy, etc. The following formula (Eq. 1.1) is often used to calculate the volume energy density (E_V, J/mm^3):

$$E_V = \frac{P}{v_s t h} \tag{1.1}$$

where, E_V is the volumetric energy density (J/mm^3); P is laser power (W); t is the layer thickness (mm); v_s is laser scan speed (mm/s); h is laser hatch space (mm).

LPBF technology has the advantages of high precision, high manufacturing density, excellent mechanical properties and saving material, which has been applied in personalized medicine, accompanying shape cooling mold, complex geometry gradient structure and functional structural parts. Through optimized LPBF process parameters and appropriate subsequent heat treatment process, the mechanical properties of metal parts such as titanium alloy, aluminum alloy, nickel alloy and iron-based alloy can reach or even better than the traditional forging counterparts.

1.3 Composition of Laser Powder Bed Fusion Equipment

Common LPBF equipment is mainly composed of an optical system, sealing and forming room (including powder laying device), control system, process software, gas circulation and purification device, and other parts.

1.3.1 *Laser and Optical System*

As the energy source of the LPBF process, the optical system is an important part of the LPBF equipment. Its stability directly determines the quality of manufacturing manufactured parts. The optical system of the common LPBF equipment consists of a laser device, beam amplifier lens or collimator lens, vibration lens and focus lens (f-θ lens), as shown in Fig. 1.2. The working principle is that the laser beam is emitted through a fiber laser, which is amplified 2–8 times by the beam expansion lens, and then projected onto the substrate at a specific position through the vibration lens and the focus lens to melt powders. At the same time, the device is also equipped

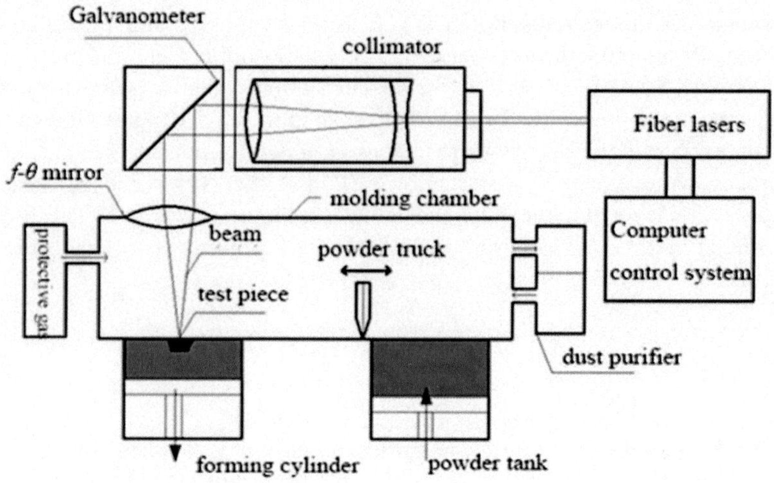

Fig. 1.2 Device of the optical path system

with a fine-tuning platform, which can be fine-tuned in the XYZ axis. Therefore, the focal length and position can be adjusted accurately to effectively avoid processing errors.

1.3.1.1 Laser

SLS technology often uses a CO_2 laser with low power, where the laser power density cannot melt the metal powders with a high melting point. In addition, the wavelength of the CO_2 laser is 10.6 μm and the laser emitted cannot be well absorbed by the metal powder. As a result, the metal parts must be sintered by mixing one metal material with another one with a low melting point.

Fiber laser is the most suitable energy source for LPBF manufacturing to melt metals due to more advantages compared with the traditional YAG, CO_2. In terms of the lasers, unique advantages of fiber laser are as follows:

(1) Higher laser focusing accuracy: the fiber laser is easy to focus to have a high-quality beam spot with a diameter of 30–100 μm. Therefore, it can obtain higher processing accuracy and higher input energy, so that almost all metal materials can be instantly melted.
(2) Good beam quality: the BPP value of the fiber laser can reach 1 mm × mard.
(3) High photoelectric conversion efficiency: the fiber laser photoelectric conversion efficiency reaches 33%. However, the traditional Nd: YAG laser only has 3%. Therefore, the fiber laser can greatly reduce power loss and operating costs.
(4) Better power stability: the traditional Nd: YAG laser has a typical power stability value of 5% compared to the fiber laser with a value of 1%. Therefore, the process can obtain a more stable power output.

(5) High reliability: the fiber laser is an all-fiber mechanism with no maintenance in the use process, thereby having a longer service life, lower maintenance cost and other advantages, which make the fiber laser very suitable for the LPBF process.

Due to the above excellent performance parameters, the fiber laser is widely used in LPBF process. At present, the main laser brands used by LPBF equipment manufacturers are IPG, SPI, Trunpf, etc. And, the Chinese laser brands mainly include Wuhan Raycus, MAX, Zhongke Meiman, Beijing Guoke and so on. At present, in the field of LPBF, the laser power used by various manufacturers is 100, 200, 400, and 500 W. Moreover, some manufacturers use lasers with 700 and 1000 W for manufacturing large-size parts.

1.3.1.2 Galvanometer Scanning System

1. Galvanometer

In the early stage of rapid manufacturing, the developed equipment adopts mechanical X/Y axis mobile scanning. Its response speed is very slow with a large error, which is gradually unable to meet the needs of rapid manufacturing technology. With the development of the technology, the scanning mirror system driven by the high-speed servo motor has been successfully applied in rapid manufacturing and has rapidly become the standard configuration of the rapid manufacturing system. This is because the galvanometer scanning system has several advantages compared with mechanical scanning.

(1) The lens deflection with a small angle can achieve the effect of mechanical scanning and large movement, using the space combination of the two lenses to achieve a large-range scanning with a more compact structure.
(2) The rotational inertia of the lens deflection is very low. The laser scanning delay can be reduced significantly by combining with computer control and a high-speed servo motor, improving the dynamic response speed of the system, and having higher efficiency.
(3) At present, the principle error of the vibration scope system can be made up for by the computer-controlled programming and adjustment mode with higher accuracy.

The galvanometer of the system consists of two galvanometers (mirror, scanning motor) and a servo circuit. The mirror is mounted on the main shaft of the scanning motor and deflects to rotate the mirror: the scanning motor deflects at a limited angle, integrated with a sensor to determine the real-time rotation angle; the servo circuit accepts the drive voltage signal to control the deflection of the scanning motor.

The working principle of the galvanometer is shown in Fig. 1.3. After the laser beam enters the vibration lens, it is first projected onto the mirror deflected along the X axis, then reflected onto the mirror rotating along the Y axis, and finally projected into the XOY working plane. Using the combination of two mirror deflection angles

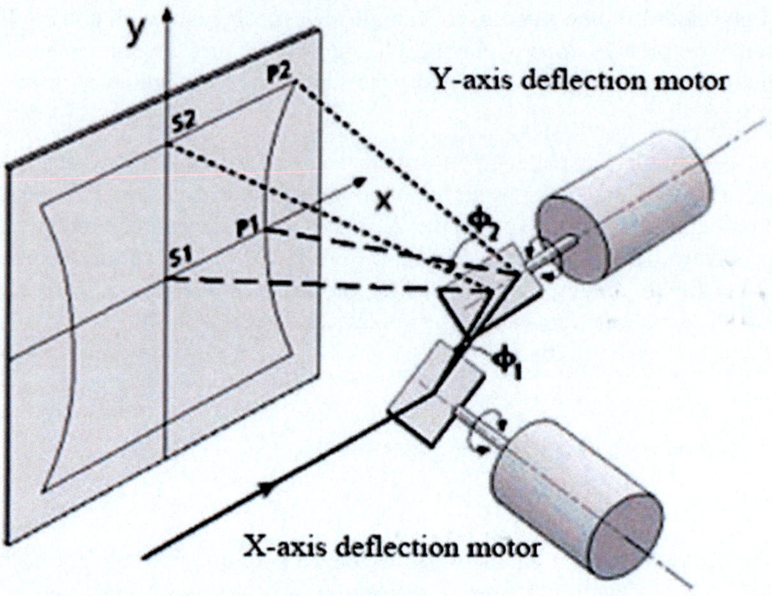

Fig. 1.3 Working principle of the galvanometer scanning system

enables the scanning of any position in the entire field of view. The scanning motor driving the mirror deflection is a special swing motor, which cannot be rotated like the ordinary motor. With mechanical torsion spring or reset torque applied on the rotor, the electromagnetic torque and the reset torque are the same, and it is also called ammeter scanning.

Based on the working principle of the galvanometer scanning system, the relationship between the coordinates (x, y) of the laser in the working plane and the angle of the two galvanometer reflectors Φ_1 and Φ_2 can be expressed as:

$$\dot{y} = d \times \tan\Phi_2 \tag{1.2}$$

$$x = \left(\sqrt{d^2 + y^2} + e\right) \times \tan\Phi_1 \tag{1.3}$$

where, d is the distance of the optical path scanned to the center of the working area by galvanometer (m); Φ_1 is galvanometer reflector angle (rad); Φ_2 is galvanometer reflector angle (rad); e is the distance between the rotating axes between the two galvanometer mirrors (m); x is coordinates; y is coordinates.

Equations (1.2) and (1.3) are transformed to obtain:

$$\left(\frac{x}{\tan\Phi_1} - 2\right)^2 - y^2 = d^2 \tag{1.4}$$

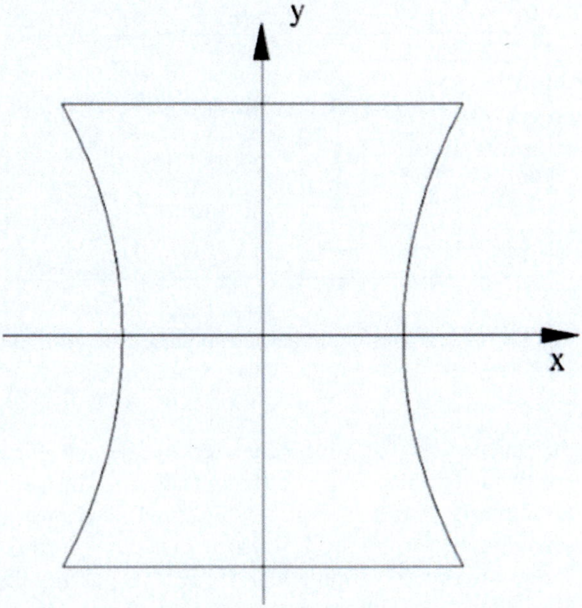

Fig. 1.4 Pincushion distortion

When Φ_1 is constant, the above equation describes a non-circumferentially symmetric hyperbola, as shown in Fig. 1.4. Therefore, from the scanning principle, the x–y two-dimensional galvanometer scanning system has inevitable deformation. The deflection angle of the galvanometer and the coordinates of the scanning point are nonlinear mappings. In addition, if the galvanometer deflection is controlled according to the conventional linear mapping algorithm strategy, pincushion distortion will occur.

To measure the magnitude of the deformation, the string height of the hyperbola was used to define the determined pillow-shaped distortion deformation amount ε. When Φ_1 is unchanged and Φ_2 changes from 0 to Φ_2, the deformation is:

$$\varepsilon = x - x_0 = d\tan\theta_1\left(\frac{1}{\cos\theta_2} - 1\right) \tag{1.5}$$

It can be seen from Eq. (1.5) that when the distance between the galvanometer scanning head and the working plane is unchanged, the distortion deformation is only related to the size of the deflection angle Φ_1 and Φ_2 and increases with the increase of Φ_1 and Φ_2, which means the distortion deformation is the smallest in the center of the working plane, and the distortion deformation is larger when scanning the edge of the working plane.

2. Field lens (f-θ lens)

The lens working near the focus surface of the objective lens is called a field lens, also known as a scanning focus lens and plane focus lens. Its main function is to overcome

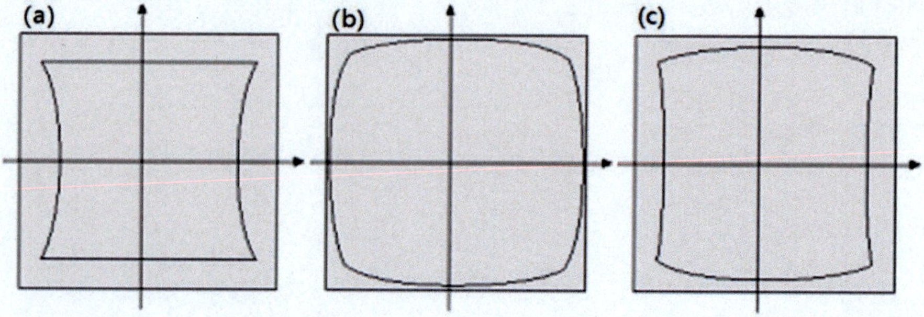

Fig. 1.5 f-θ lens correction principle: **a** pincushion deformity; **b** barrel deformity; **c** pincushion shape and barrel shape superposition

the pincushion distortion generated by the sweeping appeal tracing scope so that the focal spot has consistent focal properties in the scanning range. The correction principle is shown in Fig. 1.5. Figure 1.5a shows the distortion produced by the scanning mirror. Figure 1.5b is the distortion produced by the f-θ lens. Figure 1.5c shows the superposition effect of the laser passing through the scanning mirror and the f-θ lens. The pillow distortion is effectively improved after correction.

In addition, the f-θ lens can gather the incident parallel beams to obtain suitable size spots, which is often used in optical systems with wavelengths of 1064 nm, 10.6 μm, 532 nm and 355 nm. The spot diameter after the beam focuses through the field lens can be obtained from Eq. (1.6)

$$d = \frac{4\lambda M^2 f}{\pi n D_0} \tag{1.6}$$

where, f is lens focal length (m); D_0 is the beam waist diameter (m) before the laser beam passes through the beam expansion; λ is optical fiber laser wavelength (rad); M is the optical beam mass factor.

3. Collimator device

In the LPBF device, the main function of the collimator is to transform the transmitted light in the optical fiber into collimated light (parallel light). The working principle is shown in Fig. 1.6. The laser emitted from the laser is converted into a beam with a smaller divergence angle after the collimator. Then, the beam enters the scanning scope. The main performance indexes to be considered when selecting the collimator are divergence angle, working distance, waist beam diameter, maximum through laser power and working wavelength.

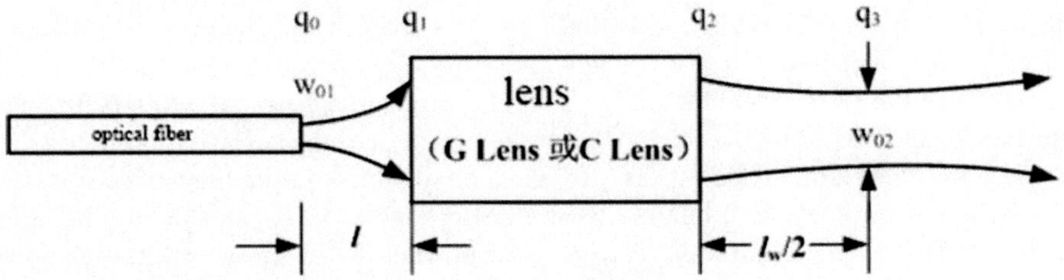

Fig. 1.6 Schematic diagram of the collimator principle

1.3.2 Powder Laying Device

In the LPBF manufacturing process, the machine structure that has the greatest impact on manufacturing quality is the powder laying device. The relative density and surface roughness of LPBFed metal parts are closely related to the quality of powder coating (surface flatness, compactness). When the powder is laid flat, the absorption of the laser by the powder bed is stable, and the flat manufacturing surface is easy to obtain after the laser melts the powder. When the powder plane is uneven, it leads to the unstable absorption of the laser by the powder bed. So, the laser focusing effect is different at different positions of the powder bed, and the laser energy absorbed by the powder bed is inconsistent due to the change in defocusing amount. Therefore, the effect of powder melting is different, so the LPBF scanning forming surface is also uneven. Finally, the relative density and surface quality of the molded parts are seriously reduced. In addition, during the LPBF manufacturing process, the manufacturing surface quality has a "feedback effect", which means that if the powder is uneven, the forming surface appears raised or concave after laser melting powders, resulting in a more uneven coating effect next time. When the powder bed is flat, the laser irradiates evenly on the surface of the entire powder bed to obtain a flat melting surface. So, the next layer of powder is easy to achieve a flat effect. LPBF completely melts the metal powder, and the powder is completely melted from about 45% of the loose density to obtain nearly 100% density. And the material shrinkage is serious, resulting in increased instability between the laser and powders. When the density of powder loosening increases, the shrinkage decreases after laser melting powders, which increases the action stability of the laser and material. Therefore, the quality of the powder laying determines the LPBF manufacturing effect to some extent. To achieve the desired manufacturing result, a flat and firm spreading effect must be obtained. At present, the powder laying method applied to LPBF technology is mainly flexible rubber strip and rigid scraper structure, two ordinary scraper structure powder spreading device collides with the forming surface during the powder laying process, or contact with the forming surface to produce great shear stress. As a result, the surface is not even after powdering, and the general scraper device has a very small effect on the compacting of the powder. The scraper pushes the powder to walk on the forming surface, and the shear force generated by the forming surface is easy to lead to the formation of thin-walled parts. As a result, precision parts with low strength are damaged.

Because the LPBF manufacturing process is prone to spheroidal, spatter, etc., the surface roughness of the formed metal parts is high, and the surface morphology is undulating. Therefore, it is necessary to find a flexible powder-laying device suitable for LPBF technology. In a flexible tooth elastic powder laying device, the flexible teeth need to have a certain strength to push the powders to the surface of the manufacturing parts. However, flexible teeth should also have good elasticity over the spatter and bulge parts without damaging the processing parts. Therefore, the flexible teeth are generally made of very thin and excellent elastic stainless steel pieces to avoid the collision caused by the rigid powder laying device and the surface bulge contact of the forming parts. The selection and design of a flexible rack are very important. It is recommended to use 316L or 304 stainless steel sheets with a thickness range of 30–100 μm that are cut by a fiber laser. The gap between teeth is 0.05–0.15 mm, tooth width is 3–5 mm, and the tooth height can be adjusted according to the actual process requirements, generally in 2–10 mm. Because the stainless steel rack is very soft, and the tooth gap of a single piece of stainless steel rack allows powder leakage, causing the powder bed surface after the powder, so, it is best to use a few pieces of stainless steel rack staggered superposition combination and flexible rack between the steel interval. In addition, a flexible rack needs to be installed before the flexible blade to avoid the flexible bar, not having enough strength to push the powder, resulting in the flexible rack's excessive bending plastic deformation and damage to the rack. The installation height difference between the scraper and the rack requires precise adjustment, and the bottom end of the scraper is 0.05–0.2 mm higher than the bottom end of the rack.

1.3.3 Control System

The LPBF control system controls the whole processing process. The quality of the control system will affect the processing speed, processing accuracy, processing efficiency, and forming quality, which are the core of the LPBF equipment. The control difficulty lies in how to coordinate the relationship between each hardware to ensure the safe and stable operation of the system. The control object of the LPBF control system mainly includes two parts: a laser optical path system and a mechanical transmission system. Through the control of these two parts, the functional requirements of the LPBF control system are realized. The control of the laser path system mainly includes the control of the laser and the vibration lens, and the control of the mechanical transmission system mainly includes the lifting action control of the working platform and the active control of the powder laying device. The hardware composition of the LPBF control system is shown in Fig. 1.7.

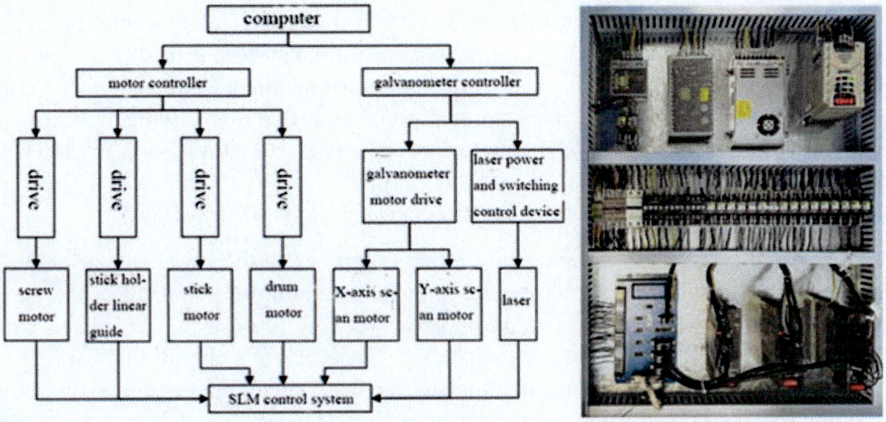

Fig. 1.7 Block diagram of control system hardware composition (left) and control circuit diagram (right)

The control system mainly includes the following functions.

(1) System initialization, status information processing, fault diagnosis, and realizing human–computer interaction function.
(2) Various control of the motor system, providing the motion control of the forming cylinder, powder cylinder, and powder laying device.
(3) Control the scanning lens, set the scan motion, scan delay, etc.
(4) Set the automatic forming parameters, such as adjusting the laser power, processing layer thickness, etc.
(5) Provide the coordinated control of the 4 servo motors of the manufacturing equipment, and complete the processing operation of the parts.

1.3.4 Gas Circulation and Purification System

In the process of LPBF manufacturing, the control of the sealed and forming indoor gas environment is very important, among which the key indicators are oxygen content, air pressure, and metal dust particle concentration. Oxygen content is directly related to the manufacturing quality of manufacturing metal parts and has a great impact on the comprehensive performance of metal parts. The manufacturing room of LPBF needs to maintain a very low oxygen content to prevent the metal parts from being oxidized, affecting the performance of the parts. During the manufacturing process, the sealed chamber should maintain a 10 kPa low positive pressure environment to ensure that the external oxygen does not penetrate the sealed chamber. The pollution and leakage of metal dust particles in the gas environment are also problems. These dust particles affect the laser irradiation, the light transmittance of the optical lens, and even the melting and solidification stability of the melting tank. Usually, metal material particle size used in the LPBF, such as 316L stainless steel, CoCrMo alloy, and titanium alloy, is only dozens of microns. Metal spatter size and powder particles are often formed during LPBF, which are easily scattered on the metal forming surface and long-term diffusion in the air, both affecting the performance of forming metal parts. There are also a serious threat to the personal safety of operators. Therefore, a gas circulation system is needed to purify the atmosphere in the manufacturing room. The structure of the gas circulation system is shown in Fig. 1.8.

The main working steps of the gas circulation system are as follows:

(1) When the computer receives the starting command of the gas cycle purification process, it will send the signal to open the vacuum pump and open the electromagnetic valve. The vacuum pump is working to push the gas mixed with oxygen into the manufacturing chamber. The protective cylinder is working to fill the inert protective gas into the manufacturing chamber, continuously dilute the oxygen concentration in the manufacturing chamber, and open the vacuum pump and push the inert protective gas to accelerate the decreasing rate of the oxygen concentration in the manufacturing chamber;
(2) The oxygen content sensor sends the detected oxygen concentration data in the manufacturing chamber to the computer in real-time, which detects whether

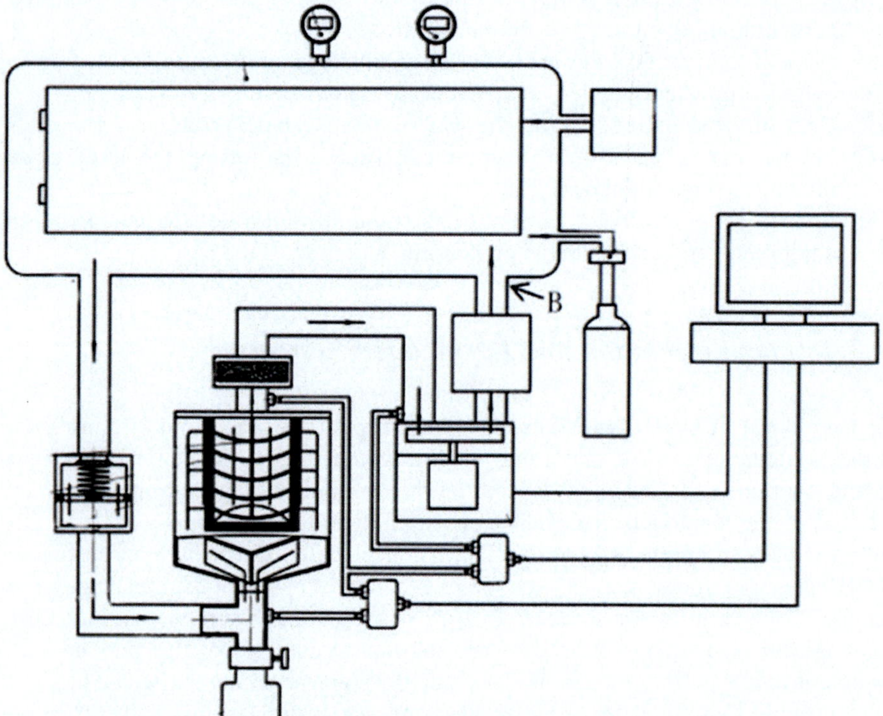

Fig. 1.8 Gas circulation system of LPBF equipment

the oxygen concentration is below 100 ppm. If the oxygen concentration is above 100 ppm, it will keep the vacuum pump pushing the inert protective gas to reduce the oxygen concentration until the oxygen concentration is below 100 ppm. In addition, it will also keep the solenoid valve open to charge the inert protective gas in the manufacturing chamber to increase the air pressure in the manufacturing chamber;

(3) The air pressure sensor sends the air pressure data in the manufacturing chamber to the computer in real-time to detect the pressure at the 10 kPa threshold. If the pressure is below this threshold, it will open the solenoid valve to maintain the supply of inert protective gas until the pressure detected by the computer reaches the threshold to maintain low positive pressure. If the air pressure is above 10 kPa, it will close the solenoid valve and disconnect the supply of inert protective gas;

(4) The airflow of metal-containing dust particles from the manufacturing chamber is evenly diverted to the fan air outlet channels through the air inlet of the gas distribution device, enters the filter box under the guidance of the cavity walls of each fan-vent channel, uniformly passes through the HEPA filter element surface and completes the metal dust in the air at the HEPA filter to filter out most of the impurities. At the same time, the electrostatic release net in close contact with the HEPA filter element can quickly release the electrostatic accumulation caused by the friction between the metal dust airflow and the HEPA filter element surface. Clean gas after one filtration, Then, goes through the activated carbon filter box to complete the secondary filtration of the gas to continue to improve the cleanliness of the airflow. The two-stage filtered clean gas is driven by a circulating gas pump through the second directional gas valve and sent back to the manufacturing room to complete a single gas cycle operation.

1.3.5 Manufacturing Process Monitoring System

In the whole LPBF process, the manufacturing quality of parts is affected by multiple factors, such as scanning speed, scanning hatch space, layer thickness, scanning path, spot compensation, laser power and density. Therefore, to obtain a high-quality part, a series of key parameters in the LPBF process must be monitored in such a complex process. Quality assurance and process monitoring have become necessary means for AM technology to improve from the model processing level to the first-class workshop manufacturing level. The molten pool of LPBF contains rich quality information, which directly determines the manufacturing quality of the parts. Therefore, the quality control of the molten pool is conducted by focusing on the shape of the molten pool and the brightness of the molten pool, etc. The main problem that quality control solves is the variability of the AM device or laser-material interaction because the latter can in turn disrupt the microstructural or macromechanical properties of the metal.

The author adopts a coaxial monitoring method combining the high-speed camera and the photodiode in the LPBF process to monitor the melting process of the 3D metal layer by layer. The coaxial real-time monitoring device is based on two detectors and a photodiode, which are distributed in the same plane. They share the same optical system with the laser to achieve coaxial monitoring through laser optics and precise positioning, which is conducive to obtaining high local resolution and fast scanning rate. For a detailed description of the system, refer to Sect. 7.4.

1.3.6 Other Important Components

Other important components in the manufacturing system mainly include (1) an oxygen content sensor, (2) a pressure sensor, (3) an intake flow sensor, and (4) a position sensor. The oxygen content sensor is mainly used for the real-time monitoring of the oxygen content in the manufacturing indoor atmosphere. The oxygen content shall be less than 100 ppm during the equipment processing process. The pressure sensor is used to monitor pressure values in real-time. In the processing process, we should ensure that the manufacturing room is in a positive pressure environment to prevent oxygen from entering the manufacturing room. The positive pressure of the manufacturing chamber is usually 10 kPa. The inlet flow sensor is used to monitor the intake speed of the inert protective gas by adjusting the intake flow rate to obtain positive pressure in the manufacturing room to ensure stable oxygen content. When the manufacturing chamber is poorly sealed, the intake flow rate will increase. However, if the increased intake flow rate still fails to stabilize the oxygen content, printing work should be stopped to check the equipment. The position sensor is mainly the manufacturing cylinder and powder cylinder substrate position sensor, and powder truck position sensor, The manufacturing cylinder and the powder cylinder substrate position sensor are used to monitor the position of the manufacturing cylinder inner substrate and the reserves of material in the powder cylinder. The powder truck position sensor is to monitor the real-time position of the powder car.

1.4 The Latest Research Progress of Laser Powder Bed Fusion Equipment

1.4.1 Commercial Equipment and Company Profile

EOS (Germany) is one of the global leaders in the field of metal AM [14]. In 2003, the company released the DMLS EOSINT M270, which is the most common type of metal-forming machine at present. In 2011, the EOSINT M280 began to be sold. At present, EOS company's main machine model is DMLS EOSINT M290. In 2016, the M400-4 model was released, which increased productivity by four lasers with a 400 mm × 400 mm × 400 mm manufacturing volume. Each of the four 400 W lasers has a build area of 250 mm × 250 mm (with 50 mm overlap) and can manufacture four parts simultaneously.

Concept Laser was acquired as a 75% stake by General Electric (GE) for $599 million in 2016. The company's equipment includes M1, M2, M3, Mlab, X-1000 R, and other models, of which the X-1000R has a maximum forming size of 630 mm × 400 mm × 500 mm. The formable products of the equipment are still the largest in all the SLM equipment [15].

3D systems [16] began working with MTT to sell SLM equipment in North America in 2008. In addition, Japan Matsuura Machinery (MATSU-URA) developed the hybrid SLM machine Avance-25 (SLM + machining) in 2010, in which, machining was performed after several layers were built by SLM to improve the surface finish.

Realizer GmbH is also one of the first German companies to develop and sell metal AM equipment. In 2017, well-known tool maker DMG MORI acquired a 50.1% stake in ReaLizer GmbH. From this, DMG MORI acquired AM technology from ReaLizerGmbH. Meanwhile, in addition to ReaLizer GmbH's headquarters in Borchen (Germany), DMG MORI's factory in Bielefeld will also be a manufacturing and assembly point for ReaLizer GmbH's SLM equipment.

The team of Yang Yongqiang at the South China University of Technology in China has been developing laser selection melting equipment since 2002 and has independently developed a series of equipment such as DiMetal240 (2004), DiMetal280 (2007), DiMetal100 (2012), DiMetal50 (2016) and other serial equipment, and the equipment has achieved commercialization. In addition, Xi'an Bright Laser Technologies Co., Ltd. [17], Farsoon High-Technology Co., Ltd., E-Plus-3D Additive Technology Co., Ltd. [18], Huake 3D Technology Co., Ltd. [19], Jiangsu Yongnian Laser Forming Technology Co., Ltd., Guangzhou Laseradd Additive Technology Co., Ltd. [20], Guangzhou Raytool Laser Technology Co., Ltd. and other companies also have released their commercialization equipment. The main models of LPBF equipment are shown in Table 1.1 [14–23].

As can be seen from Table 1.1, the forming format of the current mainstream LPBF equipment is generally below 300 mm × 300 mm, and the size is relatively small. This is mainly limited by the optical system of LPBF equipment. The output end of the scanning optical system is mainly composed of the galvanometer and the f-θ field lens. When the scanning area is too large, it is difficult for the f-θ field lens to compensate the focus to the forming plane for the edge position. The uniformity of the laser in the whole forming amplitude can not be guaranteed, which seriously affects the forming quality. Therefore, the forming size of the equipment is greatly limited. Due to the small forming size, the current general LPBF equipment can not meet the demand for many large and complex parts in the fields of automobile, mold, aerospace, nuclear power, etc.

On the other hand, due to the defects, such as molten pool spatter, powder adhesion and spherical effect, formed during the LPBF process, the surface roughness of the processed parts still needs to be improved (5–30 μm). At the same time, the

Table 1.1 The main model of LPBF equipment

Equipment manufacturers	Device model	Forming dimensions
EOS GmbH	EOS M080	ϕ 80 × 95
	EOS M100	ϕ 100 × 95
	EOS M290	250 × 250 × 325
	EOS M400	400 × 400 × 400
Concept Laser GmbH	M1	250 × 250 × 250
	M2	250 × 250 × 280
	M3	300 × 350 × 300
	Mlab	50(90) × 50(90) × 80
	X-1000R	630 × 400 × 500
SLM Solutions GmbH	SLM 125	125 × 125 × 125
	SLM 280	280 × 280 × 365
	SLM 500	500 × 280 × 365
Realizer GmbH	SLM 50	70 × 70 × 80
	SLM 100	125 × 125 × 200
	SLM 250	250 × 250 × 300
	SLM 300	300 × 300 × 300
3D Systems	ProX DMP 100	100 × 100 × 90
	ProX DMP 200	140 × 140 × 115
	ProX DMP 300	250 × 250 × 330
	DMP Flex 350	275 × 275 × 420
Xi'an Bright Laser Technologies Co., Ltd.	BLT-S210	105 × 105 × 200
	BLT-S310	250 × 250 × 400
	BLT-S320	250 × 250 × 400
	BLT-S400	400 × 250 × 400
Farsoon High-Technology Co., Ltd.	FS412M	425 × 425 × 420
	FS301M	305 × 305 × 400
	FS271M	275 × 275 × 320
	FS121M	120 × 120 × 100
	FS121M-E	120 × 120 × 100
Huake 3D Technology Co., Ltd.	HK M125	125 × 125 × 150
	HK M280	280 × 280 × 300
E-Plus-3D Additive Technology Co., Ltd.	EP-M100T	120 × 120 × 80
	EP-M150	ϕ150 × 120
	EP-M250	262 × 262 × 350
	EP-M450	455 × 455 × 500
	EP-M650	650 × 650 × 650

(continued)

Table 1.1 (continued)

Equipment manufacturers	Device model	Forming dimensions
Jiangsu Yongnian Laser Forming Technology Co., Ltd.	YLM-328	300 × 300 × 328
	YLM-300	ϕ 300 × 300
	YLM-T150	ϕ 150 × 100
Guangzhou Laseradd Additive Technology Co., Ltd.	DiMetal-50	ϕ 50 × 50
	DiMetal-100	100 × 100 × 120
	DiMetal-280	250 × 250 × 300
	DiMetal-500	500 × 250 × 300

dimension accuracy of LPBF technology is not high enough (±50 μm) due to the inherent diameter of the laser spot, the extremely high cooling gradient of the molten pool, the thermal deformation caused by heat accumulation, the step-by-step effect of software layered slicing and equipment errors. At present, the accuracy and surface roughness of LPBF equipment can not directly meet the application needs of various industrial fields, and it is necessary to improve the post-processing of sandblasting, CNC machining, abrasive particle flow, electrochemical polishing, etc.

In addition, although many studies have proposed to design the optimization of structures with different objectives based on the AM processing degrees of freedom, these structural designs mainly rely on the density and spatial distribution of a single material to improve the performance, but rarely directly use the spatial distribution of different materials to achieve the optimization of parts. Multi-material LPBF processing also faces some technical challenges, such as multi-material powder laying structure, mixing and separation technology of different material powder added into the powder bed simultaneously, and forming process of melting between any two or multiple materials under the action of the laser. At present, the LPBF processing methods or equipment for multiple materials still lack fully mature solutions.

Facing the current problems, the development of large-size, high-precision, and multi-material LPBF equipment will be an important development direction in the future.

1.4.2 Development Status of Large-Size Laser Powder Bed Fusion Equipment

LPBF equipment suppliers have developed a series of equipment for large-size parts forming, and have successful application cases. At present, there are three main technical schemes: long focal length f-θ field lens, moving galvanometer, and multi-laser multi-galvanometer splice forming.

(1) Long focal length f-θ field lens

The principle of this scheme is shown in Fig. 1.9. The size of the forming projection surface is affected by the extreme motion angle and focal length of the galvanometer, where L and L′ are the side length of the scanning range, f and f′ are the corresponding focal length, and the geometric relationship is as follows:

$$L'/f' = L/f \tag{1.7}$$

where, L is the side length of the scan range (m); L′ is the side length of the scan range (m); f is the focal length (m); f' is the focal length (m).

Because the scanning side length L is directly proportional to the focal length f, the long focal length f-θ field lens is used to increase the L. The M400 of the German EOS company uses this scheme. The maximum forming size of M400 is 400 mm × 400 mm × 400 mm, and a 1000 W laser is used. Because the long focal length increases the spot after focusing, a higher-power laser is needed to compensate for the loss of laser power density. However, larger spots will reduce the precision and surface roughness of the parts, so this technology still has some limitations.

(2) Moving galvanometer

The galvanometer can be used to scan more areas to achieve a wide range of processing by moving the scanning galvanometer as a whole. The principle is shown in Fig. 1.10. This forming scheme adopts the strategy of scanning in different regions, which moves to the next area and continues to scan after completing the scanning

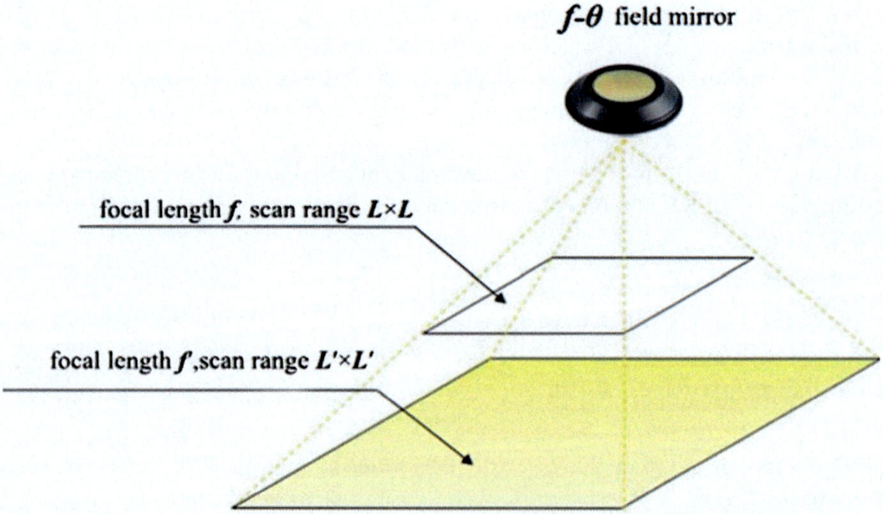

Fig. 1.9 Principles of expanding the forming region with long focal length f-θ field mirror

range. Concept Laser X-1000R uses this scheme with a forming size of 630 mm × 400 mm × 500 mm.

The shortcomings of this scheme are also obvious:

(1) It is necessary to install linear motors and guides as a motion system, and the whole machine system is more complex.
(2) Only a single laser single galvanometer is used, and the efficiency is still low when forming large-size components.
(3) Due to the low processing efficiency, the end time of the parts in the processing of different partitions is very different, and the temperature change of the entire forming surface is uneven, which is not conducive to the stress control of the parts.

(3) Multi-laser and multi-galvanometer splicing and forming

The principle of this scheme is shown in Fig. 1.11, It is equivalent to adding a galvanometer and laser to cover a larger forming area in the scanning range of the above subregions, which can not only expand the forming size but also improve the processing efficiency. Therefore, it is the mainstream scheme for the development of large-size LPBF equipment, but with the increase in the number of scanning systems, the corresponding difficulty of the system control, software development, scanning lap area quality control, and other problems will also increase rapidly.

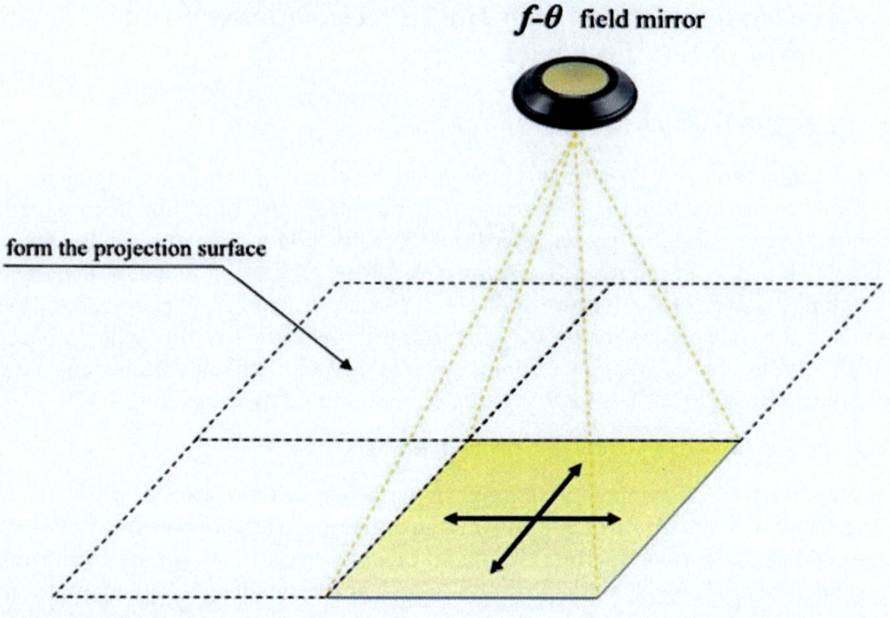

Fig. 1.10 Principles of moving galvanometer scanning

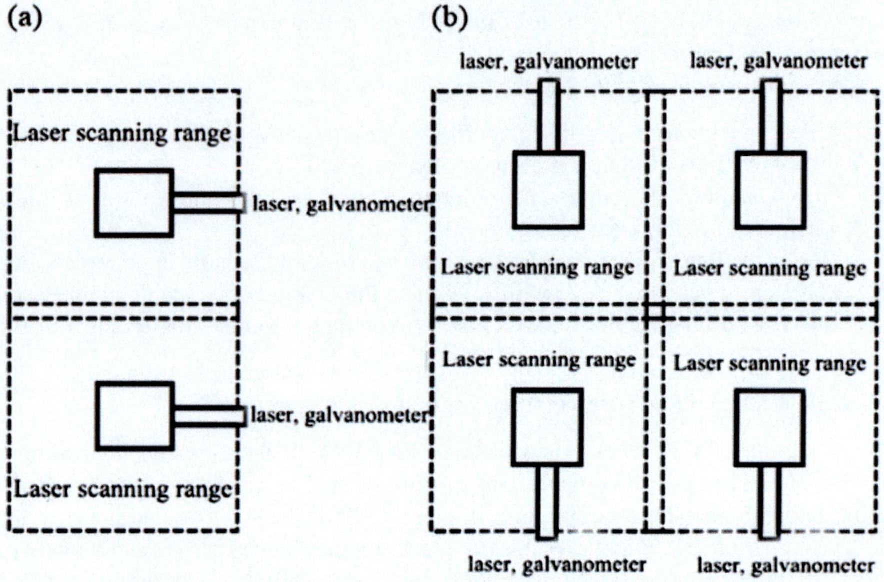

Fig. 1.11 Principle of multi-beam scanning: **a** double galvanometer and double field group; **b** four galvanometer and four field group

1.4.3 Development Status of High-Precision Laser Powder Bed Fusion Equipment

1. Short focal length f-θ field lens and low-power laser

For the direct forming process, to improve the precision of forming components, a fine and stable laser spot is inseparable. Therefore, the use of a short focal length f-θ field mirror, and low-power fiber laser, combined with a small layer thickness (below 30 μm), can improve the accuracy of LPBF processing to a certain extent, which is suitable for the formation of complex fine structures. Many vendors have adopted this scheme, as shown in Table 1.2. The deficiency of this scheme is the small forming size and the low forming efficiency. In addition, the forming accuracy (±50 μm, Ra = 5 μm) is still not compared with subtractive machining.

2. Hybrid additive and subtractive manufacturing

To improve the surface quality and machining accuracy of AM parts, people have long begun to grind, polish, and other subtractive mechanical processing of parts after AM, and to further improve the efficiency, the concept of hybrid manufacturing was put forward [24]. Numerical control (CNC) processing technology was used to reduce the material manufacturing to achieve good dimensional accuracy and surface quality of AMed complex high-performance parts. Some high-end CNC machine tool

Table 1.2 Key parameters of high-precision LPBF equipment

Vendor	Model	Forming size/mm	Laser power/W
EOS	M100	φ 100 × 95	200
Concept Laser	Mlab cusing	90 × 90 × 80	100
3D systems	ProX DMP 100	100 × 100 × 80	100
Techgine Laser Technology (shanghai) Co., Ltd.	TZ-SLM120	Φ120 × 100	200/500
Xi'an Bright Laser Technologies Co., Ltd.	BLT-A100	100 × 100 × 100	200
Hunan Farsoon High-Technology Co., Ltd.	FS121M	120 × 120 × 100	200
Guangzhou Laseradd Additive Technology Co., Ltd.	DiMetal-50	Φ 50 × 50	75

manufacturers have launched commercial hybrid additive and subtractive manufacturing equipment with their technical advantages. The existing hybrid additive and subtractive manufacturing solutions are mainly divided into two categories, one is the combination of laser cladding deposition technology and CNC processing technology, such as LASERTEC 4300 3D hybrid of German Demaggie and HYBRID HSTM1500 of Hamuel Reichenbacher. Another one is the combination of LPBF technology and CNC processing technology, such as Japan's Matsuura company's LUMEX Avance-25, Sodick's OPM250L, and OPL350L.

For the LPBF equipment manufactured by hybrid additive and subtractive manufacturing, a significant technical difficulty is to avoid metal chips generated by subtractive processing from contaminating the powder bed [25]. If the two processes of additive and subtractive are more completely separated, this problem can be easily solved, but the processing efficiency is low, which is the same as the efficiency of using subtractive equipment such as a machining center after AM process. If the two processes of additive and subtractive are alternated, the processing efficiency and the processing accuracy of the internal structure of the parts can be guaranteed, but the problem of mixing metal chips and powders is more serious.

1.4.4 Multi-material Laser Powder Bed Fusion Equipment

Multi-material LPBF technology is an AM technology that uses a variety of powder materials (at least one of which is a metal material) to prepare a single complex functional component with a variety of material structures. At present, the research on the laser melting technology of multi-material powder bed is still in the research stage, and the commercial equipment is relatively rare, which can be mainly divided into the following two technical stages.

1. Multiple materials in the build direction

According to the powder spreading principle of LPBF equipment, it is easier to achieve multi-material distribution in the Z-axis direction of the part because only the replacement of supplied powder material on a specific number of processing layers is needed. Meanwhile, the powder laying system maintains a stable laying to the powder bed. After the interface processing is completed, the subsequent processing of the number of layers is equivalent to a single material AM. The DiMetal-300 equipment of Guangzhou Laseradd Additive Technology Co., Ltd. uses a four-funnel multi-material system to achieve the forming of four materials in the Z-axis direction, as shown in Figs. 1.12 and 1.13.

2. Multiple materials on the XY plane

To freely arrange a variety of materials in different layers or different areas within the same layer, the conventional LPBF powder supply system and powder laying system need to be improved. At present, there is no mature commercial equipment, but the researchers have put forward and tested some schemes.

Chao et al. [28]. of the University of Manchester developed an LPBF system combining the powder bed display, point-by-point selective vacuum absorption, and point-by-point dry powder delivery (shown in Fig. 1.14), developed a special CAD data preparation program for LPBF, and successfully prepared 316L/Inconel 718 and 316L/Cu10Sn samples, as shown in Fig. 1.15. The samples obtained have significantly different interlayer distributions and good metallurgy at the material interface. However, some defects were still found in the region deposited by ultrasound, such as pores and cracks.

Wu et al. [29] have developed a new multi-material LPBF system based on the principle of multi-funnel powder supply + flexible cleaning and powder recovery. Based on the principle of multi-funnel quantitative powder supply and flexible cleaning and powder recovery, the system can realize multi-material AM on demand between

Fig. 1.12 CuSn/18Ni300 bimetallic porous structure. Reprinted with permission from a work by Chen et al. [26, 27]

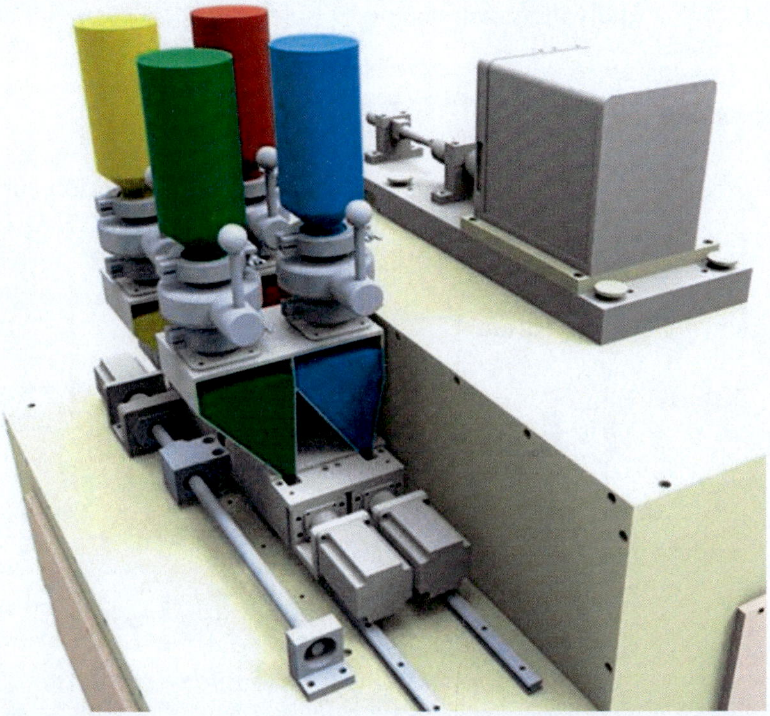

Fig. 1.13 Four-funnel multi-material system of DiMetal 300 equipment of Guangzhou Laseradd Additive Technology Co., Ltd. Reprinted with permission from a work by Chen et al. [26, 27]

different layers or in different areas within the same layer. To guarantee the performance of the system, the AM of the heterogeneous material parts was carried out, and the CuSn10/4340 steel heterogeneous material parts were successfully manufactured [30, 31], as shown in Fig. 1.16.

Multi-material metal AM technology opens a door to a new field for the manufacturing industry, providing a technological means to break through current limitations in fields such as aerospace, biomedicine, and nuclear power installations. At present, multi-material LPBF equipment also faces technical challenges such as data processing, precise preset of heterogenous materials, and avoiding powder pollution avoidance. In addition, the multi-material metal AM technology in the energy input and interface forming quality, material compatibility and interface combination, material-structure integration design, and other directions also need in-depth research and discussion.

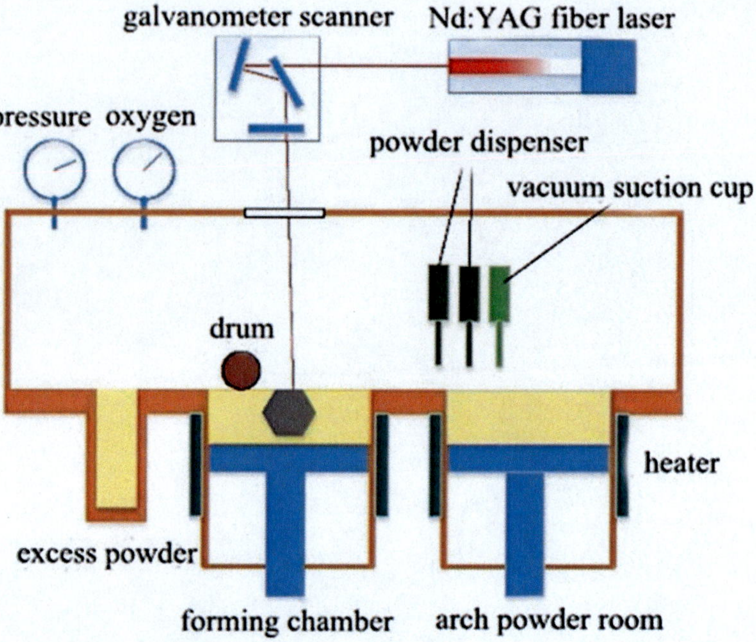

Fig. 1.14 Multi-material LPBF system with a vacuum suction system is proposed by Chao et al. Reprinted with permission from a work by Chao et al. [28]

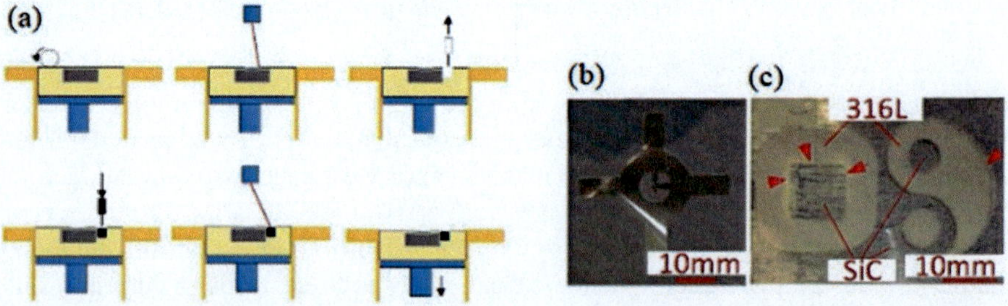

Fig. 1.15 Working schematic diagram and physical diagram of multi-material LPBF system with vacuum powder suction system proposed by Chao et al.: **a** multi-material LPBF process; **b** vacuum suction powder; **c** laser melting SiC powder layer selective powder deposition 316L box shape and semi-yin and Yang pattern. Reprinted with permission from a work by Chao et al. [28]

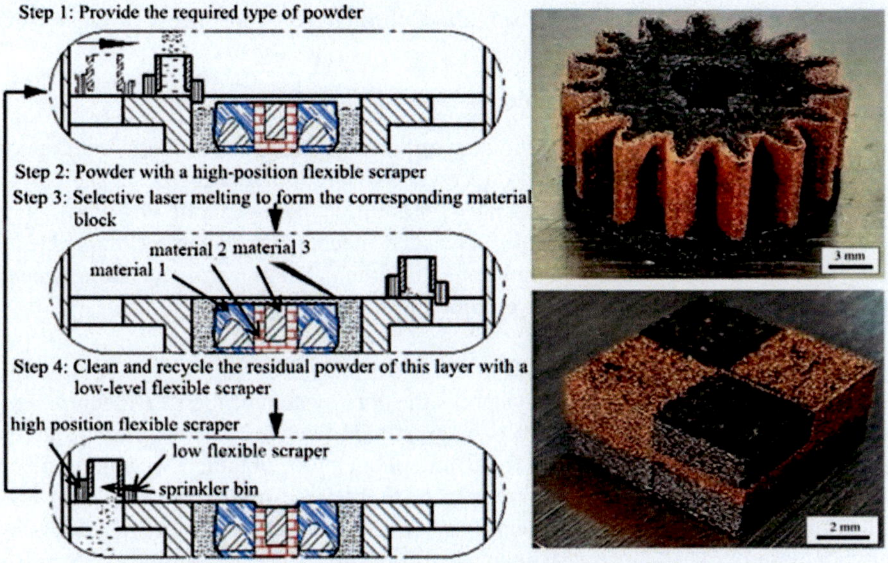

Fig. 1.16 Schematic diagram of multi-funnel powder supply + flexible cleaning and recovery powder and the CuSn10/4340 steel multi-material parts made by the equipment. Reprinted with permission from a work by Wu et al. [28]

1.5 Laser Powder Bed Fusion Crossover Technology

The crossover technology of LPBF includes material engineering technology, computer programming technology, mechanical engineering technology, optical technology, automation control technology, forming monitoring and detection technology, computer simulation technology, personalized biomedical technology, and new material technology [32].

1. Material engineering technology

At present, the materials that can be formed by LPBF technology are limited. The development of powder materials for the characterization and testing of formed parts is closely related to material engineering technology and depends on the progress and development of material engineering technology.

2. Computer programming technology

To use LPBF technology to manufacture physical parts, the first step is to digitize the product. At present, there are still few modeling methods for the AM industry upstream and many technical problems lie in the functionality and versatility of the control software. The data processing and equipment control software developed by different companies has not formed a unified data format and process standards, etc. Although there are great similarities between the control software of various AM equipment, it is not compatible. Similar modules are difficult to share, and the process requirements can still not be realized through software. These problems rely on computer programming techniques to solve.

3. Mechanical engineering technology

In the field of AM, due to the unique properties given by its process, some parts with complex geometries can be produced at a low manufacturing cost. The design of such parts is often difficult, especially for additive parts generated by topological optimization methods, which require advanced mechanical design techniques [33]. In addition, AM technology and traditional mechanical engineering technology need to be continuously integrated to overcome the shortcomings of AM itself.

4. Optical technology

As above mentioned, the optical system is the core system of the LPBF equipment, and the related optical technology is of great significance for the forming quality and efficiency. For example, fine focusing spots can be obtained with good beam quality, which directly affects the processing accuracy and minimum processing size. The multi-beam galvanometer scanning system can effectively improve processing efficiency and expand the processing range. In addition, the laser melting experiments of different wavelength lasers on highly reflective materials (such as copper and its alloys) show that the laser of specific wavelengths (blue-green) has better adaptability to copper alloy materials that are difficult to form with conventional lasers with a 1024 nm wavelength, which broadens the choice of materials.

5. Automation control technology

LPBF equipment is a precision automation product consisting of optical, mechanical, electrical, and gas. Multiple systems cooperate with a precise operation to ensure processing and manufacturing, which is inseparable from reliable automation control technology.

6. Monitoring and detection technology

LPBF process coupling is strong, and the detection methods for its process monitoring still need to be improved. Carrying out inspection technology research for LPBF technology is an inevitable requirement to explore its forming mechanism, understand its defect formation, evolution and scientific definition, improve LPBF forming quality, and finally realize process quality control and quality traceability. The detection technology for LPBF technology can be divided into two categories: online detection and offline detection. Among them, online detection has high real-time characteristics and can provide timely feedback to the control system. Offline testing is usually highly accurate, facilitating comprehensive quality inspection, and is an irreplaceable benchmark or supplement to online testing.

In terms of online detection, at present, most researchers have adopted high-speed CCD and infrared imaging devices with coaxial/paraxial in-situ architecture to obtain rich visible and infrared information of the LPBF process, developed statistical descriptors of the molten pool, melting track, spatter, plume, and temperature field distribution. Then, the correlation between related descriptors and LPBF forming quality was studied to achieve more results. A small number of scholars have carried out classification and identification research on the forming quality of a single melt track based on the sound signal information source and the radiation intensity signal of the molten pool collected by the photodiode.

In terms of offline detection, in addition to traditional material testing and analysis methods, micro-CT and laser-induced breakdown spectroscopy provide efficient and new tools for three-dimensional characterization and composition analysis of defects in LPBFed parts.

At present, LPBF process control is open-loop control based on parameter optimization. Online process monitoring and feedback control strategies are the core of the final LPBF closed-loop control. Researchers have conducted a lot of studies on it based on machine learning model prediction and traditional statistical process control (SPC). Among them, common machine learning models mainly include K-means clustering analysis, support vector machines (SVM), deep belief network (DBN), convolutional neural network (CNN), etc., which are mainly used for the extraction of statistical descriptors of LPBF processes. The research results of statistical process control are mainly applied to the analysis of the relationship between features themselves, as well as the analysis of the relationship and the generation of control charts between the feature and LPBF forming quality. In general, the integration of online and offline detection, simulation analysis application, intelligent algorithm application, and real-time improvement are the development trends of LPBF monitoring and detection technology.

7. Computer simulation technology

Computer simulation technology has been developed for many years, and it is a powerful assistant for scientific research experimental design analysis by simulating and calculating preset scenes and boundary conditions through classical formula models. Quality assurance is a key goal of metal AM. However, LPBF is a typical multi-scale, multi-physics coupling process, and the shape quality is affected by many factors. At the same time, due to the extremely small laser focusing spot, high laser spot energy density, fast scanning speed and short cooling time, a large number of physical and chemical reactions occur rapidly in different places of the powder bed, which make it difficult to completely reveal the whole process of laser melting at a certain point even if online monitoring technology is used. All of these difficulties highlight the importance of computer simulation technology.

The multi-scale simulation includes multi-scale analysis from macroscopic scale (such as overall thermal deformation and inherent stress deformation of parts) to mesoscopic scale (such as molten pool, spatter phenomenon) to microscopic scale (such as structural evolution, grain orientation). Multiphysics simulation includes multiphysics analysis such as forming temperature field, wind field (shielding gas), melt flow field (molten pool fluid), motion field (powder laying process), and solid stress and deformation field of AM structures, and multiphysics effects permeate every stage of metal AM forming. Figure 1.17 shows a schematic diagram of the airflow-spatter interaction during LPBF [34].

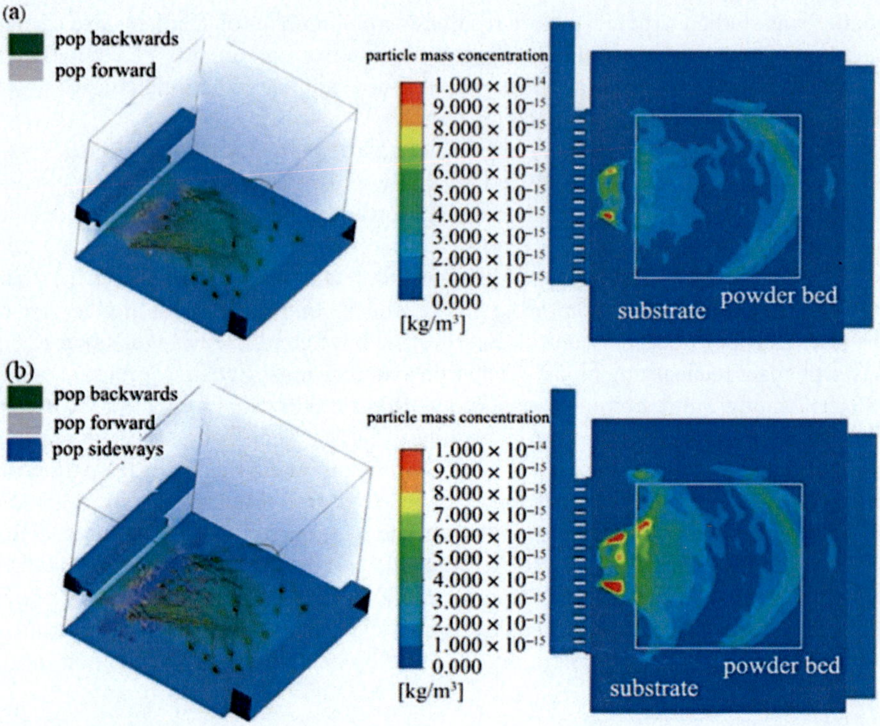

Fig. 1.17 Simulation diagram of spatter interaction during LPBF process: **a** spatter trajectory and concentration diagram of the front and rear spatter model; **b** spatter trajectory and concentration map of the three-dimensional space spatter model. Reprinted with permission from a work by Zhang et al. [34]

During the formulation of AM process, the use of CAE simulation analysis technology to obtain product performance characteristics and processing risk identification in advance is an important way to solve the quality problem of AM process. In this way, the probability of forming failures of physical parts can be reduced, and the corresponding cost losses can be avoided. In addition, AM of metal parts facilitates the evolution of design methods and design modifications. The process design process and experience can be better cured. The utilization rate of the machine and the efficiency of product processing are improved, and the process repeatability and quality can be guaranteed. If the microscopic metallographic structure and characteristic prediction can also be achieved through simulation, the simulation will significantly accelerate the development of new equipment, new processes, and new material parameter sets, and reduce R&D costs and cycles.

8. Personalized biomedical technology

The combination of LPBF technology and personalized medicine technology fully reflects the advantages of AM technology's unique free design and individual preparation, and rapid prototyping of complex internal and spatial structures. As shown in

Fig. 1.18, medical CT data is used to obtain the human bone structure and extract the size of the bone units. Then, CAD modeling techniques are used to design elements with different structural porosities and combine gradient structural models. Gradient structure samples were prepared by LPBF technology to meet the needs of cell growth for different pore sizes, matching mechanical properties such as strength and modulus. It is found that gradient structure specimens have higher strength when there is the same structural porosity in the porous structures. Therefore, the combination of biomimetic structure design, metal implants and LPBF technology for human tissue engineering reconstruction and personalized medicine has important research significance and development prospects.

9. Multi-field coupling technology

LPBF technology is coupled with magnetic fields and sound fields to obtain multi-field coupling technology. For example, adding a magnetic field to the machine-forming platform can regulate the solidification structure of the molten pool, affecting the grain size and orientation distribution. The main principle is that when convection and mass transfer effects occur in the molten pool, the magnetic induction lines in the magnetic field are cut so that the liquid metal is subjected to the Lorentz force. Therefore, by adjusting the direction of the magnetic field and the strength of the magnetic field, the microscopic movement of liquid metal can be regulated. In addition, some researchers add high-intensity ultrasonic waves during the manufacturing process to regulate microstructure and mechanical properties. RMIT University Todaro et al. [35] added high-intensity ultrasonic waves under the forming substrate in the process

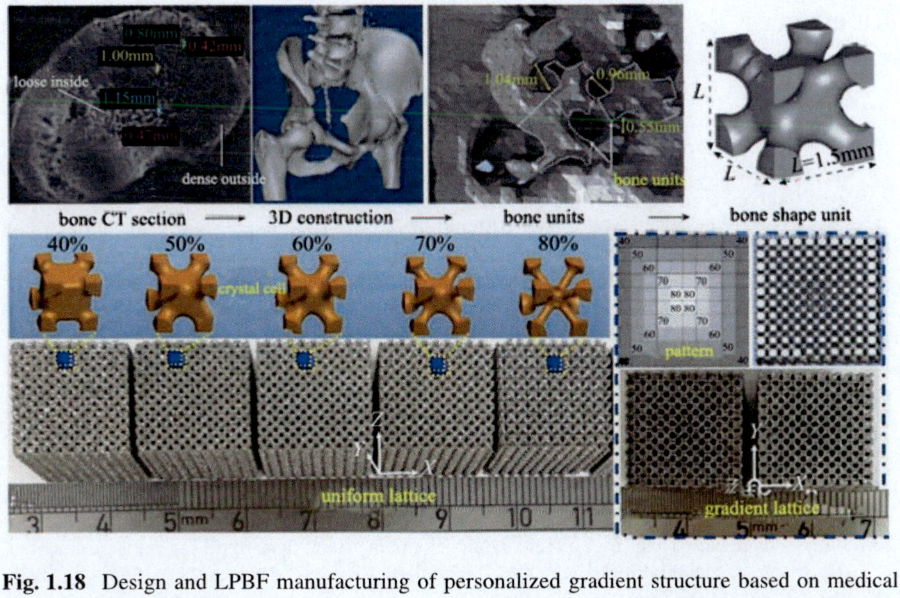

Fig. 1.18 Design and LPBF manufacturing of personalized gradient structure based on medical CT TiNi alloy implant

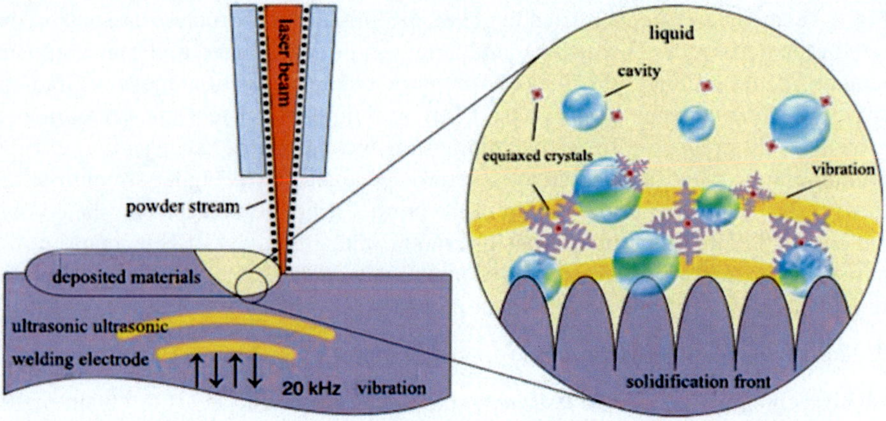

Fig. 1.19 Adding high-intensity ultrasound during AM to regulate tissue growth. Reprinted with permission from a work by Todaro et al. [35]

of laser AM Ti6Al4V, which significantly refined the structure. The original columnar crystal was transformed into an equiaxed crystal, and the mechanical properties were improved. As shown in Fig. 1.19, the main mechanism of ultrasonic regulation of microstructure is that the cavitation generated by ultrasonic waves in the molten pool promotes the growth of equiaxed crystals.

References

1. Beaman JJ (1997) Historical perspective, Chapter 3 in JTEC/WTEC panel report on rapid prototyping in Europe and Japan. WETC Hyper-Librarian
2. Bourell DL, Beaman JL, Leu MC et al (2009) A brief history of additive manufacturing and the 2009 roadmap for additive manufacturing: looking back and looking ahead, RapidTech. In: Workshop on rapid technologies, US-Turkey. Istanbul, Turkey
3. Chua CK, Leong KF, Lim CS (2010) Rapid prototyping: principles and applications (with companion CD-ROM). World Scientific Publishing Company
4. Nakagawa T (1979) Blanking tool by stacked bainite steel plates. Press Technique, pp 93–101
5. Pham D, Dimov SS (2012) Rapid manufacturing: the technologies and applications of rapid prototyping and rapid tooling. Springer, Berlin
6. Pipes A (1982) Plotting the progress of CAD/CAM: falling hardware costs and improved software are making CAD/CAM systems more attractive. Data Process 24(10):19–21
7. Wohlers T, Gornet T (2014) History of additive manufacturing. Wohlors report, vol 24, p 118
8. Mueller B (2012) Additive manufacturing technologies—rapid prototyping to direct digital manufacturing. Assem Autom 32(2):1501–1755
9. Shellabear M, Nyrhilä O (2004) DMLS-Development history and state of the art. ESO GmbH Electro Optical Systems, Erlangen
10. Song C, Weng C, Yang Y et al (2017) Development status and trend for the equipment of selective laser melted. Mech Electr Eng Technol 46(10):1–5
11. Wang Z, Huang W, Zeng X (2019) Status and prospect of selective laser melting machines. J Netshape Form Eng 11(4):21–28

12. Buchbinder D, Schleifenbaum H, Heidrich S et al (2011) High power selective laser melting (HP SLM) of aluminum parts. Phys Procedia 12:271–278
13. Bartkowiak K, Ullrich S, Frick T et al (2011) New developments of laser processing aluminium alloys via additive manufacturing technique. Phys Procedia 12:393–401
14. EOS. Metal 3D Printer_DMLS Printer_Additive Manufacturing Systems [EB/OL]. [2020-05-11]. https://www.eos.info/systems_solutions/metal/systems_equipment
15. GE additive. Additive Manufacturing Machines_GE Additive [EB/OL]. [2020-06-12]. https://www.ge.com/additive/additive-manufacturing/machines
16. 3D SYSTEMS metal 3D printers-3D systems [EB/OL]. [2020-06-20]. https://www.3dsystems.com/3d-printers/metal
17. Xi'an Bright Laser Technologies Co., Ltd. Products & services [EB/OL]. [2020-06-20]. http://www.xa-blt.com/home/product/index
18. Beijing E-Plus-3D Additive Technology Co., Ltd. [EB/OL]. http://www.eplus3d.com
19. Wuhan Huake 3D Technology Co., Ltd. [EB/OL]. [2020-06-30]. http://www.huake3d.com/product_detail.asp?Product_ID=48&Product_ParentID=10
20. Guangzhou Laseradd Additive Technology Co., Ltd. [EB/OL]. http://www.laseradd.com/products.aspx?TypeId=74&FId=t3:74:3
21. SLM. High quality industrial metal 3D printers_SLM Solutions [EB/OL]. [2020-06-20]. https://www.slm-solutions.com/products/machines/selectivelasermeltingmachines
22. Aniwaa. Get the right 3D printer or 3D scanner [EB/OL]. [2020-06-20]. https://www.aniwaa.com
23. Cloud 3D printing/3D printing equipment [EB/OL]. [2020-07-03]. http://www.cloud-3dp.com/hardware_goods.php?act=list
24. Flynn JM, Shokrani A, Newman ST et al (2016) Hybrid additive and subtractive machine tools—research and industrial developments. Int J Mach Tools Manuf 101:79–101
25. Hansel A, Mori M, Fujishima M et al (2016) Study on consistently optimum deposition conditions of typical metal material using additive/subtractive hybrid machine tool. Procedia CIRP 46:579–582
26. Chen J, Yang Y, Song C et al (2019) Interfacial microstructure and mechanical properties of 316L/CuSn10 multi-material bimetallic structure fabricated by selective laser melting. Mater Sci Eng A 752:75–85
27. Zhang M, Yang Y, Wang D et al (2019) Microstructure and mechanical properties of CuSn/18Ni300 bimetallic porous structures manufactured by selective laser melting. Mater Des 165:107583
28. Wei C, Li L, Zhang X et al (2018) 3D printing of multiple metallic materials via modified selective laser melting. CIRP Ann 67(1):245–248
29. Wu W, Yang Y, Mao G et al (2019) Design and implementation of SLM system for additive manufacturing of heterogeneous material parts. Manuf Technol Mach Tool 10:13
30. Wu W, Yang Y, Mao G et al (2019) Free manufacturing of heterogeneous materials part by selective laser melting. Opt Precis Eng 27(3):517–526
31. Wu W, Lin W, Liao M et al (2019) Micro-structure analysis of CuSn10 copper alloy/4340 steel heterogeneous material part made by selective laser melting. Mod Manuf Technol Equip (9):152–154
32. Li N, Huang S, Zhang G et al (2019) Progress in additive manufacturing on new materials: a review. J Mater Sci Technol 35(2):242–269
33. Bi G, Sun CN, Chen H et al (2014) Microstructure and tensile properties of superalloy IN100 fabricated by micro-laser aided additive manufacturing. Mater Des 60:401–408
34. Zhang X, Cheng B, Tuffile C (2020) Simulation study of the spatter removal process and optimization design of gas flow system in laser powder bed fusion. Addit Manuf 32:101049
35. Todaro CJ, Easton MA, Qiu D et al (2020) Grain structure control during metal 3D printing by high-intensity ultrasound. Nat Commun 11(1):1–9. Zhang K, Liu T, Liao W et al (2018) Photodiode data collection and processing of molten pool of alumina parts produced through selective laser melting. Optik 156:487–497

Chapter 2
Factors Affecting the Manufacturing Quality of Laser Powder Bed Fusion

2.1 Classification of Factors Affecting Manufacturing Quality

In the study of laser powder bed fusion (LPBF) technology, researchers found that about 130 factors will affect the final manufacturing quality of LPBFed parts. The following factors play a decisive role, which are materials (material category, particle size distribution, fluidity, etc.), laser and optical system (laser type, laser mode, wavelength, power, spot size, etc.), scanning strategy (scanning speed, scanning strategy, scanning hatch space, slice thickness, etc.), the external environment (humidity, oxygen content, etc.), machinery (whether the metal powder layer is uniform or not, the accuracy of motor movement of the manufacturing cylinder, motor stability of the powder laying device, etc.), geometric characteristics (addition of support, different shapes, different placement, etc.), as shown in Fig. 2.1.

The manufacturing quality can be evaluated by measuring the density, hardness, dimensional accuracy, strength, surface roughness and internal residual stress of the manufactured parts.

Because the material has a high melting point, which is prone to thermal demanufacture, and the manufacturing process is accompanied by spattering and balling, the process control of the LPBF manufacturing process is difficult. The key technologies that need to be solved in the LPBF manufacturing process can be found in Sect. 1.2.

2.2 Effect of Process Parameters on Manufacturing Quality

The LPBF manufacturing process includes laser energy absorption and transfer, material heating melting and cooling solidification, microstructure evolution (including porosity and phase transmanufacture), flow caused by surface tension in the molten pool, material evaporation, chemical reaction, and other physical phenomena. When the laser melts the powder, the gas volume in the powder will

© National Defense Industry Press 2024

D. Wang et al., *Laser Powder Bed Fusion of Additive Manufacturing Technology*, Additive Manufacturing Technology, https://doi.org/10.1007/978-981-99-5513-8_2

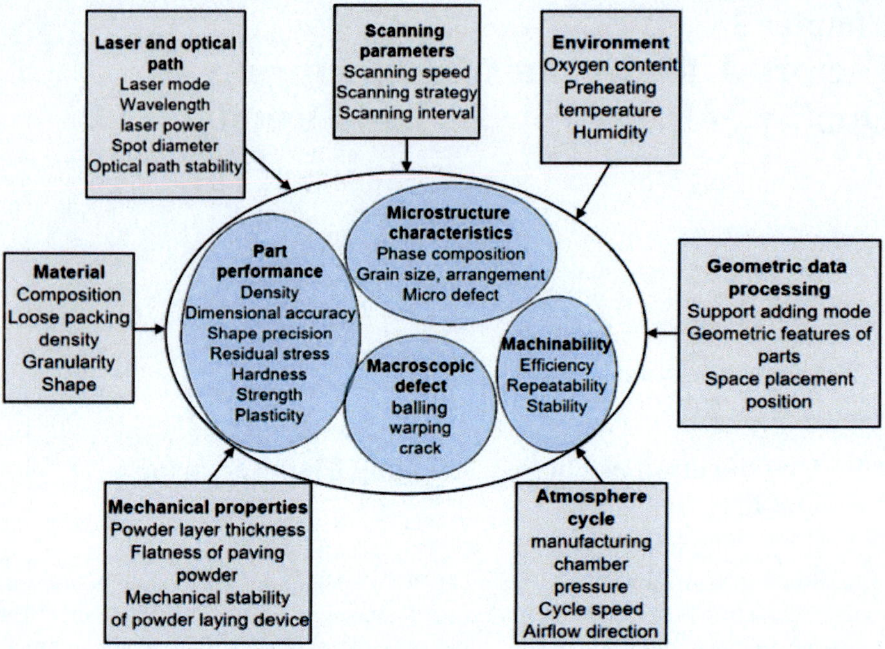

Fig. 2.1 Factors affecting the quality of additive manufacturing (AM) parts

be greatly reduced, and the density in the laser action zone will greatly increase accordingly. It is generally believed that the physical phenomena (melting or vaporization, wetting, oxidation) related to the process take the heat as the main line. Other problems, such as warping, stomata, balling, etc., are directly or indirectly related to energy input. Therefore, understanding the heat input and transmanufacture process in the LPBF manufacturing process is very important to optimize the LPBF manufacturing process and improve the quality of LPBF manufacturing metal parts.

2.2.1 Effect of Optical Parameters on Density

During LPBF manufacturing, the optical parameters which have a great influence on the manufacturing quality include laser power, laser scanning speed, scanning hatch space, defocusing, etc. In the manufacturing process, the influence of optical parameters on manufacturing quality is mutual and interdependent, so the influence of optical parameters on manufacturing quality should be comprehensively considered. Therefore, energy density is introduced to comprehensively evaluate the influence of optical parameters on manufacturing quality.

2.2.1.1 Effect of Laser Power on Relative Density

The laser power mainly affects the energy density in the laser action region. The higher the laser power, the higher the laser energy density in the laser action region. The more sufficient the melting of the material under the same conditions, the less likely to appear the bad phenomena of powder inclusion, and the melting depth gradually increases. However, too-high laser power will cause too-high laser energy density in the laser action zone, which is easy to produce or aggravate the violent vaporization or spatter phenomenon of powder materials, resulting in the manufacture of porous structure and uneven surface, and even causing warping and demanufacture and other defects.

As shown in Fig. 2.2, the surface pore micrographs of 316L stainless steel samples manufactured in LPBF under different laser power are presented. Except for different laser powers, the other process parameters of the two samples are the same. It can be seen from Fig. 2.2 that the surface of the manufactured sample is relatively smooth under the action of a high-power laser, and the scanning track itself does not split. The scanning tracks are neatly arranged, the lapping effect is good, and there are almost no pores (Fig. 2.2a). However, the surface of the sample manufactured at low power is rough, the scanning channel is interrupted, the wettability of the melt is poor, and a large number of pores appear between discontinuous melts (Fig. 2.2b). The main reason is that with the decrease of power, the temperature of the melt decreases, the wettability becomes worse, the spreadability and fluidity of the melt become worse, and a large number of pores that are not filled by the melt are manufactured. Therefore, increasing the laser power is beneficial to the densification of the LPBF manufacturing process.

The effect of laser power on the sample density is also applicable to other materials. In the LPBF manufacturing of titanium alloy (Ti-6Al-4V), the sample density increases with the increase of laser power at a certain scanning speed. However, when the laser power exceeds a certain threshold, the density of the sample decreases. It

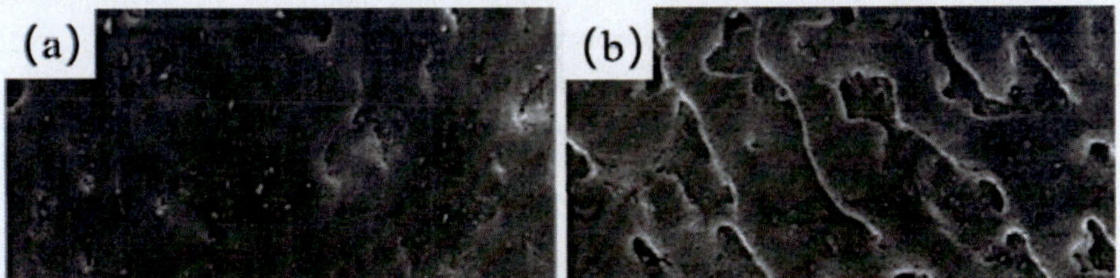

Fig. 2.2 Surface porosity of manufactured specimen under two kinds of laser power: **a** high power, **b** low power

is mainly because the laser power is too large, the energy input value is too large, the surface area of the molten pool is increased, the cooling and solidification time is longer, and the tendency of oxidation and balling is increased. At the same time, it causes the vaporization of the material and the instability of the molten pool, which increases the number of pores in the manufactured parts and reduces the density.

2.2.1.2 Effect of Scanning Speed on Density

Figure 2.3 shows the surface pores of 316L samples at different scanning speeds. It can be seen that the scanning speed has a great influence on porosity. When the scanning speed is low, although there are some small pores in the polishing cross-section of the manufactured sample, there are no macropores and the relative density is higher (Fig. 2.3a). When the scanning speed increases, the number and size of pores increase, and the corresponding density decreases (Fig. 2.3b). As the scanning speed continues to increase, when the reaction time between the laser and the powder is not long enough to melt the powder, pores with large diameters will be manufactured, resulting in a large number of pores in the parts [1] (Fig. 2.3c).

The effect of scanning speed on density is mainly reflected in the wetting and spreading ability of the melt. When the scanning speed is small, the time for the laser to stay on the powder surface is relatively prolonged, so that the molten pool has sufficient time to heat exchange with the surrounding powder, and the melt has better wetting and spreading ability. The density of the sample is higher when the pores between solid phases are filled under the action of surface tension and capillary force. However, if the scanning speed is too small, it is easy to cause too much local liquid phase, and the spatter of the liquid ball occurs under the impact of a high-energy laser. When the liquid ball falls back to the surface, it is already solidified. Therefore, some balls that are larger than the diameter of alloy particles are manufactured on the surface of the cooled molten pool, which leads to balling defects and discontinuity of a new layer of the molten pool. Pores, inclusions and other defects around the ball reduce the density of LPBF samples. At the same time, if the laser beam stays on the powder for a too long time, the surrounding powders are easy to be absorbed

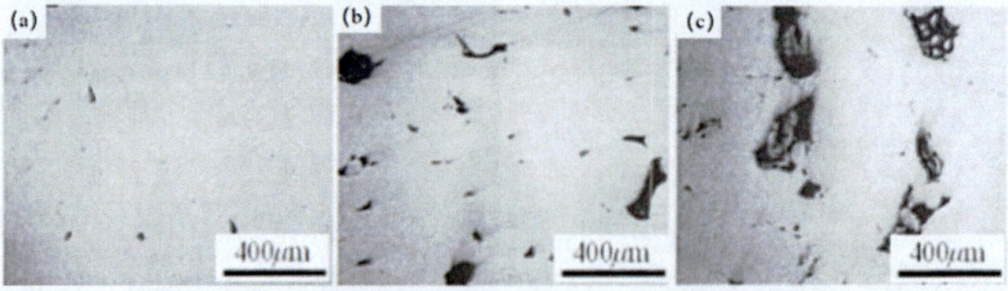

Fig. 2.3 Surface porosity of manufactured specimen at different scanning speeds: **a** low scan speed, **b** medium scanning speed, **c** high scanning speed

into the molten pool. When the laser beam moves to the next position, the amount of powders is not enough, resulting in the manufacture of a pit. When the scanning speed is too high, the input energy is not enough to melt the powders, and the wetting and spreading ability becomes worse, resulting in the existence of unmelted powder in the sample and more pores, which greatly reduces the density.

2.2.1.3 Effect of Hatch Space on Density

The hatch space is the distance between two adjacent scanning lines, as shown in Fig. 2.4. When the scanning hatch space *L* is greater than the width of the molten pool *D*, there is no superposition between the two adjacent scanning lines, and the powder between them cannot be melted, which reflects that there are independent scanning lines on the scanning surface. When the hatch space is less than the width of the molten pool, there is an overlap between the scanning lines, and the change in the superposition rate affects the morphology and density of the scanning surface. Generally speaking, the morphology of the scanning surface is the smoothest when the hatch space is equal to or slightly less than 1/2 of the molten pool width. This is because if the hatch space is too large, the combination between the adjacent scanning lines is not close, there is not enough metal filling between the peak and valley, and the surface of the scanning surface shows a wavy change. If the hatch space is too small, that is when the overlap rate is too high, the phenomenon of accumulation will occur on the scanning surface, and the surface morphology will be uneven.

As shown in Fig. 2.5, we can see that when the overlap rate is too high, the phenomenon of accumulation on the scanning surface and the uneven surface morphology can be seen. When the overlap rate is too small, the overlap between the scan lines is too little, resulting in the combination of the scan lines not being close and uneven surface morphology, showing wavy changes. There is no metal filling between the peaks and valleys.

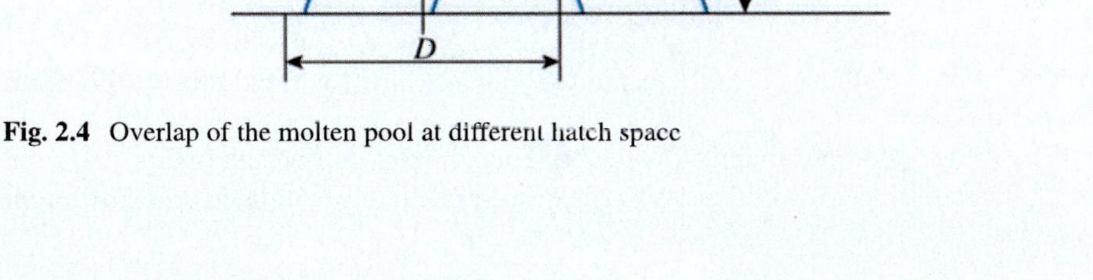

Fig. 2.4 Overlap of the molten pool at different hatch space

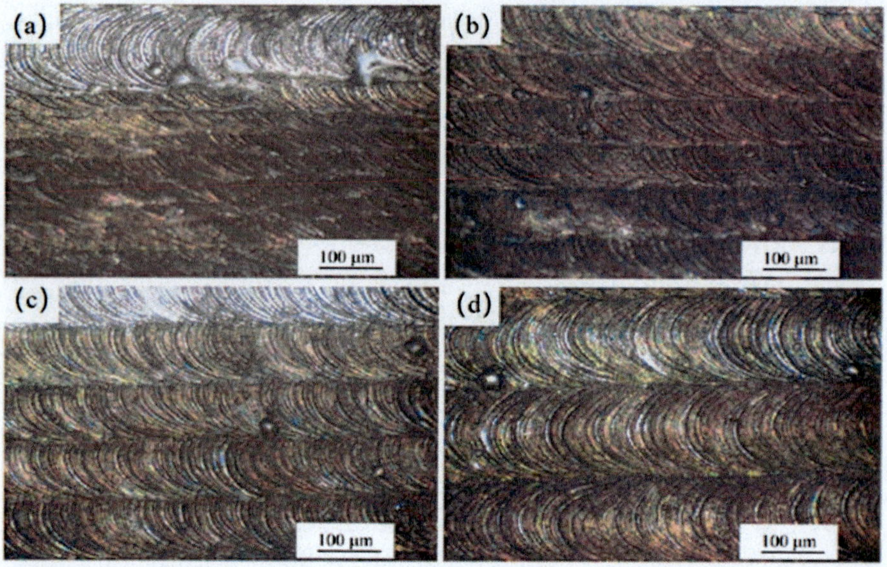

Fig. 2.5 Sample surface with different hatch space: **a**, **b** excessive overlap, **c**, **d** overlap rate is too small

2.2.1.4 Effect of Defocus on Density

The amount of defocus refers to the distance between the focus of the laser and the acting material. It is positive defocus when the focal plane is above the workpiece and negative defocus when the focal plane is below the workpiece. Under the condition that other laser process parameters remain unchanged, the focusing spot and power density can be directly affected by changing the amount of defocus. The manufacturing state of the molten pool can be significantly changed.

$$Z_R = \frac{\pi \omega_0^2}{\lambda_0} \tag{2.1}$$

where, ω_0 is the waist radius; λ_0 is the wavelength; Z_R is Rayleigh length.

$$\omega(z) = \omega_0 \sqrt{1 + \left(\frac{z}{Z_R}\right)^2} \tag{2.2}$$

where, z is the distance from the beam to the waist; $\omega(z)$ is the radius of the beam from the waist.

The value of defocus mainly affects the spot diameter. According to Eqs. (2.1) and (2.2), the radius and power density of the laser spot can be calculated under different defocus.

$$PD = \frac{P}{\pi \omega^2} \tag{2.3}$$

where, *PD* is power density; *P* is laser power; ω is spot radius.

Different defocus reflects different laser spot diameters and laser power densities. The energy density in the center of the beam with a small focus spot is higher, and the distribution of heat affected zone and energy density is more uniform, which is roughly equal to the laser spot. When the defocus value increases too much, the laser spot is too large. The energy density in the center of the spot is too low, so that the bottom of the powder layer can not be melted, resulting in the existence of unmelted powder in the manufactured sample, manufacture of pores and reduction of the density.

2.2.2 Influence of Optical Parameters on Manufacturing Accuracy

Because the laser spot is a circle rather than an abstract point in theory, there will be a laser spot size deviation on the original basis in the LPBF process, as shown in Fig. 2.6a. When manufacturing large-size parts, the influence of size deviation is small because of the large difference in magnitude between part size and spot size deviation. However, when manufacturing small-size parts, especially when they have small feature structures or personalized parts, the errors caused by laser spot size can not be ignored. Therefore, after determining the manufacturing process, it is necessary to use spot compensation or design compensation to eliminate the size deviation.

Another phenomenon that laser affects the accuracy of LPBFed parts is the laser deep penetration, as shown in Fig. 2.6b. Due to the existence of the step effect and transverse hanging structure in slicing treatment, when the laser spot is hit on a layer

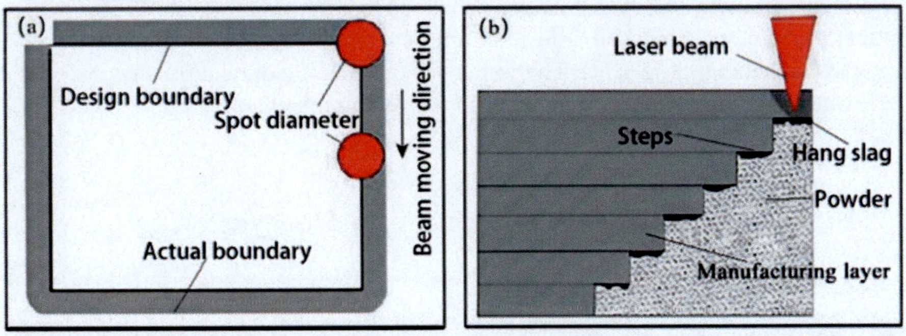

Fig. 2.6 Mechanism of the laser spot and deep penetration affecting manufacturing precision: **a** laser spot effect, **b** deep penetration effect

of powder during manufacturing inclined plane or overhanging structure, there is no solid support under the layer of powder but is supported by powders. On the one hand, the thermal conductivity of the powder is small, which leads to the continuous deepening of the molten pool. On the other hand, under the dual action of the capillary force of the powder and the gravity of the molten pool, the molten metal liquid begins to infiltrate downward, showing an increase in the depth of the molten pool.

Under the influence of the above two factors, the geometric size of the lower side of the manufactured part increases. At the same time, due to the lack of heat, a large number of hanging slag manufactured by the molten pool adhering to the powder is produced on the lower surface, which seriously affects the accuracy of the manufactured part. Due to the constraints of data and structure, the phenomenon of laser deep penetration is difficult to be eliminated. At present, the main method to reduce this influence is to optimize the placement position to weaken the "step effect" and add support to weaken the slag defect [2].

2.2.3 *Effect of Optical Parameters on Mechanical Properties*

The mechanical properties of materials reflect the ability of materials to resist demanufacture, which is related to a series of factors, such as material chemical composition, internal structure, processing methods, etc. For the same material, the main factors affecting the mechanical properties are the preparation method and internal structure of the material. Many studies have shown that the number of defects such as pores, cracks and residual stress will directly affect the mechanical properties of the parts obtained by casting or LPBF technology. Generally speaking, the smaller the porosity and residual stress of the material, the higher the density, and the better the mechanical properties of the material. The LPBF is a process in which the metal powder is melted and cooled rapidly by a high-energy laser. The metal powder needs to absorb enough laser energy to achieve complete melting to eliminate pores and other defects to the maximum extent and obtain LPBF samples with higher density. However, with the increase of input energy, it will also produce greater residual stress. Therefore, the optical parameters should be adjusted comprehensively to obtain optimal comprehensive mechanical properties. The impact of the specific optical parameters on the performance will be described in detail in the remaining chapters.

2.2.4 *Influence of Scanning Strategy on Manufacturing Quality*

The laser scan line is the smallest component of part manufacturing. To reduce stress concentration, the long scan line is divided into various shape regions in the manufacture of overlap, and then these areas are scanned in a different order (i.e. different scanning strategies are manufactured). At present, there are many kinds of AM scanning strategies in the market, but all kinds of designs are inextricably linked with the traditional welding process, among which the focus of the scanning strategy is to control the lap and reduce the stress concentration. Too much overlap in the manufacturing process of metal parts will directly lead to the concentration of thermal stress, causing excessive demanufacture of the parts or leading to cracks in the parts. At the same time, the local heat input is too large, and the internal defects of the parts increase, resulting in the decline of the mechanical properties of the parts. To avoid the above problems, the scanning strategy of metal AM has changed a lot from the initial simple strip scanning strategy to gradually evolved line scanning, arc, chessboard, island and other scanning strategies. The effects of different scanning strategies on manufacturing quality are also different, so further analysis of scanning strategies is needed.

2.2.5 *Effect of Powder Layer Thickness on Manufacturing Quality*

Because the LPBF rapid prototyping process is a laminated manufacturing process, theoretically, the thinner the powder layer thickness, the higher the manufacturing accuracy. However, the use of thin layer thickness will also greatly increase the manufacturing time and reduce the manufacturing efficiency, thus leading to an increase in the cost of the manufacturing process. If the powder layer is too thick, the powder material can not be completely melted, which will lead to poor interlayer bonding and defects in the manufactured parts after accumulation.

Figure 2.7 shows the surface and internal morphology when the powder layer thickness increases from 0.02 to 0.05 mm. It can be seen that the surface morphology is very good when the powder layer is suitable, the surface is relatively flat and almost no voids can be seen. The internal gaps are very few. However, with the increase of powder layer thickness, gaps gradually appear, the molten channel begins to break, and the surface becomes uneven. This is because when the thickness of the powder layer is small, the thickness of the laser-melted powder is relatively small. On the one hand, the complete melting of the powder can be realized. On the other hand, sufficient heat can further fully remelt the upper part of the previously solidified layer to make it flatter. The joint action of the two aspects makes the whole manufacturing surface flatter. With the increase of the thickness of the powder layer, the energy needed for powder melting also increases, and the energy provided by the laser gradually becomes relatively inadequate. When the energy can not maintain the melting of the whole powder under the action of slight disturbance and surface tension, the molten track will break off and there is a gap between the front and back, which can be proved by the appearance of the gap in the middle of the molten channel in Fig. 2.7c.

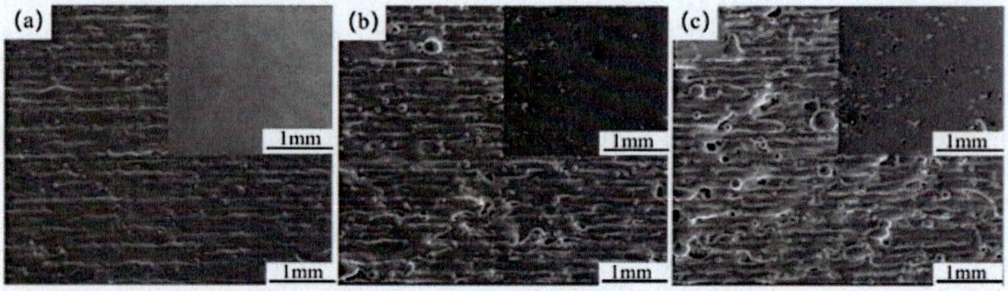

Fig. 2.7 Surface and internal morphology of specimens with different powder layer thicknesses: **a** 0.02 mm, **b** 0.035 mm, **c** 0.05 mm

In the actual processing, the thickness of the powder layer should be designed according to the laser power provided by the equipment laser to avoid holes, faults, and other defects caused by insufficient energy input.

2.3 Influence of Powder Laying Device on Manufacturing Quality

2.3.1 Structure of Powder Laying Device

Powder laying is an important step in the LPBF process, and the powder laying device is also one of the key components of L-PBF manufacturing equipment. There are two common ways of laying powder: the hopper type of powder feeding from top to bottom and the cylinder type of powder feeding from bottom to top [3] (Fig. 2.8). The powder laying device, as the part in the powder laying system that is in contact with the powder, must be able to make the powder flat, compact, and evenly on the manufacturing cylinder. Only by rapid prototyping on this basis, the manufactured parts with high density and high precision can be obtained and the manufacturing process can be ensured smoothly. In the actual processing process, although the thickness of the powder layer should be equal everywhere in theory and the plane after powder melting should also be in the same horizontal plane, the surface of the scanning area is not a complete plane due to the change of instantaneous conditions in the machining process. Sometimes, due to the deterioration of processing conditions, the protruding part of the surface of the manufactured parts may even far exceed the thickness of the powder, which may cause a collision between the powder-laying device and the solidified parts. Therefore, the powder-laying device not only needs to obtain a flat, uniform and compact powder layer but also needs to ensure that it will not damage the manufactured parts.

With the development of technology, a variety of flexible powder-laying devices gradually replace rigid powder-laying rollers. In the author's laboratory, a kind of rigid paving powder scraper with a pre-compaction function and a flexible paving

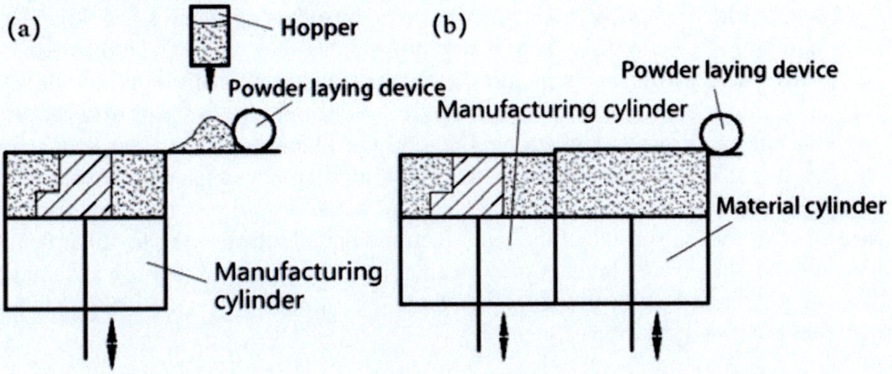

Fig. 2.8 Schematic diagram of two powder spreading methods: **a** hopper type, **b** cylinder type (double cylinder)

brush has been designed and applied to different types of experiments and material manufacturing, and good manufacturing results have been obtained.

The principle of the rigid powder laying scraper with pre-compaction function is that the motor drives the bracket to move between the powder laying cylinder and the manufacturing cylinder, thus driving the scraper and the pressing plate to promote the powder laying action. The scraper assembled in front of the pressing plate (along the powder laying direction, as shown in Fig. 2.9) first flattens the powder accumulated in front of the powder laying device along the powder laying direction and obtains the powder of uniform thickness h_g on the manufacturing cylinder plane. Then, under the push of the pressing plate with a certain radian, the powder layer is pushed from the thickness h_g to the thickness h_y, so that the powder layer thickness reaches the process set thickness h and the powder layer density is increased at the same time [4].

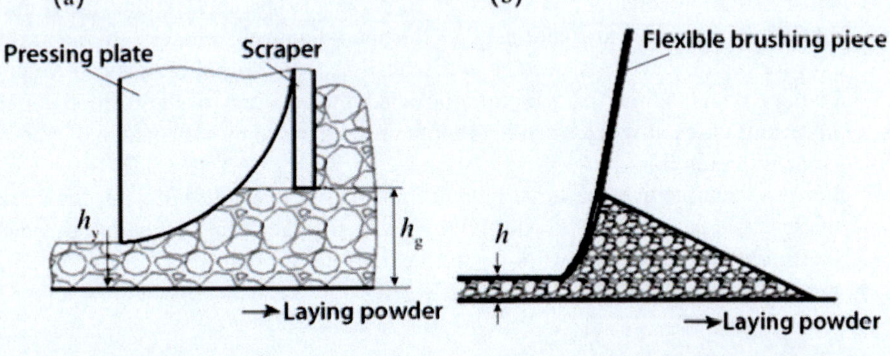

Fig. 2.9 Design schematic diagram of two powder spreading devices: **a** powder spreading scraper with pre-compaction function, **b** flexible paving and painting

The principle of a flexible paving brush is that the ultra-thin stainless steel sheet of 0.01–0.06 mm is used to manufacture a painting group to promote the movement of the powders. The brush is manufactured as follows: a laser is used to cut the stainless steel sheet into a brush piece of equal width 3–5 mm. There is a gap between the two adjacent brush pieces, which depends on the diameter of the laser spot used when cutting. The biggest feature of the flexible brush powder laying device is to use the elasticity of a thin metal sheet to promote the powder. At the same time, it can cross the uneven surface of the manufactured parts in the powder laying process to ensure that the powder layer is smooth and uniform, which will not cause damage to the manufactured parts. In theory, the more the number of powder brushes used, the better the powder laying effect.

2.3.2 Selection of Powder Laying Device

When choosing a rigid or flexible powder-laying device, it needs to be judged according to the structural characteristics of the manufactured parts. The main selection principles are as follows:

(1) For the parts with complex and changeable shapes on the scanning plane, the manufacturing surface fluctuates greatly, and the edges sometimes warp slightly, so the flexible paving brush device has better adaptability than the rigid powder laying device.

(2) In the case of using the pre-compaction powder laying device, when there is friction between the powder laying device and the surface of the formed parts, the whole powder laying device will be slightly raised, which will cause the rigid powder laying device to vibrate, resulting in a whole row of wavy powder laying surface, and there is a vicious circle effect. On the other hand, the flexible paving brush is in the shape of a tooth sheet, and its influence range is only the width of the brush piece when it passes the bump portion.

(3) Most of the personalized precision parts have a drape surface, and the drape surface is easy to warp seriously in the manufacturing process. If the rigid powder-laying device is used, it will cause a collision between the powder-laying device and the formed parts to reduce the accuracy of the formed parts. In serious cases, it will cause the manufactured parts to be damaged and cannot be manufactured.

(4) For the manufactured parts with slice thickness between 25–35 μm, the slice thickness is close to the powder laying limit, whether using a flexible powder laying device or rigid powder laying device, there is not much difference in powder layer density. And it has little effect on the density of manufactured parts.

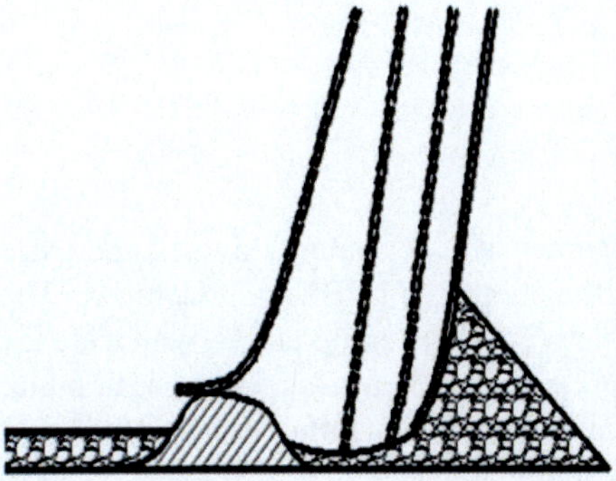
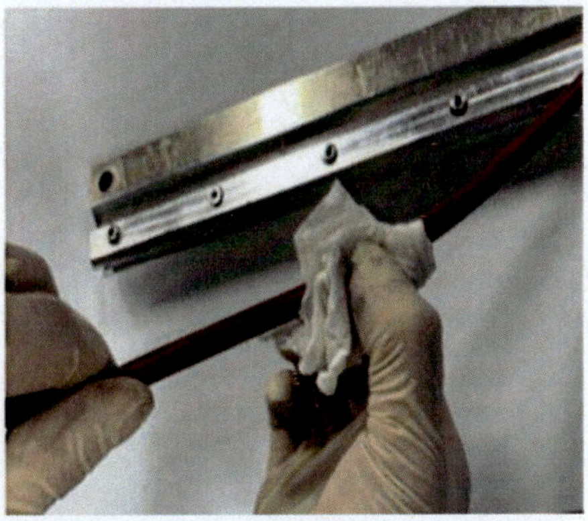

Fig. 2.10 Powder defects of flexible paving (left) and silica gel scraper (right)

(5) In the overall structure, the powder laying device with pre-compaction function has a simple structure. The main parts (scraper and pressing plate) are relatively easy to process, while the flexible paving brush manufacturing requirements are more stringent and maintenance is more time-consuming.

Although the adaptability of the flexible paving brush is high, there are also some shortcomings. As shown in Fig. 2.10, the brush is demanufactured when pushing the powders. When there is always powder in front, the demanufacture is more stable and the powder is flatter. The greater the demanufacture after the obstacle, the greater the spring-back potential energy. After crossing the obstacle, the quality of the powder is not enough to prevent the steel sheet from springing back, so the brush will bounce back to the original shape and the powder will be flattened again. However, the transition zone will manufacture a "pit" lower than the normal height of the powder layer. When the potential energy of the steel sheet is very large, it is possible to eject forward and raise the powder.

To improve the deficiency of rigid brush, the mainstream equipment uses the silica gel strip shown in Fig. 2.10 instead of a rigid brush to complete the powder laying work. The use of a silica gel strip can further simplify the structure of the powder laying device. At the same time, it is easier to clamp, which can not only ensure that the powder laying is smooth, uniform and compact but also avoid a rigid collision with the parts in the powder laying process because of its flexibility. Therefore, using a silica gel strip as a scraper is a better solution at present.

2.4 The Influence of Atmosphere on Manufacturing Quality

2.4.1 *Effect of Oxygen Content*

Oxidation is a common problem in the LPBF process. A continuous high-temperature liquid-phase molten pool is manufactured in the process of LPBF manufacturing. The liquid phase metal in the molten pool is very active at high temperatures, which is easy to react with oxygen to manufacture metal oxides, or even burn (such as pure titanium powder). Therefore, the atmospheric environment of the manufacturing cavity must be strictly protected in the process of LPBF manufacturing.

The effect of oxygen content on the quality of LPBF manufacturing can be explained by the interfacial thermodynamics. The molten pool manufactured in LPBF is the process that liquid phase metal wets solidified solid metal. The three-phase equilibrium diagram of the liquid phase, solid phase and gas phase is shown in Fig. 2.11.

The wettability of liquid metal to solid metal is mainly determined by its contact angle and liquid phase viscosity. It can be seen that the liquid metal can well wet the solid interface when the contact angle is less than 90°. And when the contact angle is greater than 90°, the liquid metal will contract to spherical, as shown in Fig. 2.12.

In the process of the development of the molten pool, the free energy of the system develops to the lowest direction. When the molten pool develops to the surface of metal oxide, surface-free energy is much smaller than that of the liquid metal and gas phase. Therefore, because of the trend of system-free energy, liquid metal is difficult

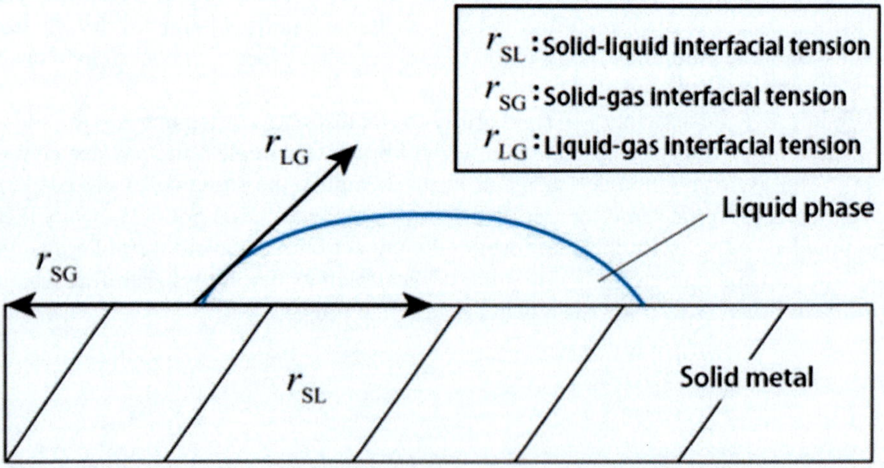

Fig. 2.11 Three-phase equilibrium diagram

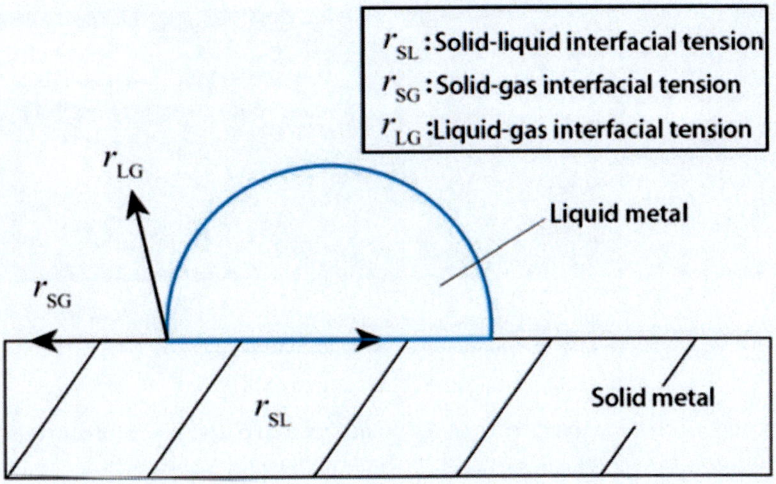

Fig. 2.12 Balling occurs

to wet metal oxide under general laser power. However, spherical surface free energy is the lowest, which leads to balling effect, as shown in Fig. 2.13.

Because LPBF is a layered manufacturing process, the small balls produced by metal oxides will seriously affect the bonding between the scanning lines, and even delamination will occur when the oxidation is serious due to the too small bonding force between the upper and lower layers. And the joint of the two layers will be broken under the action of cumulative thermal stress, which greatly reduces the

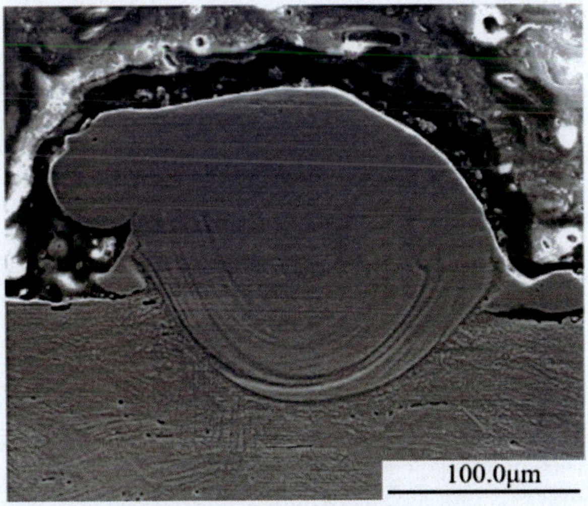

Fig. 2.13 Balling phenomenon

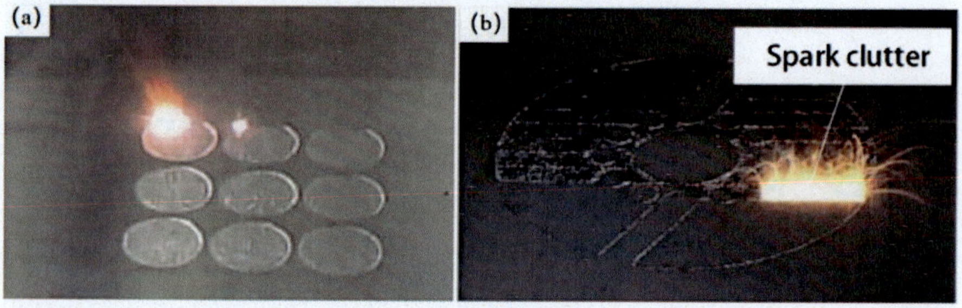

Fig. 2.14 LPBF manufacturing sparks: **a** laser and powder when the stability of the spark, **b** the spark when a laser interacts with a powder in an unstable manner

density and mechanical properties of the manufactured parts. Therefore, in the manufacturing process, it is necessary to input the appropriate protective gas to avoid oxidation.

Figure 2.14 shows the instant sparks produced by laser and powder under different oxygen content during the manufacturing of LPBF. Figure 2.14a shows the sparks during the stable interaction between laser and powder. It can be seen that the sparks diverge in a straight line. Figure 2.14b shows the phenomenon of spatter or serious oxidation when the laser interacts with the powder. It can be found that the spark no longer presents a straight line and becomes messier, and a lot of spatter material falls into the manufactured surface, which will have varying degrees of influence on the density and surface roughness after continuous accumulation.

To ensure the low oxygen content in the manufacturing chamber, the inert protective gas needs to be continuously injected during the manufacturing process to ensure the manufacture of a positive pressure environment in the manufacturing chamber. At the same time, the atmosphere circulation filtration system is used to filter the atmosphere of the manufacturing chamber to blow away the plume in time, so the powder bed is very clean after laser scanning, and there are few soot and spatter particles.

When there is no gas protection and circulating purification, there is no gas flow in the manufacturing chamber. As a result, the manufactured gas is scattered near the manufacturing area, resulting in an obvious layer of soot and spatter particles on the powder bed [5], as shown in Fig. 2.15.

2.4.2 *Effects of Different Protective Gases*

To prevent metal products from oxidation in the process of LPBF manufacturing, it is common and necessary to be protected by inert gas. The main inert gases used in the process of LPBF are nitrogen, argon and helium. Different inert gases also have different impacts on manufactured quality. For example, the use of nitrogen in the manufacturing process of Ti6-Al4-V, in which titanium will react with nitrogen to

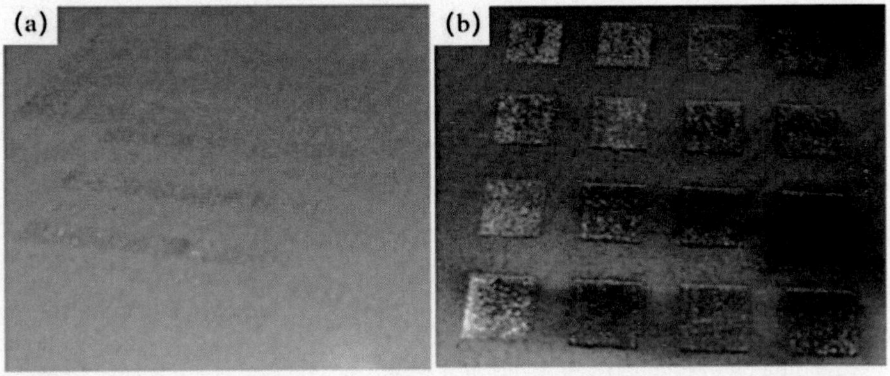

Fig. 2.15 The surface of the powder bed after laser scanning: **a** powder bed surface with gas protection and circulation purification, **b** powder bed surface without gas protection and circulation purification

produce TiN, can strengthen the hardness and strength of the sample. Therefore, the influence of protective gas on manufacturing quality also needs to be considered.

2.4.3 Atmosphere Cycle Purification

When stainless steel powder (including new powder and used powder) and pure titanium powder are manufactured under the protection of nitrogen and argon respectively, it will be found that metal plume will be produced instantly when the laser interacts with the powders. The main source of metal plume is caused by the combustion and vaporization of carbon elements, low melting point alloy elements and impurity elements in metal powder. The long-term repeated use of the powder aggravates the accumulation of metal plume, which means that although the new stainless steel powder produces a small amount of metal plume, the metal plume will pollute the powder. And the metal plume is more and more serious when the laser interacts with the powders for a long time. At present, all manufacturers or scientific research institutions are still unable to fundamentally solve the problem of metal plume. The main negative effects of metal plume include polluting the lens and polluting the powders.

When scanning at low speed, the laser energy input is large and the amount of metal plume produced is also large. The metal plume quickly makes the lens stick with a layer of metal plume particles, resulting in serious power attenuation when the laser passes through the lens. The lens will become hotter and hotter until it breaks down. The pollution of metal plume to the lens results in insufficient power of laser radiation on the surface of the powder bed and insufficient melting of powders, which leads to poor quality of manufactured parts and even inability to complete printing. Therefore, metal plume pollution has a great impact on the manufacturing efficiency and quality of LPBF. In addition, after the metal plume is produced, a small part of the metal plume is blown out of the powder bed by protective gas, and most of them still falls on the surface of the unused powders, which aggravates the pollution of the powder and affects the quality of the next printed parts.

To reduce the influence of metal plume, the mechanical design of the manufacturing chamber or protective cover can be optimized: (1) The lens is as far away from the surface of the powder bed as possible. (2) In the mechanical design, the lens is convenient to install so that when it is contaminated by metal plume, it can be quickly removed and then quickly installed after wiping to avoid too long downtime of the manufacturing machine. (3) A gas circulation filter is added to the manufacturing chamber, and after the atmosphere in the manufacturing chamber passes through the filter, the metal plume and vaporization products are filtered out, and the clean gas is obtained and then passed into the manufacturing chamber.

Figure 2.16 shows a summary of the relationship between various defects in the process of LPBF manufacturing. From the figure, it can be found that defects such as spatter, insert, metal plume and material pollution are closely related to the control of the manufacturing process, and affect the mechanical properties of the LPBF manufactured parts. Moreover, the defects influence each other. For example, the generation of metal plume leads to the attenuation of laser power, the insufficient melting of powder and the deterioration of manufacturing surface quality, which affects the stability of the LPBF manufacturing process and leads to spatter. Spatter is the main source of surface inserts, which finally leads to a great decline in the mechanical properties of LPBFed parts. Therefore, in the manufacturing process, pollutants such as spatter and metal plume should be filtered and purified through the atmospheric circulation system in time to reduce the content of impurities in the formed parts and improve the manufacturing quality.

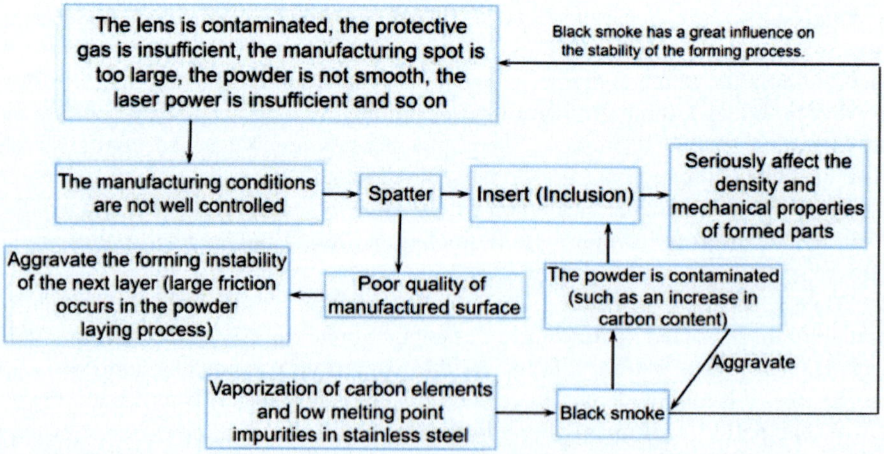

Fig. 2.16 Relationship between various defects in the manufacturing process of LPBF

2.5 Influence of Other Factors on Manufacturing Quality

2.5.1 Powder Material

2.5.1.1 Powder Properties

Metal or alloy powder materials are usually used in LPBF manufacturing. The manufacturing process is the rapid melting of powder materials under the action of a high-energy laser. According to the manufacturing characteristics of LPBF and the application of parts, the material is required to ensure the uniformity of the composition and properties of the manufactured parts. According to the manufacturing requirements, the composition, and physical and chemical properties of metal powders should be comprehensively considered to select or prepare suitable powder materials.

At present, the metal powder materials used include mixed powder, pre-alloyed powder and elemental metal powder. Among them, the mixed powder is obtained by mixing the powder particles of various components evenly by mechanical ball milling. However, in the manufacturing experiment, because the mixed powders will be separated in the process of powder transfer, especially during powder laying and in the process of scanning, the melting of the powders is a selective dynamic solidification process. And the degree of thermal action varies from place to place. Therefore, the composition of the LPBF samples made of mixed powder is uneven, which affects the performance of the LPBFed part. For the pre-alloyed powder, that is, the powder and elemental metal powder prepared by atomization, there is no uneven composition. Because the following two kinds of powders are more suitable materials for the study of LPBF technology. Figure 2.17 shows the sample manufactured by TiNi alloy mixed powder and pre-alloyed powder. It can be seen that the sample manufactured by mixed powders has defects. However, the sample manufactured by pre-alloyed powder has a better manufacturing effect without obvious defects.

Atomization methods can be divided into gas atomization and water atomization, both of which are different in powder particle morphology and oxygen content. The

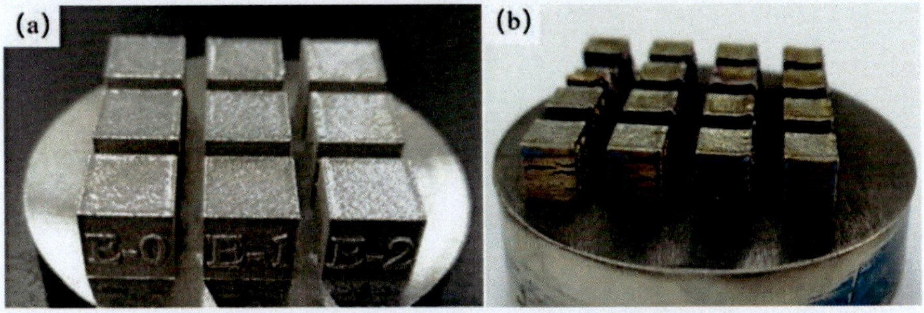

Fig. 2.17 TiNi samples manufactured with different powders: **a** mixed powder, **b** pre-alloyed powder

difference between the two micro-morphologies is shown in Fig. 2.18. The particle shape of 316L stainless steel powder by water atomization is irregular. However, 316L stainless steel powder manufactured by gas atomization is spherical. Because the morphology of the powder particles affects the fluidity of the powder, the fluidity of the powder affects the uniformity of powder laying in the processing process. If the powder is not uniform, it will lead to uneven metal melting in each part of the scanning area and uneven development of the molten pool. As a result, the microstructure of the manufactured parts is uneven, that is, the structure of some areas is dense, while gaps may appear in other areas. The effect of powder morphology on the manufacturing quality can be seen in Fig. 2.19. Therefore, compared with an irregular granular powder, the spherical granular powder is more beneficial to obtain denser LPBFed parts.

In the selection of the LPBF materials, it is also necessary to consider the relationship between the characteristics of powder materials and laser action, that is, the

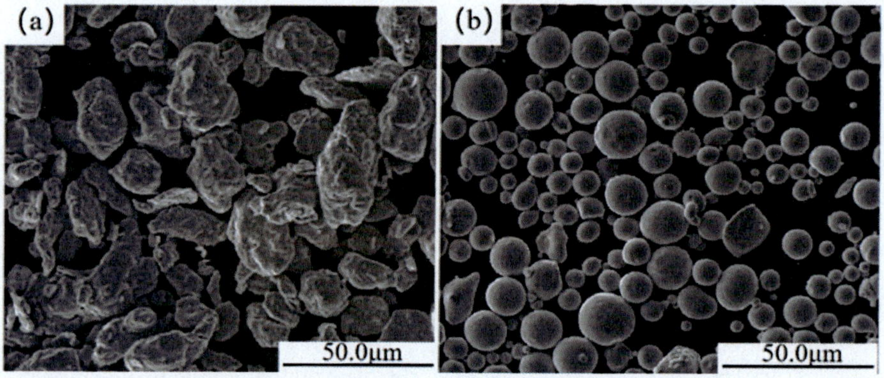

Fig. 2.18 Microstructure of 316L stainless steel powder: **a** water atomization, **b** gas atomization

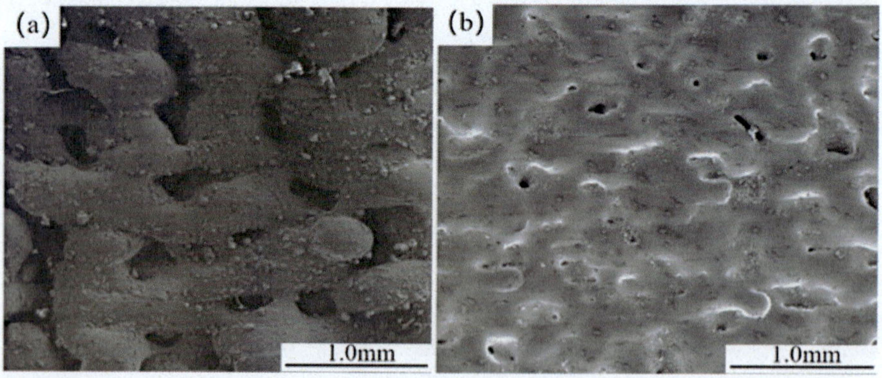

Fig. 2.19 Surface morphology of 316L stainless steel L-PBF specimen: **a** water atomization, **b** gas atomization

Fig. 2.20 Copper alloy L-PBF specimen

laser absorptivity of metal powder. The laser absorption of each component in metal powder is different. On the one hand, it is caused by the difference in their thermophysical and chemical properties. Thermophysical properties include melting point, boiling point, the heat of melting, heat of vaporization, emissivity, thermal conductivity, coefficient of thermal expansion, atomic structure, resistivity, etc., especially the resistivity of metals. On the other hand, the same metal material has different laser absorptivity for different kinds of lasers.

This mainly depends on the laser wavelength. Generally speaking, with the increase of the laser wavelength, the reflectivity of metal materials to the laser increases, that is, the absorptivity decreases. For example, when the LPBF manufacturing copper alloy powder is carried out by using a fiber laser with a wavelength of 1064 nm, the surface quality of the manufactured sample is poor due to the lack of energy adsorbability due to the high reflectivity of copper alloy to laser, as shown in Fig. 2.20.

The particle size and distribution of the powders and the particle shape of the powder determine the loose density of the powders. On the one hand, small particle size can increase bulk density. On the other hand, a single powder particle size is not conducive to improving the loose density. Therefore, the powder particle size distribution is particularly important. The experimental results show that the particle size with Gaussian distribution is the most beneficial to increase the loose density. At the same time, the particle shape of the powder is also a major influence factor, such as spherical powder, its packing density is small, so a better combination of coarse and fine powder is needed. The loose density has a direct effect on the manufacturing of LPBF. Under a large enough laser energy density, the high loose density provides enough metal for laser melting to manufacture a molten pool. The higher the loose density, the easier it is to achieve densification.

2.5.1.2 Oxygen Content of the Powder

Because the LPBF manufacturing process has strict requirements on the oxygen content, it is necessary to control the oxygen content in the powder material in addition to controlling the oxygen content in the atmosphere in the manufacturing chamber. On the one hand, the oxygen in the powder material comes from the residue of incomplete reduction in the production process. On the other hand, because the metal powder particles are small, they react with oxygen in the air to manufacture an oxide film to cover the metal powder particles. The oxides remaining in the manufactured powder oxidize the liquid metal under the action of high temperature, which increases the surface tension of the liquid molten pool, enhances the balling effect, reduces the quality of the molten pool, and affects the subsequent manufacturing, thus affecting the internal structure of the parts.

2.5.1.3 Powder Impurity

Defects such as spatter, oxidation and balling often occur in the melting process of metal powder, resulting in metal powder contamination. The powder impurities are mainly composed of some large particles, metal balling products oxidized on the powder surface, spatter products and slag dropped during the manufacturing of overhanging structure. If the powder is used repeatedly for a long time without sieving, the surface of the manufactured parts will be embedded with a lot of inclusions. The composition of these inclusions is the same as that of the waste slag produced in the steelmaking process. After analysis, the main inclusions are SiO_2, CaO, MnO, etc. The inclusions in the internal structure of the LPBF manufactured parts seriously affect the mechanical properties of the metal parts and may threaten the stable operation of the powder-laying device because the particle size of balling particles and spatters is generally several times larger than the diameter of the powder particles. However, the height of the powder-laying device and the surface of the formed part is generally only tens of microns (set layer thickness). When the powder-laying device encounters the above inclusions, mechanical failures, such as sticking, are easy to occur. Therefore, eliminating the intercalation and reducing the oxide particles in the powder are very important for the laser melting of high-quality metal parts in the powder bed. The reduction of impurities in materials can be achieved by the following methods: (1) Using single-phase powder as far as possible; (2) Stabilizing the oxygen content in the manufacturing chamber at a low level as far as possible; (3) Sieving and drying powder materials periodically.

2.5.2 *Placement of Parts*

From the principle of LPBF manufacturing, because of the special manufacturing mode, the performance of the manufactured parts will be anisotropic. The placement of the parts will also have an obvious impact on the manufacturing quality. Therefore, to ensure manufacturing quality, the placement of the manufactured parts should be considered in LPBF manufacturing.

In the laser manufacturing process, when the laser irradiates the powder layer, if there is a solid metal below the powder layer, the heat will be transferred from the molten pool to the lower structure through the solid body. And, a portion of the lower solid structure will be remelted to manufacture a solid bond. Because of the physical structure, the heat of the molten pool can be transferred effectively and the molten pool can be solidified. If the part has a cantilever part, at least a small portion of the area below the molten pool will be unmelted powders, which are much lower in thermal conductivity than solid metal. As a result, the heat from the molten pool will be retained for longer, resulting in more powder sintering around. Due to the poor thermal conductivity of the powders, the heat will gather in the powders below the cantilever part, which can not be fully melted, resulting in adhesion to the bottom surface of the cantilever area and demanufacture of the lower surface of the cantilever part or very poor surface roughness.

The emergence of the overhanging structure is usually determined by the way the parts are placed during manufacturing. When there is a large horizontal overhang part (Fig. 2.21a), not only a large amount of support is needed, but also the surface quality becomes very poor after manufacturing. And the long-line scanning may cause the bottom surface to warp seriously and break away from the support. Therefore, a large area surface should be avoided as a horizontal overhanging surface, unless it can be placed directly on the manufactured substrate without support. The method of setting the surface with a small area as the bottom surface (Fig. 2.21b) has the minimum amount of support. However, the size of the Z axis is the largest and the manufacturing time is the longest. Figure 2.21c shows an inclined placement, the α angle is generally a reliable manufacturing angle and, in special cases, the minimum manufacturing angle. Inclined placement not only needs fewer support structures but also greatly reduces the size of the Z-axis compared with the previous method. Another advantage of inclined placement is that the bottom surface of the part is supported by linear contact, while the two types shown in Fig. 2.21a, b are surface contact which is much easier to remove the support during the placement as shown in Fig. 2.21c. Inclined placement ensures the structural integrity and functional realization of parts by sacrificing the quality of unimportant parts (the surface quality is worse than vertical manufacturing), so this arrangement is generally used in the manufacturing of complex parts.

In general, the supporting structure needs to be added to the part where the horizontal angle is less than 45°, which is usually called the lower surface. And the surface roughness of the lower surface is usually rougher than that of the vertical wall and the upper surface. At this time, the lower surface of the parts is particularly important, it can not only optimize the performance of the task parts but also determine whether the manufacturing is successful or not. When the parts have a variety of placement options, they should be selected to realize the ideal self-support placement of the parts to reduce the post-processing cost as much as possible.

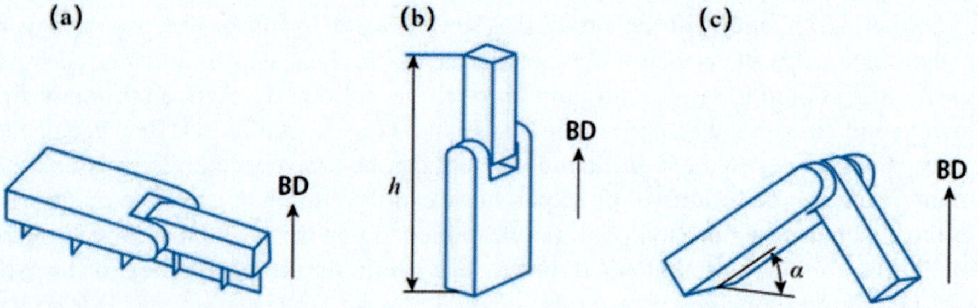

Fig. 2.21 Different placement modes of LPBF in the manufacturing process: **a** horizontally, **b** vertically, **c** set at an angle

References

1. Mai SZ (2016) Study on laser selective melting manufacturing technology and properties of personalized CoCr alloy crown fixed bridge. South China University of Technology
2. Wang YD (2015) Study on laser selective melting process, microstructure and properties of CoCrMo alloy. South China University of Technology
3. Lu JB (2011) Design optimization and process research of selective laser melting direct forming of personalized precision metal parts. South China University of Technology
4. Luo ZY (2011) Study on selective laser melting manufacturing technology and influencing factors of thin-walled parts. South China University of Technology
5. Bai YC (2018) Study on the mechanism and control of laser selective melting forming of maraging steel. South China University of Technology

Chapter 3
Study on Single-Track, Multi-track and Multi-layer Manufacturing

3.1 Foundation and Control of Single-Track Manufacturing

The performance of parts manufactured by laser powder bed fusion (LPBF) is largely based on the line surface body manufacturing method and the single-track and single-layer manufacturing quality of materials. The defects formed on the single-track directly affect the manufacturing quality of the single-layer. This defect point will also accumulate during the next layer of laser scanning. When the defect is not eliminated, it will form an internal defect in the part. Therefore, this section will introduce the basis and control of LPBF single-track manufacturing.

3.1.1 Single-Track Manufacturing Foundation

During single-track manufacturing, the micro molten pool is not solidified at a specific position fixed on the manufacturing substrate but moves with the movement of the laser beam. The position irradiated by the laser beam is always the metal in the molten state, while the position far away from the laser beam solidifies rapidly [1]. At this time, the solidification of the micro molten pool is affected not only by the surrounding powders but also by the molten pool in the front melting state and the metal solidified at the rear end. At this time, the laser power and scanning speed have the greatest influence on the single-track in the process parameters. The laser power determines the instantaneous energy value obtained by the powder, while the scanning speed determines the time for the powder to continuously obtain energy. In addition, the defocusing amount and powder layer thickness also has a great impact on the single-track manufacturing effect. When the processing plane is near the focus, the energy is concentrated, which is conducive to the formation of a thin and straight single-track scanning molten. For the material, the continuous single-track can be obtained through a reasonable configuration of laser power and scanning speed within

© National Defense Industry Press 2024

D. Wang et al., *Laser Powder Bed Fusion of Additive Manufacturing Technology*, Additive Manufacturing Technology, https://doi.org/10.1007/978-981-99-5513-8_3

a certain layer thickness. However, for the powders with different particle shapes and particle sizes, the morphology of the melting track will be different.

In the process of a single-track molten pool, the upper and lower parts of the molten pool are separated by the matrix. Under the molten substrate, the force exerted by solid particles on liquid particles is greater than that between liquid particles. The wetting angle is acute, and the shape of the molten pool is fan-shaped. Above the matrix, the wetting angle tends to be obtuse, and the shape of the molten pool tends to be spherical because the force exerted by solid powder particles on liquid particles is less than that between liquid particles.

Under the action of the above two forces, the shape of the molten pool changes and becomes oval. Figure 3.1 is a schematic diagram of the single-track manufacturing section. The molten pool is divided into upper and lower parts. And the upper part of the base (powder melting zone) includes the molten pool manufacturing zone and powder-free zone. The molten pool under the substrate belongs to the laser remelting part, which is used to ensure the bonding strength between layers during the LPBF manufacturing process. It can be seen from Fig. 3.1 that the elliptical molten pool above the matrix belongs to the powder melting and solidification manufacturing area, while the molten pool below the matrix belongs to the remelting part of the previous molten pool. Although these two parts are approximate ellipses, their causes are quite different. For the elliptical molten pool on the upper part of the matrix, the main reason is the surface tension of the liquid molten pool formed by the laser-melted powder, while the elliptical molten pool on the lower part of the matrix is mainly caused by the Gaussian energy distribution of the laser energy. The energy distribution of laser energy in the radial direction of the circular spot can be expressed as the formula:

$$e = \frac{4P\sqrt{d^2 - 4y^2}}{\pi d^2} \quad \left(y \le \frac{1}{2}d\right) \tag{3.1}$$

where, P is laser power; d *is* laser spot diameter; e is laser energy; y is the distance from the flare to the center point, and $y \le 12d$.

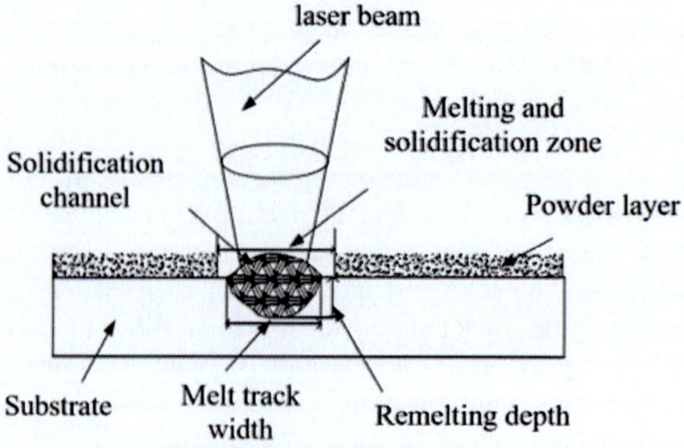

Fig. 3.1 Schematic diagram of single-track manufacturing section

3.1.2 Single-Track Manufacturing Control

3.1.2.1 Effect of Substrate Thermal Conductivity

When the laser melts the powder layer, most of the laser energy absorbed by the powder diffuses through the heat conduction of the substrate. However, only a small part is used for the melting of the powder itself. Therefore, as the main thermal conductor, the substrate material directly affects the single-track manufacturing quality. At the same time, different substrate materials and manufacturing materials have different bonding properties. A good bonding property is conducive to enhancing the deformation resistance in the manufacturing process and avoiding the defects such as warping deformation or cracking caused by stress concentration during the manufacturing process. The main reasons for the difference in single-track forming quality caused by substrate materials are:

(1) The thermal conductivity of substrate materials is different. The thermal conductivity is too high and the heat diffusion rate is too fast, resulting in that the residual heat is not enough to completely melt the original powder to form an intermittent single-track. With the influence of the surface tension of the liquid metal, spheroidized particles are formed;
(2) Different wettability of liquid raw materials to substrate materials. The better the wettability, the better the spread of the molten liquid on the substrate. The worse the wettability, the more obvious the spheroidization phenomenon.

Figure 3.2 shows the effect of printing 316L stainless steel material on the aluminum substrate. Due to the large thermal conductivity of aluminum and fast heat dissipation, the melting path discontinuity occurs during the manufacturing process. At the same time, the bonding performance of 316L material and the aluminum substrate is poor, resulting in warping. Therefore, when selecting the formed substrate, the matching between the substrate material and the formed material should be considered to avoid printing failure.

3.1.2.2 Single-Track Morphology

In addition to material factors, the factors that affect the manufacturing quality of single-track are mainly laser power and scanning speed. Figure 3.3 shows the shape of a single-track and its corresponding section under different energy input conditions, and the energy input value decreases gradually from left to right. It can be seen from the figure that under the condition of high energy density input (Fig. 3.3a), the shape of the melting track is regular and continuous, and the width of the melting track is the largest. However, there is a large range of powder-free zone near the melting track, and the range of the powder-free zone is closely related to the P/v ratio. After the power density is reduced, the melting track morphology shown in Fig. 3.3b can be obtained. Compared with the first one, the width of the melting track is significantly

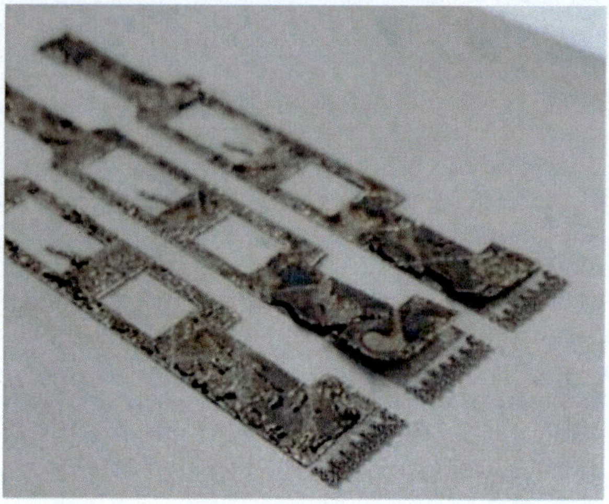

Fig. 3.2 Effect of 316L stainless steel material printed on the aluminum substrate

reduced, and the most important thing is that the powders around the melting track are still in the original position. However, when the energy density is insufficient, the morphology of the melting track is shown in Fig. 3.3c, d. The shape of the melting track becomes irregular and discontinuous, and spheroidization occurs.

In single-track manufacturing, high laser energy input is easy to obtain smooth and continuous single-track, but high laser power density makes the material vaporize, which reduces the material quality in the molten pool, and blows away the powder around the molten pool. When the next molten pool is scanned, there is not enough powder to ensure the fullness of the molten pool. With large energy input, the width and depth of the molten pool are large. The width of the molten pool of the first melting track shape shown in Fig. 3.3a is 2–3 times the spot diameter, and the wetting angle of the molten pool is acute. When the laser energy input decreases, the molten pool cannot reach the vaporization point, the molten pool is continuous and the surrounding powder is not vaporized and blown away. Under this condition, the width of the molten pool formed is 1–2 times the spot diameter, and the wetting angle of the molten pool becomes larger. As the energy input continues to decrease, only the energy in and around the spot center is enough to melt the metal powders. The high scanning speed makes the cooling speed of the molten pool faster. The shape of the molten pool is intermittent occasionally. At this time, the width and depth of the molten pool are small, the wetting angle is more than 90°, and the width of the molten pool is 1/2 to 4/5 of the spot diameter. When the laser energy is not enough to melt more powder, that is, when the scanning speed is too high, the molten path is more discontinuous, the fracture is serious, the wettability with the matrix is poor, and even spheroidization occurs. Therefore, the smaller the wetting angle of the molten pool, the more conducive to manufacturing high-density parts, that is, when the single-track shape and section are shown in Fig. 3.3a or b, it is more suitable for manufacturing high-density metal parts.

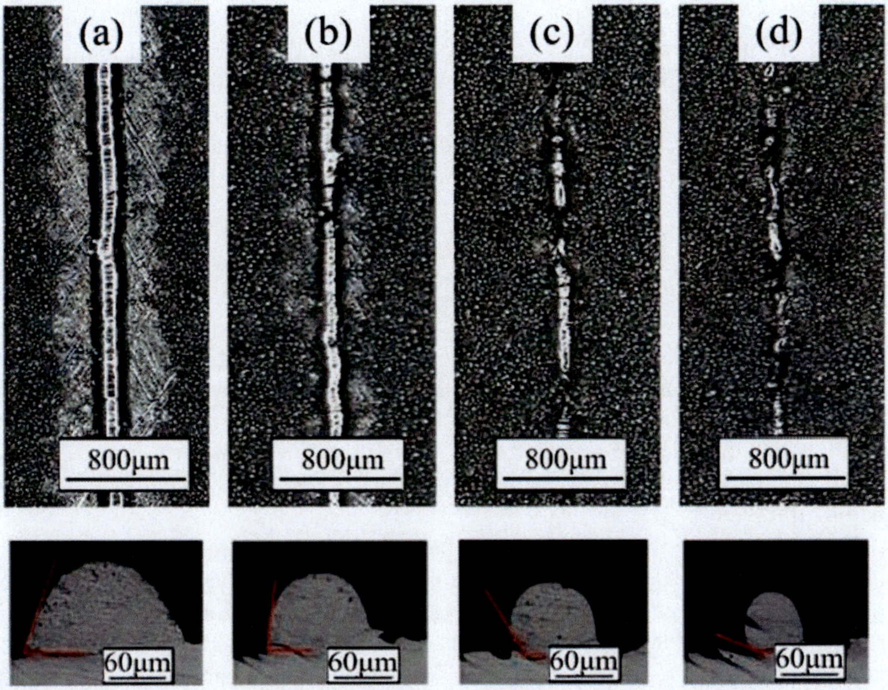

Fig. 3.3 Four typical single-track patterns and their corresponding sections: **a** continuous and regular melting track shape; **b** continuous and regular melting track with smaller width; **c** irregular and discontinuous melting track shape; **d** irregular discontinuities and spheroidized melting track morphology

3.1.2.3 Molten Shape Control

Theoretically, for a specific material, the principle of heat conservation can be used to judge whether enough energy can be obtained to completely melt the powder under a certain laser input parameter, that is

$$\frac{\alpha P}{v} = \left(c_p \times \Delta T + \Delta H\right) \times \rho \times V \tag{3.2}$$

where, α is laser absorption of powder; P is laser power; v is laser scanning speed; c_p is material heat capacity; ΔT is the material can reach the temperature required for complete melting; ΔH is latent heat of crystallization of materials; ρ is loose density of powder; V is the volume of material.

Equation (3.2) indicates whether the material has absorbed enough heat for powder melting. However, in the actual calculation process, it is necessary to consider the absorption of the material to the laser α value and the material mass m of the molten pool. The absorptivity of the material is closely related to the density of the powder, the shape of the powder, and the flatness of the powder surface. It is also related to the emission wavelength of the laser. As for the mass m of the solidified melting track in the single-track molten pool, because the width and height of the melting track are determined by processing parameters, such as processing layer thickness, laser power, and scanning speed, with the change of the above parameters, the mass m of the melting tracks per unit area is different, that is, m is not a quantitative, but a variable. Therefore, Eq. (3.3) is often used for calculating energy input value in actual processing ψ_0,

$$\psi_0 = \frac{4P}{\pi d^2 v} \tag{3.3}$$

where, P is laser power; d is spot diameter; v is laser scanning speed.

When using different kinds of lasers (the laser mode is generally selected as the basic mode) and different focusing spot diameters for LPBF manufacturing, it can be judged whether the LPBF manufacturing of high-density metal parts can be carried out by calculating whether the value of ψ_0 reaches the energy input of the above melting track.

The molten metal in a single-track will adsorb the surrounding powders and suck the surrounding powders into the molten pool, but some of the powders cannot be melted and remains granular and bonded to the solidified melting track, as shown in Fig. 3.4a. These partially melted powder particles will gather and stick together after multi-layer manufacturing. As shown in Fig. 3.4b, more unmelted powders stick to the wall. Powder adhesion will make the wall thickness larger, which is not conducive to the manufacturing of precision parts. Therefore, in the actual processing process, the energy input should be optimized according to the adhesion of powder particles on the surface of the melting track when the melting track is continuous to obtain a smooth and continuous melting track [2].

In the process of LPBF, the parameters of manufacturing continuously molten in single-track scanning are generally selected as the process parameters of surface,

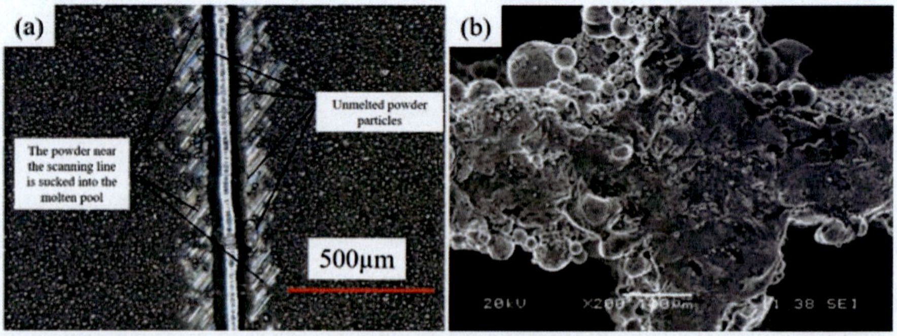

Fig. 3.4 Adsorption of laser on surrounding powder: **a** the molten pool adsorbs the surrounding powder and the unmelted powder; **b** unfused powder accumulates

body and part manufacturing. However, the status of a single scan line does not fully represent the status of multi-layer processing. The main reasons are as follows:

(1) The shrinkage of the powder results in a change in the thickness of the powder during processing;
(2) The uneven powder thickness or overall deviation (greater than or less than) from the set value results in the change of the position of the focus in the powder (each layer, each track), which is also the reason why the single scanning track under the same parameters is smooth and continuous, but some are twisted;
(3) The optimal parameters of single-track manufacturing may not be the optimal parameters of solid manufacturing because there are many other interference factors. Therefore, the coupling analysis of the main factors should be carried out to obtain the optimal process parameters.

3.2 Multi-track Overlapping Process

3.2.1 Overlap Between Adjacent Melting Track

In the process of single-layer manufacturing, in addition to the influence of the above two factors of laser power and scanning speed, single-layer manufacturing also depends on the periodic remelting between two adjacent melting tracks. Therefore, the overlap ratio between two tracks has an important influence on the manufacturing of the whole single-layer, which is reflected in the hatch space in the process parameters. The hatch space mainly determines the bonding quality between the two tracks. Because of the overlap of the molten beads, there is a part of the common area between adjacent molten beads. When the adjacent molten beads are formed, this area will be remelted by the laser beam once, and its microstructure will change slightly. The overlap of the molten bead can be expressed by Eq. (3.4):

$$\mu = \frac{D - s}{D} \times 100\% \tag{3.4}$$

where, μ is the overlap rate; s is the hatch space; D is the width of the molten pool.

As can be seen from Eq. (3.4), there is a negative correlation between scan spacing and overlap rate. When the width of the molten pool is constant, the overlap rate will decrease gradually with the increase of hatch space. Figure 3.5 show the schematic diagram of molten overlap in one layer. It can also be seen from the figure that the manufacturing accuracy of the upper surface of a single-layer is significantly related to the overlap rate. With the increase of the overlap rate, the distance between the top of the fusion and the bottom of the two arc-shaped fusion gradually decreases, thus improving the surface quality of each layer.

Figure 3.6 shows the molten overlap morphology at different hatch space. It can be seen that at low hatch space, the surface is flat due to the molten overlap, and it

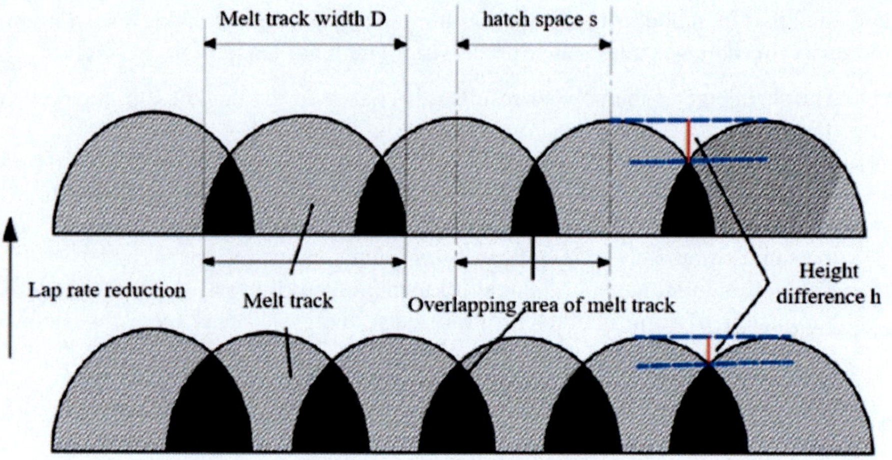

Fig. 3.5 Schematic diagram of overlapping in one layer

is almost impossible to identify the molten trace. However, some spatter particles are found attached to the surface (Fig. 3.6a). With the increase of hatch space, the surface began to be uneven and the molten traces became obvious. Especially, when the hatch space reached 0.12 mm, the independent melting track could be seen, which indicates that there was no overlap between the melting tracks at all. In the upper right corner of each figure is the morphology of the endpoints of the melting track. It can be seen that the overlap condition of the adjacent two melting tracks changes with the change of hatch space. The overlap rate decreases obviously with the increase of hatch space. When the overlap rate is negative, it indicates that there is no overlap. As shown in Fig. 3.6f, the hatch space is too large, resulting in a large gap between adjacent tracks in the same manufacturing layer, so the two adjacent tracks completely fail to bond. This situation will lead to a high internal void fraction of the parts when manufacturing the solid parts, and seriously reduce the relative density and mechanical properties of the parts.

3.2.2 *Heat Accumulation in Multi-track Overlapping*

Due to the existence of an overlap rate in multi-track scanning, the influence of the former molten pool on the heat of the latter molten pool is inevitable. Therefore, based on the energy input equation of the single-track (3.3), the heat influence factor should be added, and the Eq. (3.3) should be rewritten as

$$\psi_1 = \frac{4P}{\pi d^2 v}(1 + \beta) \tag{3.5}$$

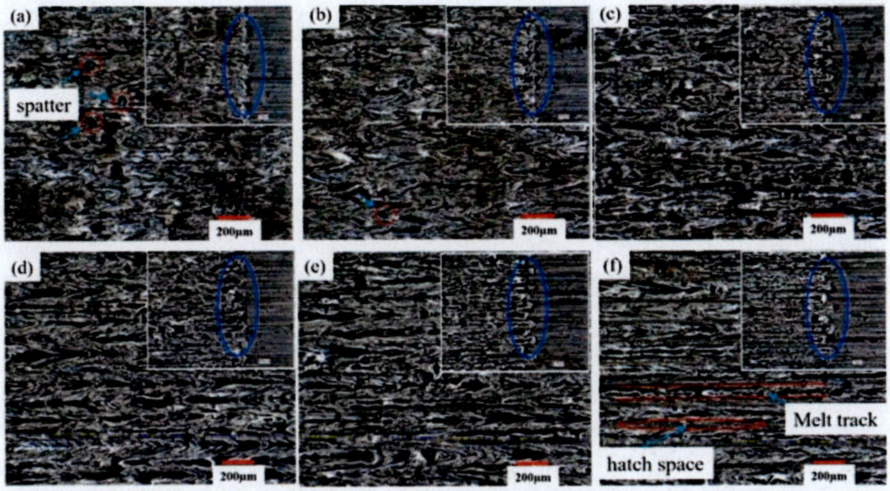

Fig. 3.6 Overlapping morphology of the melting tracks at different hatch space: **a** 0.04 mm; **b** 0.06 mm; **c** 0.08 mm; **d** 0.09 mm; **e** 0.10 mm; **f** 0.12 mm

where, β is the heat accumulation factor of the molten bead overlap, and its value is closely related to the overlap rate. However, the existence of a multi-track overlap ratio makes the former molten pool provide a wetted base for the latter, which is conducive to the smooth formation of the molten pool. Therefore, lower laser power or higher scanning speed can be used to obtain the ideal shape of the molten pool. The heat accumulation factor β of molten bead overlap is mainly related to the overlap ratio μ, so it can be defined $\beta = \chi\mu$, The corresponding χ value is different for different overlap rates, and χ is defined as the thermal influence coefficient of different overlapping rates.

Assume that the melting track shape shown in Fig. 3.3b is the ideal shape of LPBF manufacturing, while the corresponding manufacturing parameters are P_a and v_a for single-track manufacturing, and the other parameters are set as fixed values. When multi-track manufacturing is adopted, the energy input required to make the melting track equal to the width of a single-track due to the above heat effect and preheating effect is less than the energy input for manufacturing a single-track. At this time, the manufacturing parameters are P_b and v_b, so β can be expressed as as Eq. (3.6)

$$\beta = \frac{P_a v_b - P_b v_a}{P_b v_a} \tag{3.6}$$

In the actual manufacturing process, the laser power P is generally unchanged, and the energy input is adjusted by adjusting the scanning speed, so Eq. (3.6) can be simplified as Eq. (3.7):

$$\beta = \frac{v_b - v_a}{v_a} \tag{3.7}$$

For different overlap ratios μ, the thermal influence coefficient with different overlap ratios can be obtained χ, as shown in Eq. (3.8).

$$\chi = \frac{v_b - v_a}{\mu v_a} \tag{3.8}$$

3.2.3 *Single-Layer Manufacturing*

The scanning surface in LPBF is obtained by overlapping multi-track. Therefore, the overlapping and interaction between the scanning tracks have a critical impact on the quality of the formed surface. Under inappropriate parameters, the formed surface will appear yellowish brown or black, seriously affecting the manufacturing quality [3]. To obtain a dense and smooth multi-track lap, it is necessary to consider the overlapping rate between the tracks based on a single molten pool.

According to the analysis of single-track, type I (Fig. 3.7) and type II (Fig. 3.8) are suitable for LPBF manufacturing. When multi-track lapping manufacturing is carried out because there is a large range of powder-free zone near the melting track, most of the material in the next molten pool comes from the remelting of the previous molten pool, and the volume of the molten pool will decrease. Therefore, for type I lapping manufacturing, it is necessary to use a large hatch space to make the post-scanning molten pool avoid the powder-free zone as far as possible and ensure the source of powder in the molten pool. Figure 3.7 shows the surface morphology of type I lap when the overlap ratio is 30% and 10% respectively. It can be seen from the figure that, when the overlap ratio is 30%, the manufacturing surface presents an overmelting state and is uneven due to the large value of the heat accumulation factor β of the molten pool. In this case, there is a serious shortage of materials at the top of the molten pool, which is the result of laser remelting of the molten pool. It is also partially caused by the plasma recoil force generated after the molten pool vaporizes, which flattens the molten pool under the liquid. When the overlapping rate is 10%, the molten pool after solidification can be seen, and deep grooves are generated between the melting tracks. The rule that makes each molten pool continuous is that the molten pool should be full, which is conducive to the laser melting manufacturing of the powder bed. Therefore, the low overlapping rate should be adopted for the type I track.

For the type II track, the width of the molten pool has been greatly reduced, and the powder around the molten pool has not been blown away, so a higher overlap ratio is required. Figure 3.8 shows the surface morphology when the overlap ratio of the type II track is 30% and 10%, respectively. It can be seen from the figure that the surface looks not tight and the molten bead is not firmly lapped when the lapping rate is 10%; When the overlap ratio is 30%, not only the firm overlap between the molten pools is ensured, but also the formed surface is very smooth. The main reason is that the hatch space value is similar to the laser spot diameter when the overlap

Fig. 3.7 Surface morphology of different overlap ratios of type I track: **a** the overlap rate is 30%, and **b** the overlap rate is 10%

Fig. 3.8 Surface morphology of different overlap ratios of type II track: **a** the overlap rate is 30%, and **b** the overlap rate is 10%

ratio is 30%. This hatch space range can ensure the uniform distribution of laser energy input on the manufacturing surface without causing defects such as warping and insufficient melting of the formed parts.

It can be seen from the above results that the size of the laser hatch space affects the energy distribution of the laser transmitted to the powders. When the hatch space is about equal to the laser spot diameter, the laser energy can also be uniformly distributed after the molten is superposed, so the ideal lap manufacturing effect can be obtained. In the actual LPBF manufacturing process, it is necessary to consider: (1) processing efficiency; (2) residual stress accumulation caused by excessive heat input, which leads to warping, cracks, etc.; (3) the influence of impurities produced by material vaporization in the manufacturing chamber on the manufacturing process. Therefore, the authors believe that the design of process parameters of LPBF manufacturing should be optimized based on type II melting track morphology.

Under the condition of appropriate process parameters and hatch space, the surface formed in single-layer scanning has four typical morphologies, as shown in Fig. 3.9. As shown in Fig. 3.9a, it is a wavy shape, and the surface of the layer scanning block forms dense short ripples. As shown in Fig. 3.9b, the shape is relatively smooth, but some points in local areas are higher than other areas, so it is called "convex point type". As shown in Fig. 3.9c, the surface formation is very poor, and the protruding melt is densely scattered on the surface of the manufacturing surface, making the surface look like a particle strewn with particles, so it is called "loose particle type". As shown in Fig. 3.9d, the single-layer surface with good manufacturing effect is smooth without obvious bulges, ripples and melt particles. Among the four surfaces, the surface roughness in Fig. 3.9d is the lowest, while the surface roughness values in Fig. 3.9b, a and c increase in turn.

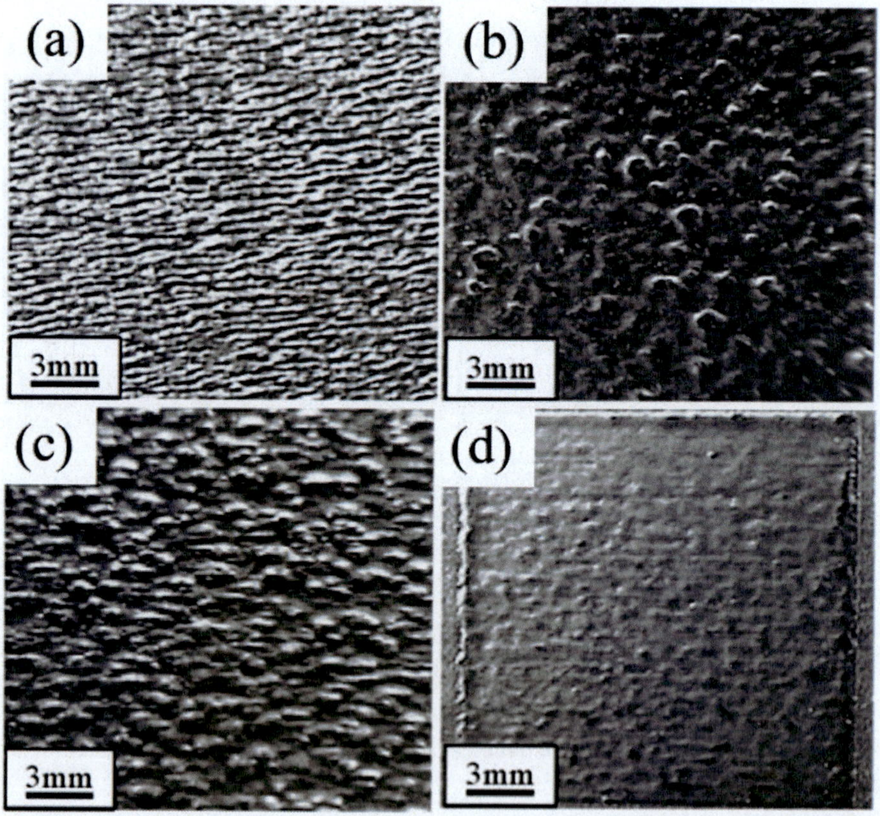

Fig. 3.9 Typical single-layer scanning surface topography: **a** corrugated type, **b** convex point type, **c** loose grain type, and **d** smooth

3.3 Multi-layer Superposition Manufacturing Process

3.3.1 Multi-layer Superimposed Energy Input Model

In the process of multi-layer manufacturing, the input energy accumulates continuously, which will affect the manufacturing quality. Therefore, the energy input model should be evaluated comprehensively. Use a Z-shaped scan policy during printing, as shown in Fig. 3.10. The solid line is the laser scanning track, and the dashed line is the laser jump track. The length and width of the manufacturing surface are denoted by L and W respectively. v_s represents the speed of the v line, namely the scanning speed mentioned previously, and the v_f represents the speed of the manufacturing direction in one layer.

Assume that the influencing factor of the scanning strategy is κ, It can be obtained that the input laser energy is as Eq. (3.9)

$$\psi_s = \kappa \times (1 + \varnothing) \times \frac{P}{d\upsilon h} \tag{3.9}$$

where, ψ_s is laser energy input size (J/mm^3); $\varnothing$ is overlap rate between adjacent melting tracks; d is focus spot diameter (mm); P is laser power; υ is laser scanning speed; h is processing layer thickness (mm).

The laser light spot is nearly circular, so most companies use the $\varnothing = \frac{d-s}{d} \times 100\%$ to calculate the overlap rate of the molten tracks, and s is the scan line distance (mm). This equation does not consider the width of the actual molten pool, so the calculated overlap rate cannot accurately represent the actual one. Therefore, it is more accurate to use melt pool width d_m instead of d to calculate lap rate, that is, $\varnothing = \frac{d_m-s}{d_m} \times 100\%$. In Eq. (3.9), only the scanning strategy influence factor κ is unknown.

The manufacturing volume of the molten pool along the scanning direction in unit time is approximately expressed as $V_1 = v_s \times d_m \times h$, and the volume of the molten pool along the manufacturing direction in unit time is $V_2 = v_f \times W \times h$.

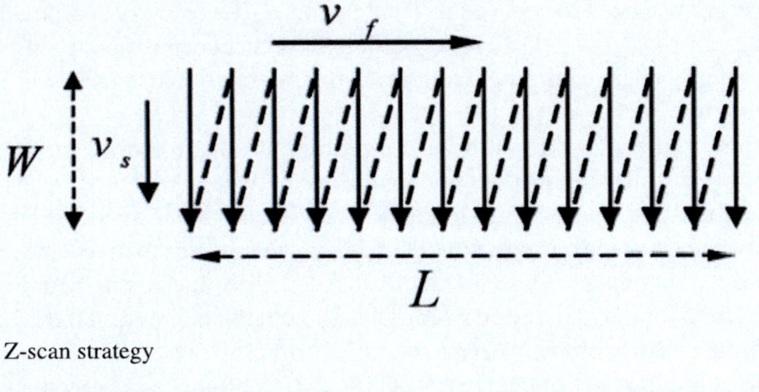

Fig. 3.10 Z-scan strategy

$$vf = \frac{s}{W + \frac{v}{v_{jump}} \times \sqrt{W^2 + s^2}} \cdot v \tag{3.10}$$

Because the laser energy absorbed by the powder per unit of time and volume is equal, that is

$$c_p v d_m h t = c_p \kappa v_f W h t \tag{3.11}$$

It can be found simultaneously

$$\kappa = \frac{v d_m}{v_f W} = \frac{d_m \times \left(1 + \frac{v}{v_{jump}} \times \sqrt{1 + \frac{s^2}{W^2}}\right)}{s} \tag{3.12}$$

By substituting Eq. (3.12) into Eq. (3.10), we can get:

$$\psi = \frac{d_m \times (1 + \frac{v}{v_{jump}} \times \sqrt{1 + \frac{s^2}{W^2}})}{s} \times (1 + \varnothing) \times \frac{P}{vdh} \tag{3.13}$$

In the actual manufacturing process, v_{jump} is generally one order of magnitude larger than v, so Eq. (3.13) can be approximately written as

$$\psi = \frac{P(d_m - s)}{vdsh} = \frac{P}{vdh} \times \frac{2d_m - s}{s} \tag{3.14}$$

From the energy input Eq. (3.14), it can be seen that the energy input is related to the laser power, hatch space, scanning speed, processing layer thickness, spot diameter, and actual molten pool width. The main difference between the energy input formula (3.14) and other energy input equations is that the actual molten pool width and the effect of energy accumulation between tracks are considered. A rough energy accumulation factor $2d_m/s - 1$ can be obtained by calculating the ratio of the actual molten pool width to the hatch space. If you need to calculate the exact energy accumulation, you need to add a coefficient factor ω. The energy accumulation factor becomes $\omega \times (2d_m/s) - 1$. Different coefficient factors correspond to different laser power, scanning speed and processing layer thickness ω. It is usually taken as 1 for simplification.

To obtain dense parts, in addition to getting sufficient energy input, it is also necessary to consider the overlapping conditions between the fusion beads to obtain good manufacturing results. The overlapping ratio makes the formed molten track and formed layer preheat the current processing layer, which leads to the increase of actual energy input. Therefore, Eq. (3.14) can be understood that for a specific material, sufficient energy input is required to completely melt the powders at the beginning of manufacturing to obtain a smooth and continuous melting track. However, the actual energy input increases with the energy accumulation effect caused by the overlap of the melting track and the superposition of multi-layer. At this time, the equivalent energy input of the laser is $(P/(v \times d \times h)) \times (2d_m/s - 1)$. The increase of effective energy input leads to the increase of internal stress during the manufacturing process, which is easy to produce defects such as warpage and cracks. Therefore, to obtain compact part manufacturing, it is necessary to reduce the initial set laser energy input, namely $P/(v \times d \times h)$ value, to $P/(v \times d \times h)$ after processing a certain layer $P/(v \times d \times h) \times (2d_m/s - 1)$.

To sum up, the conditions for manufacturing dense metal parts by LPBF are summarized as follows:

$$\begin{cases} E_1 = \frac{P}{vdh} \\ E_2 = \frac{P}{vdh} \times \frac{2d_m - s}{s} \\ \varnothing = \left(\frac{d_m - s}{d_m}\right) \times 100\% \end{cases}$$

(Sufficient energy to completely melt the powder to obtain a continuous single-track at the beginning of manufacturing)

(A dense manufacturing effect can be obtained by reducing energy after multi-layer manufacturing)

(The overlap ratio between scanning lines must meet a certain value)

3.3.2 Heat Accumulation of Multi-layer

The heat accumulation effect can not be ignored when the multi-layer manufacturing surfaces are superposed. The accumulated heat that cannot be dissipated in time preheats the subsequent manufacturing layer. For multi-layer superimposed manufacturing, the multi-track manufacturing energy can be substituted into Eq. (3.5) to obtain:

$$\psi_1 = \frac{4P}{\pi d^2 v}(1 + \beta) + E \tag{3.15}$$

where, E is the accumulated heat value of the formed layer. The value of E is related to the *P/v* value, layer scanning area and heat dissipation rate. Due to the existence of heat accumulation value E, an ideal smooth melt track can be obtained under the conditions of lower laser power and higher scanning speed. Only when the accumulated heat value of each layer is equal to the dissipated heat, the temperature of the part will not increase, and the E value will remain constant.

The approximate value of heat accumulation value E in multi-layer superposition can be calculated by the following method: according to the formation of type II single-track, select the corresponding parameters P_a, v_a, and record the width of the corresponding single-track under this parameter. Then, through the multi-layer stacking manufacturing experiment, it can be observed that the single-track on the surface of the formed part reaches the effect of the type II track shape. At this time, the required energy input will be smaller than that of single-track manufacturing. When measuring the width of the multi-track stacking and the manufacturing width of the single-track are the same, the processing parameters P_b and v_b in the multi-track stacking are recorded. The heat accumulation factor β between the melting tracks can be directly calculated in the multi-track stacking, so b is a fixed value. The value of E can be calculated by Eq. (3.16).

$$E = \frac{4P_a}{\pi d^2 v_a} - \frac{4P_b}{\pi d^2 v_b}(1+\beta) \tag{3.16}$$

When the actual laser power remains unchanged, the energy input is adjusted by scanning speed. Equation (3.16) can be simplified as:

$$E = \frac{P_a}{\pi d^2}\left(\frac{4}{v_a} - \frac{4(1+\beta)}{v_b}\right) \tag{3.17}$$

Therefore, the approximate value of heat accumulation value E can be obtained only by determining the overlap ratio and v_a, v_b under the same melting track width.

3.3.3 Multi-layer Manufacturing

In general, it is much more difficult to superpose the multi-layer manufacturing surfaces than to overlap the single-track or multi-track because the manufacturing conditions are much more complex when stacking solid parts. To ensure the stability of the manufacturing state, it is very important to control factors such as the flatness of powder spreading, the stability of effective laser power, and the oxygen content in the manufacturing chamber. Even if the above conditions are strictly controlled, the surface morphology of the formed parts will deteriorate gradually during the manufacturing process [4]. Figure 3.11 shows the morphology of different scanning layers formed by the "S" shaped orthogonal stacking scanning strategy. Figure 3.11a shows the scanning morphology of the first layer. The tracks of each melting track are continuous straight lines, and the surface is fish scale. The melting tracks are closely overlapped. The entire formed surface is dense and flat, and the boundaries on both sides of the surface are straight. Figure 3.11b shows the morphology of the second layer. The formed surface is still dense and flat, and a small amount of powder particles begin to adhere between the molten beads and on both sides of the formed surface. Figure 3.11c and d show the scanning morphology of the third layer and the fourth layer respectively. It can be seen that the powder particles adhered between surface tracks and on both sides of the manufacturing surface gradually increase, making the manufacturing surface gradually rough, and the two sides of the surface gradually present serrated boundaries. The main reason is that the formed surface quality has a positive feedback effect. After the formed surface quality gradually becomes worse, the powder surface gradually becomes uneven and the molten pool becomes irregular, which makes the formed surface rougher.

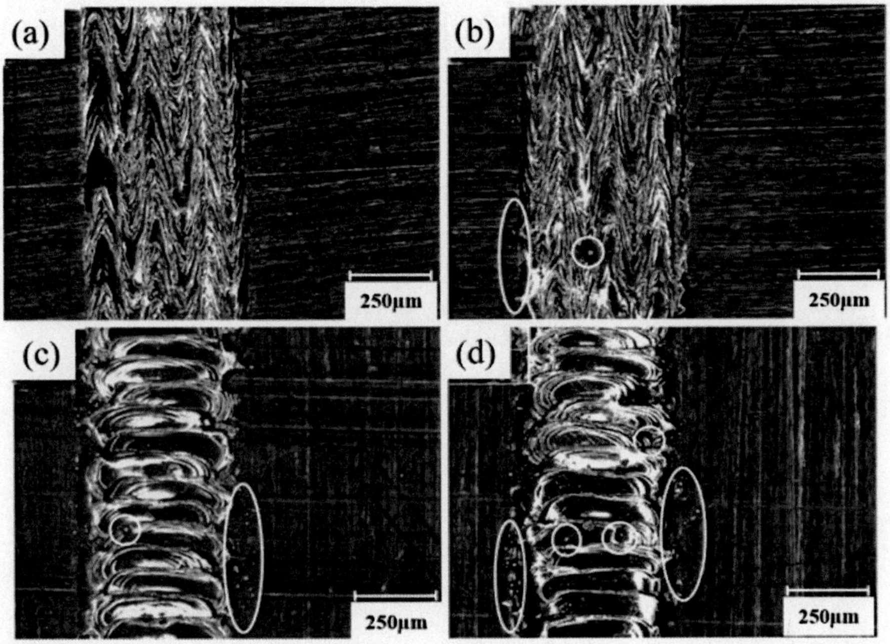

Fig. 3.11 Multi-track and multi-layer melting track morphology: **a** the first floor, **b** the second floor, **c** the third layer, and **d** the fourth floor

It can be seen from the manufacturing characteristics of a single-track that the powder around the scanning line will be adsorbed during the formation of the track, and finally, a semicircular section will be formed above the melting tracks. When multi-track are lapped to form a surface, there are small gullies at the overlap, manufacturing a lapping layer with a small height difference, which makes the thickness of the next layer of powders slightly uneven. With the increase in the number of layers, this non-uniformity gradually increases due to the accumulation of unevenness on each layer. When the thickness of the powder layer at the lower part exceeds a certain value, it is easy to cause the powder to not completely melt to adhere to the formed surface. In addition, the powders adsorbed at the edge of the layer gradually accumulate, eventually leading to the deterioration of the surface quality of the formed surface.

The fundamental reason for instability in multi-layer manufacturing is the solidification shrinkage of powder after melting. The laser melts metal powders by using a high-energy laser beam to scan the powder bed and then a solid part was accumulated in a layer-by-layer way. The metal powders used are generally 400–500 meshes, and the packing density is about 45% of the corresponding solid material. That is to say that after the powder material is heated, melted, and solidified, it will inevitably produce volume shrinkage in the X, Y, and Z axis directions during the process of changing from loose state to dense state, thus affecting the precision of parts.

Shrinkage is an inherent physical property of metal materials, which goes through three interrelated shrinkage stages from the melting state to room temperature:

(1) Liquid shrinkage: shrinkage within the temperature range from melting temperature to solidification (liquidus temperature);
(2) Solidification shrinkage: shrinkage within the temperature range from the beginning of solidification to the end of solidification (solidus temperature);
(3) Solid shrinkage: the shrinkage within the temperature range from the termination of solidification to cooling to room temperature.

Under the influence of unstable factors of solidification shrinkage, the processing layer thickness of the nth layer can be derived from Eq. (3.18):

$$h_n = h + h_{n-1}\left(1 - \frac{\rho_0}{\rho}\right) \tag{3.18}$$

where, h_n is the thickness of the n layer; h_{n-1} is the thickness of the n − 1 layer; h is the thickness of the powder layer; ρ_0 is the loose density of powder; ρ is solid density.

When the number of layers increases from n to $+\infty$, the limit of h_n is:

$$\lim_{n\to+\infty} h_n = \lim_{n\to+\infty} h\frac{1-\left(\frac{\rho_0}{\rho}\right)^n}{1-\frac{\rho_0}{\rho}} = h \times \frac{1}{\frac{\rho_0}{\rho}} = h \times \frac{\rho}{\rho_0} \tag{3.19}$$

Equation (3.19) shows that the layer thickness h_n gradually increases, but tends to a fixed limit value. If $h = 35\ \mu\text{m}$, $\rho_0 = 4.04\ \text{g/cm}^3$, $\rho = 7.98\ \text{g/cm}^3$, the actual processing layer thickness tends to be 70 μm stable. It can be seen that the actual processing layer thickness is proportional to the density of the formed part and inversely proportional to the loose packing density of the powder. The greater the thickness of the processing layer, the greater the instability of the manufacturing process, which is not conducive to the firm combination of the molten pool and the substrate, the greater the spheroidizing tendency, and ultimately the worse the quality of the formed surface.

To prevent the formation surface from becoming worse due to the increase in layer thickness, the commonly used effective methods are as follows: (1) During the manufacturing process, the manufacturing parameters are adjusted in real-time. If the scanning speed is properly increased, the surface quality will slowly improve, but the density will decrease; (2) Remelt the formed surface; (3) Special scanning strategy is adopted.

Using a special scanning strategy to eliminate the above increase in layer thickness is also conducive to improving the density of multi-layer manufacturing parts. From the process of single-track manufacturing and multi-track manufacturing, it can be known that the molten pool is elliptical after solidification. When the molten pools are overlapped, micro-grooves will be generated. And the unstable factors at the overlap joint of the molten pool are more than those inside the molten pool, which is an active zone for the generation of pores, inclusions and other defects. Many research results show that the densification effect of formed parts can be improved by using an interlaminar stagger scanning strategy. This is mainly because the interlaminar stagger scanning strategy can repair the manufacturing defects of the previous layer, and the liquid metal is easy to wet the substrate at the gully of the previous layer. Figure 3.12a shows the interlaminar stagger scanning strategy and its evolution. The lower layer tracks are filled and scanned in the valley between the two tracks of the previous layer. The solid line represents the n-th layer track, and the dotted line represents the (n + 1)-th layer track. The procedure is as follows: first, scan the solid line, then lay a certain thickness of powder on the layer, and then scan the dotted line, that is, layer n + 1. The hatch space is the same as that of the n-th layer, but the molten region is located between the n-th layer tracks. As shown in Fig. 3.12b, orthogonal scanning is performed based on Fig. 3.12a. In orthogonal scanning, the scanning methods of the layer n and the layer n + 1 are the same as those shown in Fig. 3.12a, except that the scanning directions of layer n + 2 and layer n + 3 are orthogonal to those of layer n and layer n + 1. Figure 3.12c shows the addition of border scanning based on the method shown in Fig. 3.12a. Figure 3.12d shows the addition of frame scanning based on the scanning method shown in Fig. 3.12b. Figure 3.13 shows the stacking diagram of staggered scanning between the current layer and adjacent layer.

In addition, the scanning area also has a significant impact on the manufacturing quality of multi-layer manufacturing parts. The scanning area refers to the area where the laser and powder interact. The space occupied by the part on the substrate is the scanning range. Generally, the scanning area is equal to or smaller than the scanning range. In the manufacturing process, the scanning area determines the energy input per unit area, which has a considerable impact on heat accumulation. In the manufacturing process of large-area solid parts, due to a large number of overlapping between scan lines, it is easy to produce melt convergence and spheroidization on the formed surface. Each group of thin-walled parts is usually composed of only one or several scanning lines. Even if the scanning area of the whole part is small, the spheroidization area and level of the formed surface are small due to the small manufacturing area during powder spreading. Besides, the friction resistance and collision of the powder spreading scraper are relatively weak. For multi-component mixed powders, the tendency of spheroidization is greater during large-area manufacturing. The surface is uneven during solid block manufacturing with many large particles gathering on the surface. Therefore, the powder-spreading collision caused by spheroidization cannot be ignored. In the manufacturing process of some large-scale solid parts, due to the very serious collision, it is necessary to use a flexible powder spreading device, while thin-walled parts with the same manufacturing range can use a rigid powder coating plate. The weak friction resistance and collision make it possible to use rigid powder spreading plates in the manufacturing process. The powder spreading surface can be relatively flat and the manufacturing accuracy will be higher.

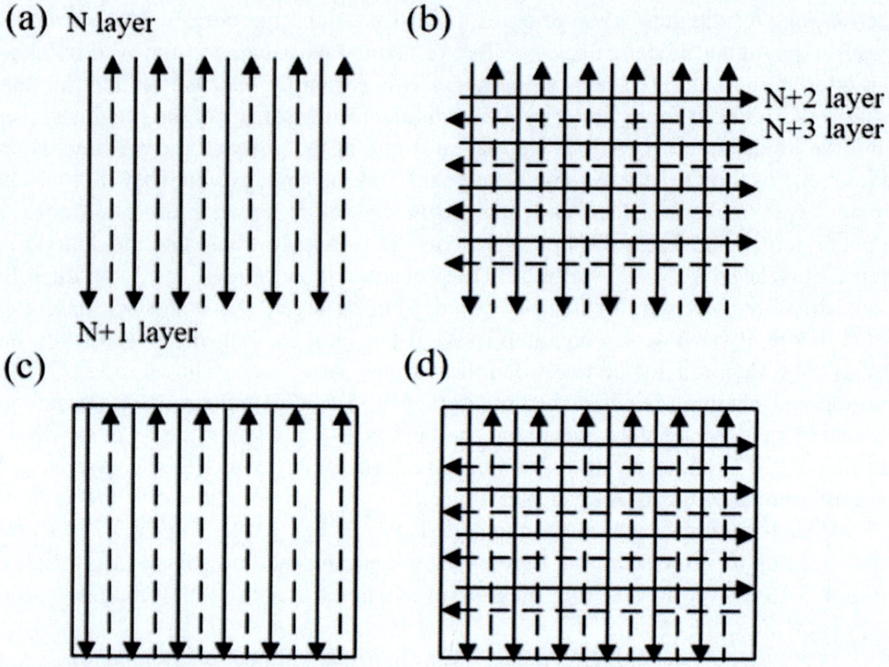

Fig. 3.12 Interlaminar stagger scanning strategy and its evolution: **a** interlayer stagger scanning strategy, **b** interlayer staggered orthogonal scanning strategy, **c** interlayer staggered border scanning strategy, and **d** interlayer staggered orthogonal border scanning strategy

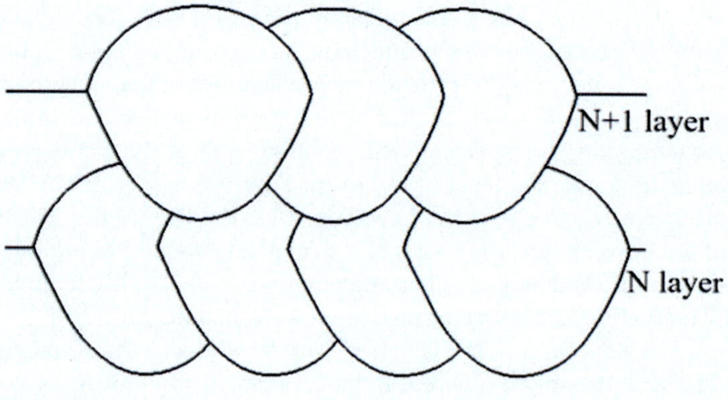

Fig. 3.13 Interlayer stagger scanning strategy overlay diagram between adjacent layers

3.3.4 Layer Thickness

Single-layer scanning manufacturing has realized the face manufacturing process of the part, and the final manufacturing of the entire 3D physical part can be achieved by melting multi-layer entities. In addition to the influence of laser power, scanning speed and hatch space, the thickness of the powder layer will play an important role in the fusion process of multi-layer entities. The thickness of the powder layer is determined by the thickness of the slice but is limited by the particle size of the powder. In principle, the thickness of the powder layer is not less than D50 of the particle size of the powder. Different powder layer thicknesses have an extremely important impact on manufacturing efficiency and quality. A large powder layer thickness means that the height of each layer is large. Therefore, the total number of layers manufacturing the whole part is reduced, which improves manufacturing efficiency. However, due to limitations such as laser power and geometric structure of the formed part, an excessive layer thickness, on the one hand, will lead to insufficient powder melting and reduce the density of the part. On the other hand, it will increase the overhang size and reduce manufacturing accuracy. Figure 3.14 shows the side morphology of the sample with the thickness of 0.02 mm and 0.06 mm respectively. It can be seen that the size of the melting track of the two is different, and the height of the side melting track of the latter is about 1.6 times the former. The size and number of voids in the latter are also large because the increase in layer thickness makes the melting track get more powders, which increases the width of melting track, but also leads to the increase in the distance between adjacent melting tracks in the manufacturing direction. In addition, some powders cannot be melted because they cannot get enough heat. As a result, some voids are formed in these powder areas.

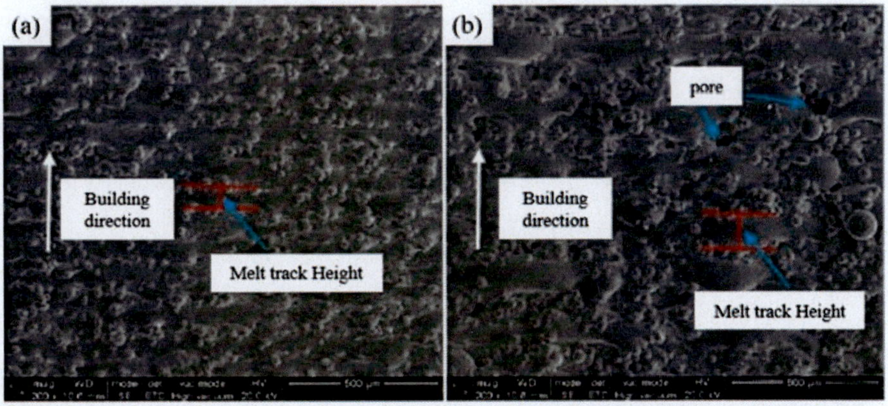

Fig. 3.14 Profile of specimen with different powder layer thickness: **a** 0.02 mm, and **b** 0.06 mm

References

1. Bai YC (2018) Research on the mechanism and properties controllability of selective laser melting of maraging steel. South China University of Technology
2. Mai SZ (2016) Study on the manufacturing processes and properties of customized Co Cr alloy crowns and fixed bridges manufactured by selective laser melting. South China University of Technology
3. Wang YD (2015) The study on process, microstructure and properties of CoCrMo alloy manufactured by selective laser melting. South China University of Technology
4. Luo ZY (2018) Study on process and effective factors of thin-wall parts manufactured by selective laser melting. South China University of Technology

Chapter 4
Unstable Factors and Types of Defects in Laser Powder Bed Fusion Process

4.1 Classification of Influencing Factors of Instability

Laser powder bed fusion (L-PBF) is a complex process. Before machining, the CAD model of the part is sliced and dispersed by professional data processing software and the necessary supporting structure is added, then the scanning path is planned, and the processed data contains contour information which can control the movement of the laser beam. Then the data is imported into the L-PBF equipment, and the computer will adjust the profile information layer by layer and control the scanning mirror to deflect, so that the laser spot can melt the metallic powder selectively and bond with the previous layers of material. However, the powder in the area that has not been irradiated by laser remains the original shape and can be recycled use. After a layer of powder is scanned, the powder supply cylinder is raised to a certain height, while the manufacturing cylinder is lower to a certain thickness. The powder is scraped from the powder supply cylinder to the manufacturing platform by laying brush, and then the newly laid metal powder is melted by laser. Repeat this until the whole manufacturing process is completed. After manufacturing, it is necessary to carry out post-processing processes such as wire cutting, removing support, polishing and so on. As shown in Fig. 4.1, the whole machining process is affected by many factors, which can be classified as follows:

(1) Prophase data. Part CAD drawing, STL file format conversion, support addition, slice layering, etc.
(2) Material properties. Material shrinkage, particle diameter, fluidity, impurities, etc.
(3) Equipment accuracy. Manufacturing cylinder lifting, powder laying system, optical path and scanning system, substrate installation plane, etc.
(4) Processing principle. Laser deep penetration, spot diameter, powder adhesion, spheroidization, etc.
(5) Process parameters. Laser power, scanning speed, hatch space, scanning strategy, powder layer thickness, etc.

© National Defense Industry Press 2024

D. Wang et al., *Laser Powder Bed Fusion of Additive Manufacturing Technology*, Additive Manufacturing Technology, https://doi.org/10.1007/978-981-99-5513-8_4

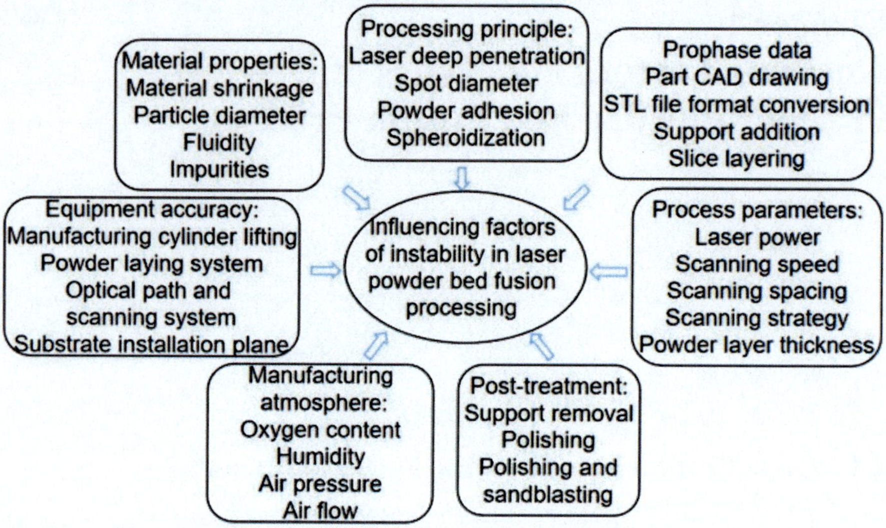

Fig. 4.1 Influencing factors of the L-PBF manufacturing process

(6) Manufacturing atmosphere. Oxygen content, humidity, air pressure, air flow, etc.
(7) Post-processing. Support removal, grinding, polishing and sandblasting, etc.

Among the above factors affecting the instability of L-PBF manufacturing process, some factors do not need to be paid special attention and in-depth study, such as the errors caused by early data processing, as long as the processing operation is carried out strictly in accordance with the requirements. This error can be regarded as an inherent error. The error of post-processing is also well known, although it is affected by the operating level and subjective influence of the operators, but the error caused by this unstable operation is usually not within the scope of the study. Some factors can be studied together because they often have interaction effects on each other, such as process parameters and processing principle, process parameters and material properties and so on.

4.2 Instability of Machining Process Caused by Stress and Deformation

4.2.1 Stress and Deformation

The L-PBF technique uses a tiny laser beam with extremely high energy density to melt the metallic powder. Under the irradiation of the laser beam, the powder undergoes a very non-uniform physical processes of rapid heating and rapid cooling

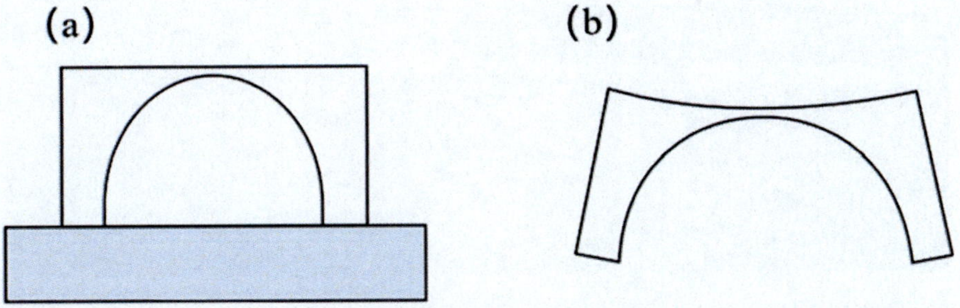

Fig. 4.2 Schematic diagram of residual stress deformation: **a** before cutting, **b** after cutting

in a very short time, and the corresponding molten pool and the surrounding material are heated, melted, solidified and cooled at a very high speed, resulting in volume shrinkage deformation, but it is limited by the material in the surrounding cold area, thus resulting in residual stress in the parts. Thus it can be seen that the generation of residual stress in the process of L-PBF manufacturing process is inevitable, but the deformation caused by the excessive residual stress not only affects the dimensional accuracy, shape accuracy and mechanical strength of the manufactured parts, it will also cause the part to crack in the layer, thus threatening the manufacturing process. At the same time, the problem of residual stress is not easy to be detected experimentally, as shown in Fig. 4.2, if there is more residual stress inside the parts, no abnormality can be observed before the manufactured part is separated from the substrate due to the close combination between the manufactured part and the substrate. However, obvious deformation will be observed after cutting the manufactured part from the substrate.

The methods used by researchers to measure residual stress include drilling method, neutron diffraction method and X-ray diffraction method, etc. The drilling method is to drill a hole in the sample by mechanical method, and to calculate the residual stress by measuring the strain, which not only produces a large processing stress, but also needs to stick a strain gauge near the drill hole, so it requires higher surface quality and size of the sample.

However, neutron diffraction and X-ray diffraction can not measure the internal stress of parts. There are always some deficiencies in the current measurement methods, so that the research work on residual stress of additive manufactured parts can not be perfect. Therefore, some scholars establish mathematical models to predict the residual stress in the additive manufacturing (AM) process, while others established finite element models to simulate the thermal process of the L-PBF manufacturing process, the stress distribution in the manufacturing process is obtained by thermodynamic coupling method, and the validity of the model is verified by measuring residual stress, which can characterize the residual stress in the AM processes to a certain extent.

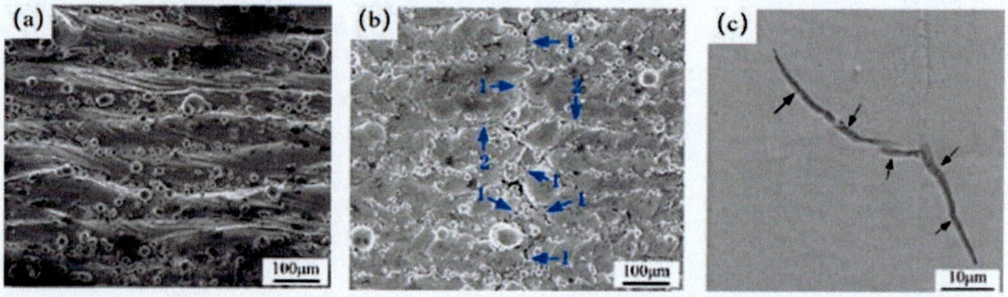

Fig. 4.3 SEM image of 316L stainless steel parts: **a** the top view, **b** the side view, **c** microcrack

4.2.2 Relationship Between Microstructure and Residual Stress

Figure 4.3a, b show the top and side SEM images of the 316L part fabricated by L-PBF, the top SEM image shows that the parts are well overlapped and undulated, and there are grooves between the adjacent melt tracks, while the tracks are slightly raised. There are no obvious holes and the density of the parts is high. The side SEM image shows that there is a crack about 500 μm in length on the side of the part, the No. 1 arrow refers to the intra-layer crack and the No. 2 arrow refers to the interlayer crack. Figure 4.3c shows a polished SEM image of the side, with a crack length of about 30 μm. The cause of this kind of crack is the high-speed movement of the laser beam with high energy density, which leads to the melting and solidification of the material in a very short time, resulting in a great temperature gradient inside the part, manufacturing cracks to release thermal stress.

As shown in Fig. 4.4, the temperature field, stress field and microstructure of the L-PBF affects each other, the direction of the arrow in the figure indicates the influence of one factor on another, in which the dotted line arrow indicates a weak influence, while the solid line arrow indicates a strong influence. It is worth emphasizing that the factor of microstructure transformation should be considered when analyzing the residual stress of L-PBF process, because the transformation of microstructure not only determines the chemical composition of the material, but also determines the heating process of the material.

4.2.3 Stress Distribution and Evolution

4.2.3.1 Vertical Stress Distribution

In order to study the characteristics of residual stress along the height direction, the test samples shown in Fig. 4.5 are designed and fabricated by L-PBF. The heights of the samples are 12 mm, 16 mm and 20 mm, and the thickness and width are fixed with 1 mm and 30 mm respectively. Process parameters: laser power 150 W,

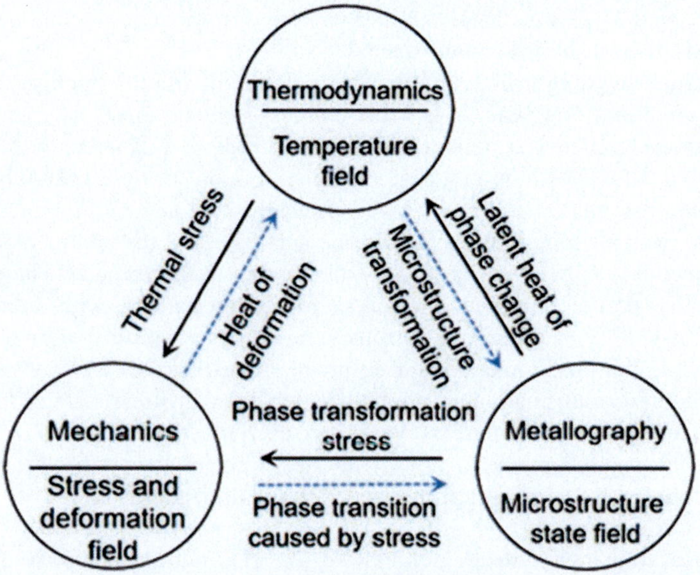

Fig. 4.4 Decomposition and interaction of temperature field, stress field, deformation field and microstructure

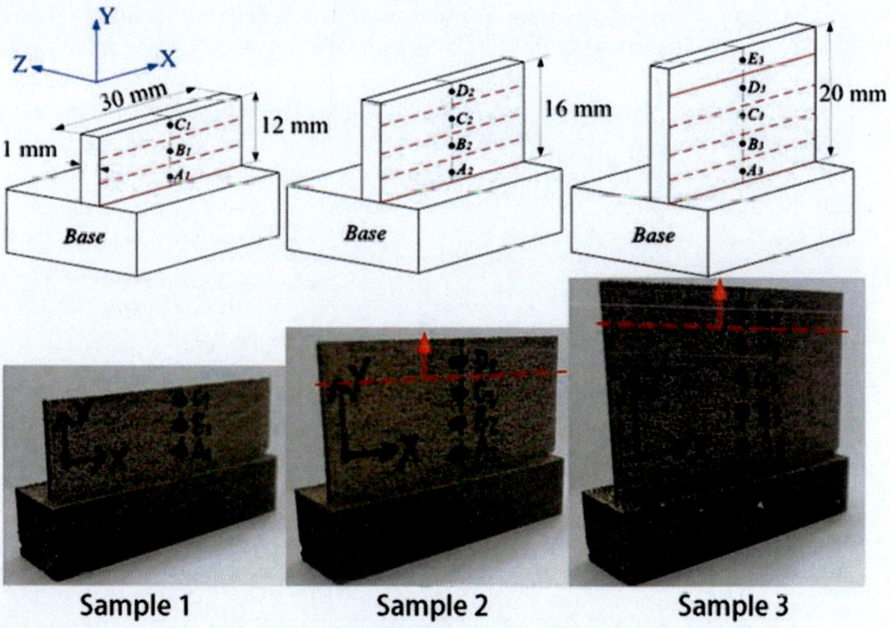

Fig. 4.5 Test sample for residual stress measurement in height direction

scanning speed 400 mm/s, hatch space 0.08 mm, powder layer thickness 0.04 mm, "Z-shaped" X-axis unidirectional scanning strategy. The scanning direction is X-axis direction, the height direction is Y-axis direction, and the thickness direction is Z-axis direction. The material is 316L stainless steel atomized spherical powder, and its physical parameters: tensile strength is 485 MPa, yield strength is 170 MPa, elongation is 30%, elastic modulus is 196 GPa, Poisson's ratio is 0.294. In order to prevent the stress change caused by wire cutting, the abutment was fabricated before the sample was manufactured, and then the substrate and the abutment were heat treated together, kept for 2 h at 450 °C and cooled in the furnace. The sample is manufactured on the abutment, and the sample is cut by wire with the abutment. Because the surface of the sample is rough and oxides are produced on the surface of the sample during the manufacturing process, these oxides will absorb X-rays, so the measured results can not accurately reflect the real stress state of the sample surface. Therefore, the sample is electrolytically polished with saturated sodium chloride solution and polished 1 min. After measurement, the surface roughness *Ra* of the sample is 8.5 μm, which meets the national standard GB/T7704-2017 for stress measurement by X-ray.

The measurements of residual stress were carried out on the X-ray diffraction equipment D8ADVANCE, experimental conditions: copper target, input wavelength of 0.15418 nm, Ni filter, tube pressure of 40 kV, pipe flow of 40 mA, scanning step length of 0.02°, scanning speed of 0.001°/s, slit DS = 1°; RS = 8 mm (corresponding to LynxExe array detector). The residual stress along the height direction is shown in Fig. 4.6, the residual stress of the part belongs to low horizontal stress in the range of −90 to 110 MPa.

The material near the substrate is the residual compressive stress, with the increase of the number of manufacturing layers, the residual compressive stress reaches the maximum in the middle of the sample, and then gradually decreases until it is transformed into residual tensile stress. The maximum compressive stress of the three

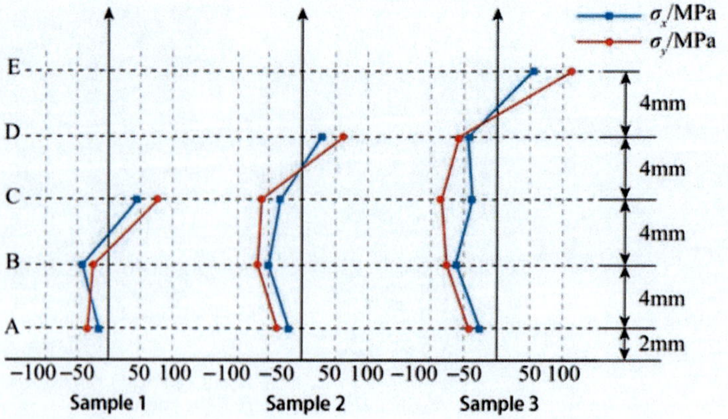

Fig. 4.6 Residual stress in height direction

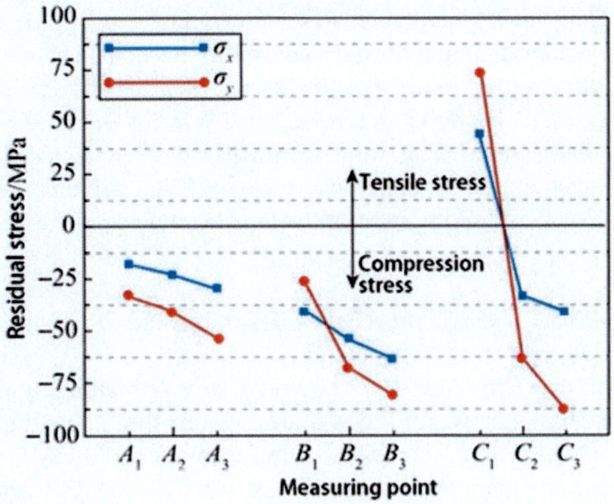

Fig. 4.7 Comparison of residual stresses of specimens with different heights

sample: the X-axis directional stress (σ_x) is −41 MPa, −54 MPa, −64 MPa respectively, and the Y-axis directional stress (σ_y) is −33 MPa, −68 MPa, −88 MPa respectively. The maximum tensile stress is at the top of the sample, σ_x is 44 MPa, 31 MPa, 54 MPa respectively, σ_y is 73 MPa, 61 MPa, 114 MPa respectively. This change trend is due to the fact that the material near the substrate solidifies and cools first, and then the cooling shrinkage of the scanning layer causes the lower layer to bear the compression stress, so that the compression stress of the material near the substrate is larger. With the increase of the number of manufacturing layers, the molten pool is far away from the substrate, the path of heat transfer becomes longer, the internal heat of the formed part continues to accumulate, and the heat becomes relatively uniform, the subsequent thermal cycle tempers the manufactured layer and releases the stress to some extent.

In order to study the effect of scanning layers on thermal stress, the stresses of three samples were compared, Fig. 4.7 shows the comparison between sample 1 and sample 2, relative to the compressive stress (σ_x and σ_y) at A_2 point of sample 1 and sample 2, 5 and 8 MPa are increased, 13 and 42 MPa are increased at point B_2, and the transformation from tensile stress to compressive stress at C_2 point. When the layer of point D_2 is scanned by laser, the layer of point A_2 and B_2 is heated to a higher temperature (but does not exceed the plasticizing point), while the layer of point C_2 is heated beyond the plasticizing point, thus causing stress release in the layer of point C_2. The thermal expansion of the layer of point A_2 and B_2 will be limited by the material below it, so the layer of point A_2 and B_2 will produce compressive stress. In the cooling stage, the shrinkage of the layer where the D_2 point is located is limited by the materials below it (including the layers of A_2, B_2 and C_2), therefore, residual tensile stress occurs in the layer of D_2 point, the residual compressive stress occurs in the layer of C_2 point, and the compressive stress of the layer at A_2 and B_2 points increases. The distance between point A and point D is 12 mm, and the distance between point B and point D is 8 mm, the subsequent thermal cycle at point

B is more remarkable, so the change of stress value is greater. The change of σ_x is smaller than that of σ_y, because the subsequent thermal cycle in Y direction (height direction) is more significant than that in horizontal direction. Similarly, when the layer of E_3 point is scanned, the interior of the part experiences the same stress change trend. This shows that when the new material is piled on the manufacturing layer, the subsequent thermal cycle (STC) will increase the compressive stress in the lower part and transform the tensile stress in the upper part into compressive stress.

4.2.3.2 Horizontal Stress Distribution

In order to study the horizontal stress distribution of L-PBF-fabricated parts, the samples shown in Fig. 4.8 are manufactured, and the test points are marked *m* and *n*, where *m* represents the sample number, *m* takes 1, 2, 3, 4, 5 and 6, *n* to represent the test point number, and *n* takes A, B, C, D, E, F and G. The width of all samples is 6 mm, the thickness is 1 mm, and the distance between the test points along the length of the scan line is 6 mm. The process parameters are as follow: laser power of 150 W, hatch space of 0.08 mm, powder layer thickness of 0.04 mm, and "Z-shaped" X unidirectional scanning strategy. The lengths of samples 1, 2 and 3 are 42 mm, 30 mm and 18 mm, respectively, and the scanning speed is 400 mm/s, which is used to analyze the effect of scanning line length on stress distribution; The lengths of samples 3, 4 and 5 are all 18 mm, but the scanning speed is 400, 200 and 800 mm/s, which is used to study the response of the energy input ($\psi = P/\upsilon$) on residual stress; Sample 6 is 42 mm in length, but is divided into three segments to analyze the effect of segmented scanning on stress.

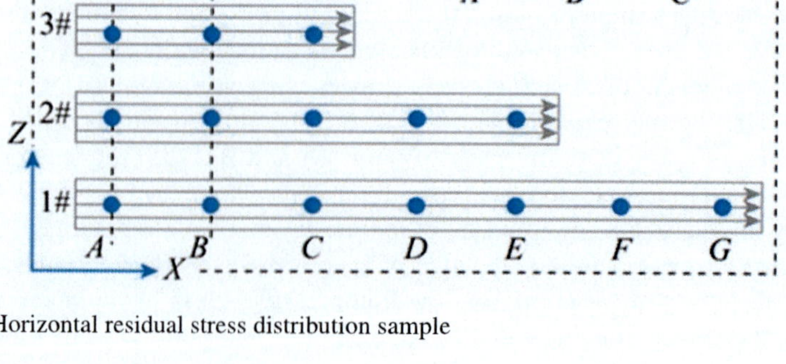

Fig. 4.8 Horizontal residual stress distribution sample

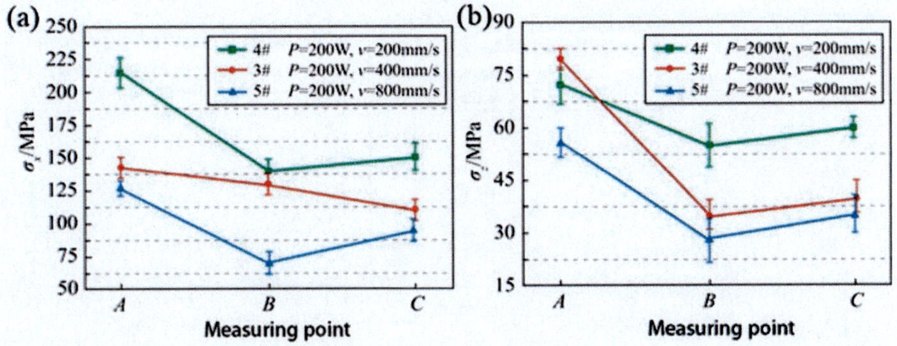

Fig. 4.9 Relation curves between energy input and residual stress: **a** along scanning direction, **b** vertical scanning direction

1. Effect of energy input on stress distribution

Figure 4.9 shows the relationship between residual stress and energy input. According to the formula $\psi = P/\upsilon$, the energy input of sample 4 is twice as much as that of sample 3 and 4 times as much as that of sample 5. The residual stress of sample 4 is the largest, followed by that of sample 3 and that of sample 5 is the smallest. This shows that the greater the energy input, the greater the residual stress in the part. The reason can be attributed to the larger the energy input, the larger the molten pool, and the greater the volume shrinkage of the molten pool after solidification, thus manufacturing a greater thermal stress.

Comparing the curves in Fig. 4.9a, b, it is found that σ_x is much larger than σ_z, the range of σ_x of sample 4 is 137–212 MPa, the range of σ_z is 53–71 MPa, the range of σ_x of sample 3 is 112–142 MPa, the range of σ_z is 35–80 MPa, the range of σ_x of sample 5 is 70–125 MPa, the range of σ_z is 30–55 MPa. This is because in the case of high-speed laser scanning, the heat-affected zone and the molten channel along the scanning direction (X) show a slender shape, and the temperature gradient in the X direction is much larger than that in the vertical scanning direction (Z), while perpendicular to the scanning direction (Z direction) mainly depends on heat conduction, so the temperature change is small. In the L-PBF process, large temperature gradient will lead to large residual stress.

2. Effect of scan line length on stress distribution

Figure 4.10 shows the relationship between the residual stress and the scan line length. Consistent with the results of the previous analysis, the upper surface of the part is tensile stress, and the longer the scan line, the greater the residual stress. The analysis shows that when the laser scans the track, the shrinkage in the layer is mainly the longitudinal shrinkage of the melting track, but when the scanning line is too long, the shrinkage compensation of the melting track is not sufficient, which leads to large residual stress. In addition, the range of σ_x of the three samples is

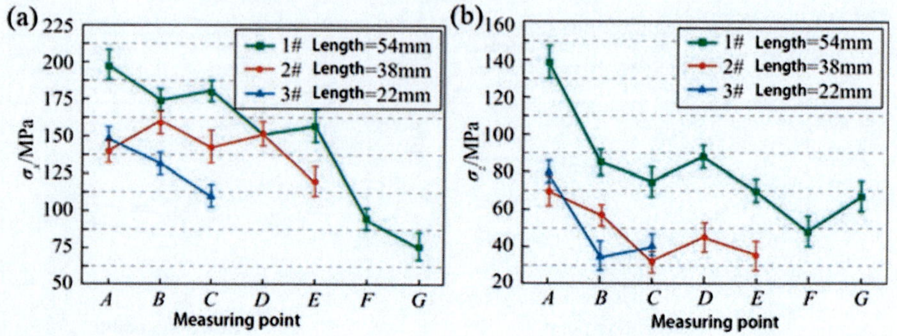

Fig. 4.10 Relation curves between scanning line length and residual stress: **a** along scanning direction, **b** vertical scanning direction

80–180 MPa, the range of σ_z is 30–80 MPa (except the starting point 1_A), and the σ_x of the three samples is larger than the corresponding σ_z respectively.

The peaks of σ_x and σ_z appear at the beginning of the track, and then decrease gradually along the scanning direction, but their changing trends are different. The decrease of σ_x is very small at the starting position, but decreases sharply in the second half of the scan line, while σ_z decreases sharply at the beginning of the scan line, and there is no significant change in the second half of the scan line. This is because at the beginning of the scan line, the molten channel is surrounded by the surrounding powder, and because the thermal conductivity of the powder is much lower than that of the corresponding solid material, the heat of the molten channel is difficult to conduct, and a large temperature gradient is produced in the molten pool, thus causing greater thermal stress. As the laser continues to move forward, the heat is gradually transmitted, the temperature gradient of the melting track decreases and tends to dynamic equilibrium, and the thermal stress decreases gradually.

3. Effect of sectional scanning on stress distribution

Figure 4.11 shows the stress comparison between sample 6 and sample 1. Sample 1 uses long line scanning, while sample 6 uses short line scanning, it can be found that the average value of σ_x of sample 6 is about 22% lower than that of sample 1, while σ_z is reduced by about 17%. It is obvious that the influence of scan line on σ_x is greater than σ_z. At the test points 1_A, 1_B, 1_C, 1_D and 1_E, σ_x generally decreases more than 50 MPa, this is because the short scan line length means that the interval time between adjacent track is short, resulting in a decrease in the temperature gradient in the X direction, thus reducing the thermal stress. This shows that short-line scanning can reduce the residual stress of L-PBFed parts, reduce the risk of warping deformation and cracking, and improve manufacturing quality.

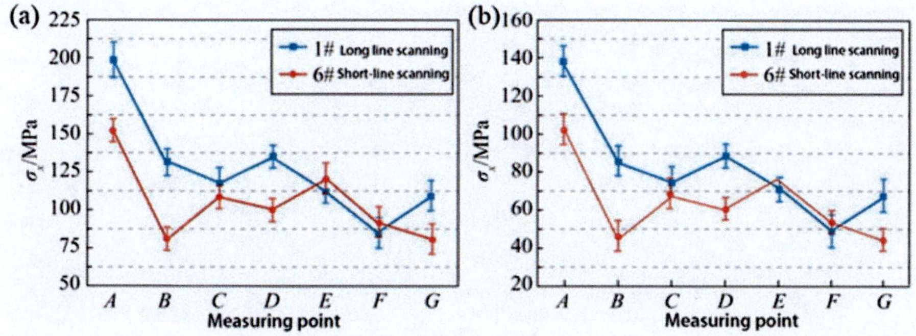

Fig. 4.11 Relationship curves between sectional scanning and residual stress: **a** along scanning direction, **b** vertical scanning direction

4.3 Unstable Manufacturing Process Caused by CAD Model Design

4.3.1 The Overhanging Structure

In theory, LPBF technique can form metal parts with arbitrary complex shape, but it cannot perfectly form all geometric features, especially the overhanging structure. As shown in Fig. 4.12, the overhanging structure makes the local shape accuracy and size accuracy of the LPBF parts cannot meet the requirements, which may lead to the scrapping of the formed parts or even the failure of the forming process in serious cases. For the processing of overhanging surfaces, the main method is to add a large number of supports to ensure the stability at present, and then remove these supports, surface grinding, sand blasting and shot peening to ensure the forming quality of the overhanging surfaces. However, compared with the LPBF vertical forming surface, overhanging surface or small angle inclined surface, the forming quality is always unsatisfactory. In a few cases, the overhanging surface is machined after LPBF process is completed. However, when the formed parts are fine and complex or the overhanging surface is inside the part, the means of adding supports or subsequent machining are no longer appropriate. Therefore, it is of great significance to improve LPBF process and expand the application scope if the overhanging surface can be directly formed completely without adding supports, or the overhanging surface can be avoided or minimized in the design stage.

1. Experimental method

The influence of inclined angle, energy input (scanning speed and laser power), stress accumulation superposition and scanning vector length on the forming of overhanging structure is analyzed by designing forming experiments. The relationship between key process parameters and critical forming angle of overhanging surface formed by LPBF is discussed, so as to provide design basis for forming overhanging

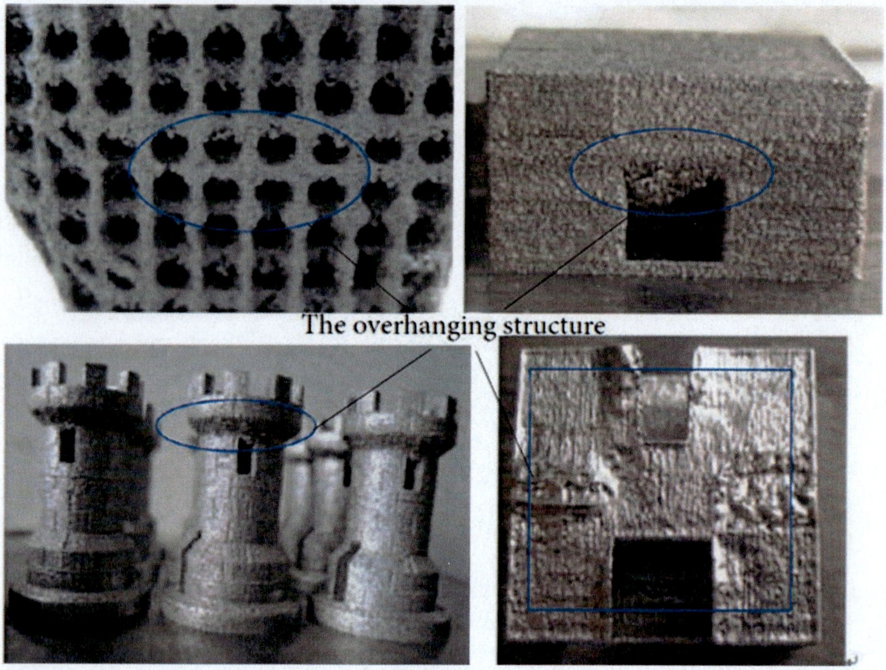

Fig. 4.12 Various defects occurred in the LPBF-fabricated overhanging structure

structure to avoid inherent geometric restrictions and optimize the quality of formed overhanging surface, LPBF technique can be used to fabricate parts with complex shape with low risk.

Figure 4.13 shows the overhanging structure in LPBF process, and is the schematic diagram of arbitrary curved surface part after layering [1]. Among them, the *ab* and *cd* sections are the overhanging surfaces during the forming process, and will form an overhanging part without self-support in slicing. The length S of the overhanging part between layers can be obtained from the following formula:

$$S = h \times \mathrm{ctg}\theta \tag{4.1}$$

where h is slice thickness, θ is the angle between the profile of the slice layer and the horizontal plane.

According to formula 4.1, the value of S is related closely to the thickness h and inclination angle θ: the higher the h or the smaller the θ, the greater S is. At present, the layer thickness used in LPBF is generally determined by the powder particle size of the raw material, and the optimized layer thickness range is 20–50 μm. In this experiment, the layer thickness h is set to 35 μm. So the value of S is mainly related to the inclination angle θ. Inclination angle θ_1 of section *ab* is obviously greater than the inclination angle θ_2 of *cd* segment. Therefore, defects are more likely to

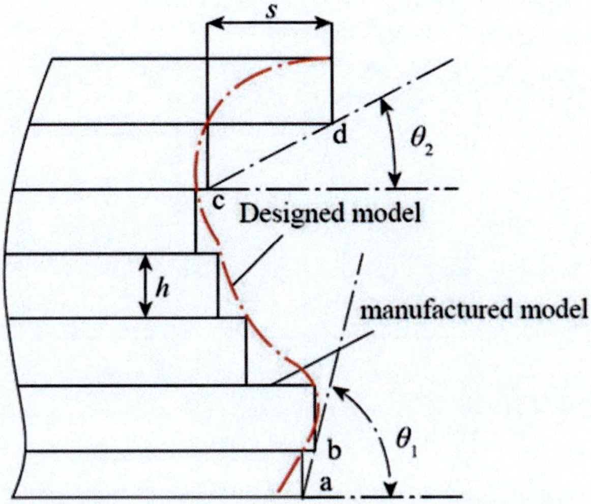

Fig. 4.13 Slice principle model of the overhanging surface

occur in section *cd* during the forming process. during the LPBF process, there is a critical inclined angle. The so-called critical inclined angle means that when the inclined angle is less than a certain value, the overhanging surface will collapse, which affects continuous processing. The two common defects of the overhanging surface in LPBF forming are dross formation and warping deformation.

As shown in Fig. 4.14, inclined planes with different inclination angles θ are designed and fabricated at the scanning speed of 200 mm/s and 600 mm/s respectively, the influences of the size of θ and scanning speed on the forming quality of the overhanging surface are discussed. The laser power is fixed at 150 W, the hatch space is 0.08 mm, and the slice thickness is 35 μm. The number of processing layers is 100, and the size of each processing layer is 10 mm × 5 mm, the XY direction interlaminar staggered scanning strategy is used.

In order to obtain the corresponding relationship between the scanning speed, laser power and the critical inclined angle, based on the above, the inclined angle θ is reduced from 50° to 25°, the scanning speed of 200–1200 mm/s is used, and the laser power is controlled to 120 W, 150 W and 180 W respectively. The critical inclined angle of the overhanging surface at different power and scanning speed is discussed. Considering the serious defects of the overhanging surface, caused by too small inclination angle or too large energy input during the experiment, hinders the smooth progress of the experiment, the overhanging surface with serious defects shall be stopped in time according to the defect degree of the overhanging surface during the processing to ensure the completion of the experiment. Through the real-time photography of the forming process, the warping deformation trend of the overhanging surface is observed, and the influence of the stress accumulation on the forming of the overhanging surface is analyzed.

The internal stress generated in LPBF process is the external force that causes the warping deformation of the overhanging surface. The internal stress due to different

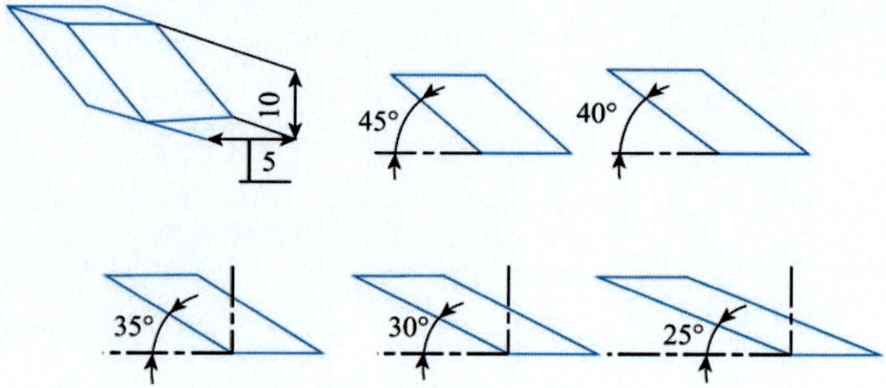

Fig. 4.14 Design of overhanging surface with different inclination angles

scanning vector lengths is also different. In the experiments, the overhanging surfaces with scanning vector lengths of 20 and 80 mm were designed, and the bottom surface was supported with the same density to verify the influence of different scanning vector lengths on the forming of overhanging surfaces.

2. Influence of inclination angle on forming of overhanging surface

Figure 4.15 shows the comparison of the effect of the overhanging surface when the scanning speed is 200 mm/s and 600 mm/s respectively, and the forming inclination angle θ decreases from 45° to 25°. It can be seen from Fig. 4.15a that when the scanning speed is 600 mm/s, only the overhanging surface with $\theta = 25°$ collapse slightly, but the forming process can continue, the overhanging surface with $\theta \geq 30°$ can be well formed. When the scanning speed is 200 mm/s (Fig. 4.15b), the deformation is serious with θ increasing from 25° to 40°, and the warping deformation of the sample is still serious even if θ is 45°. The formed upper surface area is getting smaller and the dross on the overhanging surface is getting more. It can be seen from Fig. 4.15c that the warping deformation of the overhanging surfaces at scanning speed of 200 mm/s is more serious than that at the scanning speed of 600 mm/s.

From the above results, it can be seen that the inclination angle θ and scanning speed have great influences on the forming quality of overhanging surface. According to Fig. 4.13, a smaller θ means a larger overhanging part S between layers. When S is larger than the spot diameter, the laser focused spot completely falls on the powder support area, resulting in a large volume of molten pool and subsidence into the powder. In order to stabilize the overhanging surface, S must be smaller than the diameter of spot, so that the laser spot is mostly scanned in the solid support area. So when the layer thickness $h = 35$ μm, according to formula (4.1), S must be less than the spot diameter which is 70 μm. Theoretically $\theta = 27°$ is the minimum inclined angle of the overhanging structure formed by LPBF. When S is smaller than the spot diameter, $\theta > 45°$ is a reliable forming angle for the overhanging structure of LPBF. In general, the minimum forming angle corresponds to a low energy input (low laser

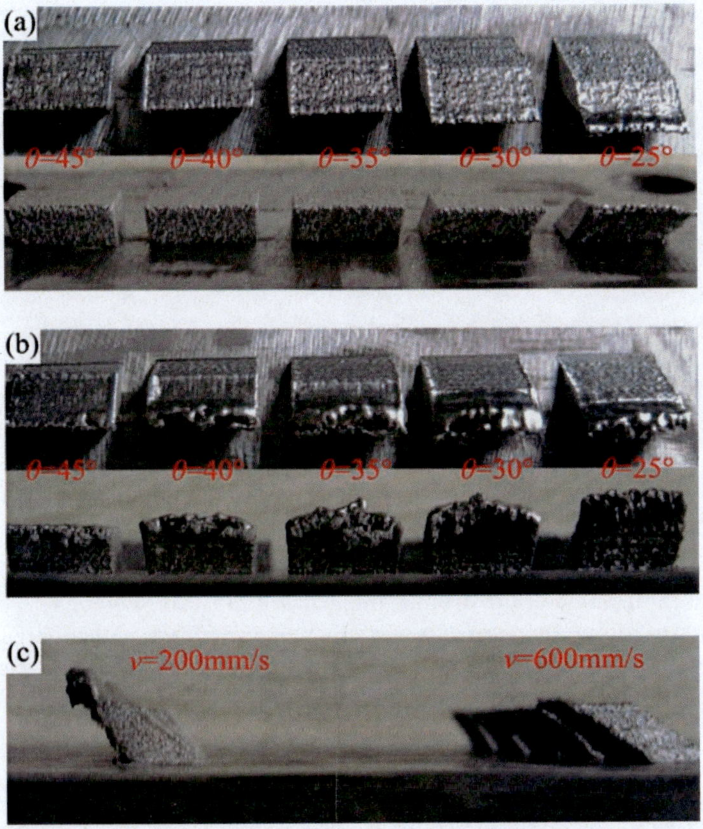

Fig. 4.15 Forming experiments of overhanging surface with different inclined angles: **a** scanning speed v = 600 mm/s; **b** scanning speed v = 200 mm/s; **c** side view of both speeds

power or high scanning speed), while the reliable forming angle corresponds to a high energy input (high laser power or low scanning speed). The above results are in good agreement with the experimental results in Fig. 4.15. When $v = 600$ mm/s, $\theta = 25°$, overhanging surface is deformed, and overhanging structure is well formed when $\theta = 30°$, indicating that the minimum forming angle is between 25° and 30°, which is consistent with the minimum forming Angle $\theta = 27°$ calculated by theory. When $v = 200$ mm/s, the heat input is about 3 times than that of $v = 600$ mm/s, resulting in a rapid increase of internal stress in the LPBF process. Therefore, in order to obtain the ideal forming quality, it is required to increase the inclination angle of the overhanging surface when $v = 200$ mm/s, that is, the minimum forming angle becomes larger. As illustrated in Fig. 4.15b, when $v = 200$ mm/s, $\theta \leq 40°$, the overhanging surface has serious warping deformation and slag hanging, and there is also a small amount of slag hanging on the overhanging surface even if $\theta = 45°$, indicating that the minimum forming angle is slightly greater than 45° when $v = 200$ mm/s, which is consistent with the above theoretical analysis $\theta > 45°$ is the

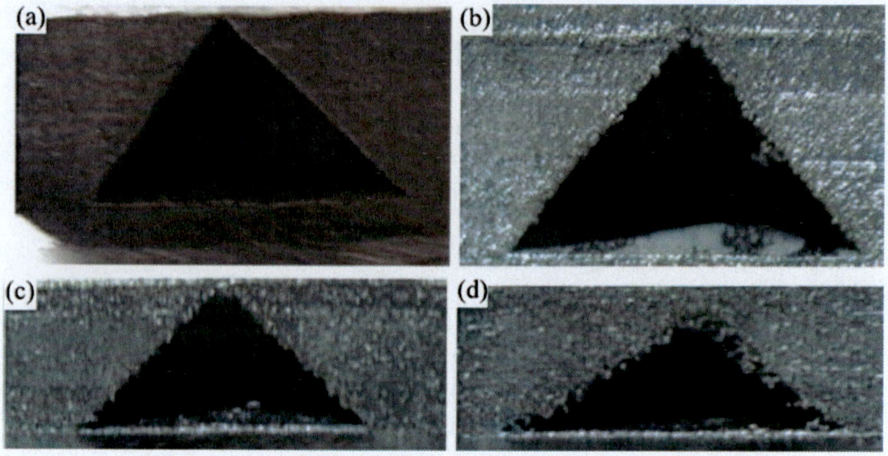

Fig. 4.16 Comparison of forming experiments of overhanging structures with different inclined angles

reliable forming angle of the overhanging surface in LPBF. Figure 4.16 shows the comparison of forming experiments of overhanging structures with different inclined angles under the above conditions. It can be found that the amount of slag hanging under the overhanging surface getting more as the inclined angle gradually decreases, verifying the above analysis conclusions.

3. Influence of scanning speed on forming quality of overhanging surface

Compared with laser sintering forming, the forming speed of LPBF is generally much lower, so as to obtain enough energy input to completely melt the current powder and partially melt the previous melting track and formed layer. However, the smaller the scanning speed is, the longer the laser heating time is. The greater the temperature difference between the current layer and the previous layer as well as the upper and lower parts of the current layer, leading to more serious deformation of the part. In Fig. 4.15, when $\theta = 25°, 30°, 35°, 40°, 45°$, and $v = 200$ mm/s, $v = 600$ mm/s, it is found that the deformation of the sample when $v = 200$ mm/s is much more serious than that when $v = 600$ mm/s at the same inclined angle, indicating that greater internal stress will be generated at low scanning speed, and the corresponding minimum forming angle should also be improved. Combining the analysis of the influence of inclined angle θ on overhanging structure, it is known that the minimum forming angle and scanning speed are mutually restricted under the same laser power. When the inclination angle θ of the overhanging surface is fixed at low level, the scanning speed must be increased to reduce the warping tendency of the overhanging structure; when it is necessary to scan at a low speed to obtain a denser formed sample, the inclined angle must be increased in design in case of unable to add supports. Although the amount of warping decreases with the increase of scanning speed, it cannot be improved by continuously increasing the scanning speed. Because the decrease of energy density caused by excessive scanning speed is bound to reduce the scanning penetration, affecting the combination between the upper and lower layers, and easily lead to interlayer cracking.

Figure 4.17 shows the comparison of the effect of the overhanging surface with different scanning speed increasing from 200 to 1200 mm/s and the forming inclined angle θ increasing from 25° to 50°, when the laser power is 180 W, so as to verify the above analysis conclusion. In order to make the experiment process stable and ensure that the number of processing layers is as large as possible, the forming status must be monitored in real time during the experiment, the processing files with serious forming defects must be stopped to ensure the smooth processing of other files. When the scanning speed is 200 mm/s, warp begins at the 15th layer, and the processing was stopped at 55th layer because of serious fracture caused by warping. When the scanning speed is 400 mm/s, processing was stopped at the 75th layer. When the scanning speed is 600 mm/s, processing was stopped at the 85th layers and when the scanning speed is 800, 1000 and 1200 mm/s, processing was stopped at the 115th layer. Figure 4.17b shows the side view of the samples. It can be seen from the above results that θ and scanning speed (energy input) have obvious influence on the forming quality of the overhanging surface. Moreover, as illustrated in Fig. 4.17, the sequence of warping deformation starts to occur from the top right (small inclined angle θ, low scanning speed) and slowly moves to the lower left as the number of processing layers accumulates (The inclined angle θ increases and the scanning speed increases).

4. The effect of laser power on the forming quality of the hanging surface

Compared with Figs. 4.17 and 4.15, it is also found that at the same scanning speed and inclined angle, the deformation tendency of the overhanging surface formed by the laser power of 180 W is larger than that of 150 W. In Fig. 4.15, when the scanning speed is 600 mm/s and the laser power is 150 W, the overhanging surfaces with $\theta \geq 30°$ can be formed well; while in Fig. 4.17, when the scanning speed is 600 mm/s and laser power is 180 W, the processing of the overhanging surface stops at the

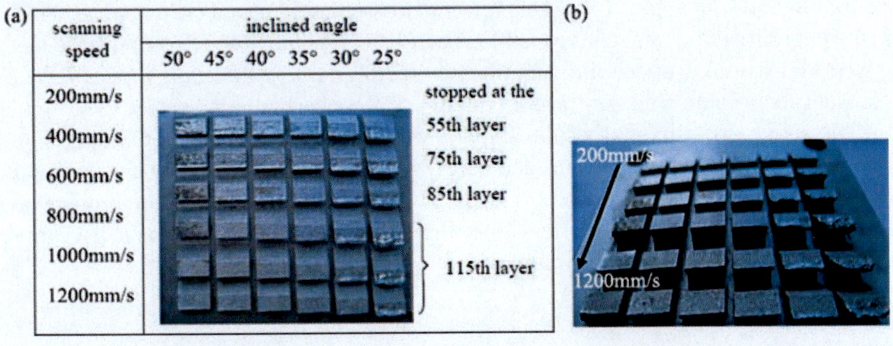

Fig. 4.17 Experiments of overhanging surfaces formed at different θ when the scanning speed increased from 200 to 1200 mm/s: **a** front view; **b** side view

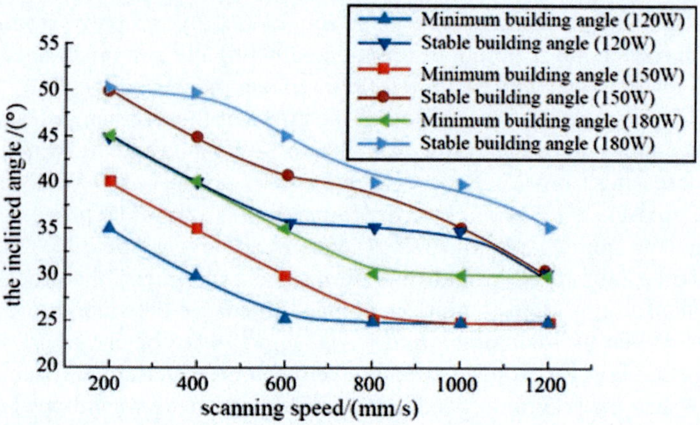

Fig. 4.18 Relationship between critical building angle and different scanning speeds, laser power

85th layer, and the overhanging surface with $\theta = 30°$ has been warped seriously. The above results illustrate that the laser power also has a great influence on the forming quality of the overhanging surface.

Figure 4.18 shows the manufacturing results of the overhanging surfaces at different scanning speeds and inclined angles θ when the laser power is fixed at 120, 150 and 180 W. It can be seen that the scanning speed and laser power restrict the selection of the inclined angle. When the laser power increases, the parameter range shifts to the upper right, that is, the minimum building angle and the stable building angle become larger. Each power corresponds to two curves, namely the minimum building angle (thin line) and the stable building angle (thick line).

Figure 4.19 shows when P = 150 W the curves of minimum building angle and stable building angle divides the parameter range into three areas: non building area, drape coverage area and stable building area. It can also be seen from Fig. 4.19 that the laser power and scanning speed have equal effects on the overhanging surface in some cases. For example, the stable building angle at P = 120 W is the same as the minimum building angle at P = 180 W when the scanning speed is 200 and 400 mm/s (the experimental angle change is 5°). Therefore, it is feasible to combine the laser power and scanning speed into an influence factor, namely the energy input *P*/*V*, to consider its influence on the forming quality of the overhanging surface.

The above experimental results and analysis show that the critical building angle of building the overhanging surface safely is closely related to the energy input. Under certain energy input conditions, the building process is more dangerous when the building inclined angle is less than the critical inclined angle, and it is more difficult to obtain the ideal forming quality of the overhanging surface.

5. Influence of stress accumulation on the quality of overhanging surface

Figure 4.20 shows the laser processing state in different time periods when the laser power is 180 W. There is no obvious wrapping in the first 10 layers, and warping

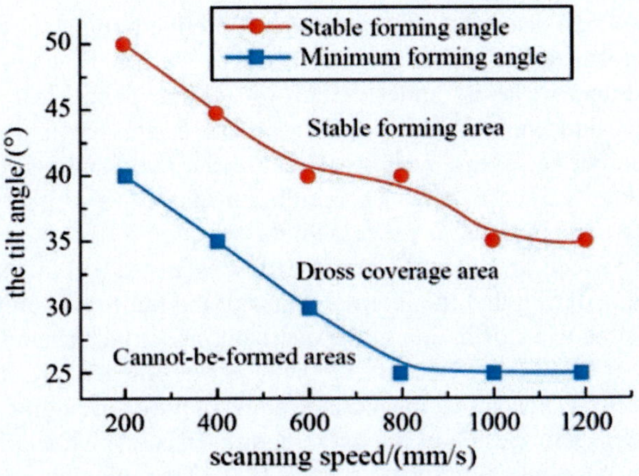

Fig. 4.19 Curves of critical building angle (including minimum forming angle and stable forming angle) divide building into three areas (P = 150 W)

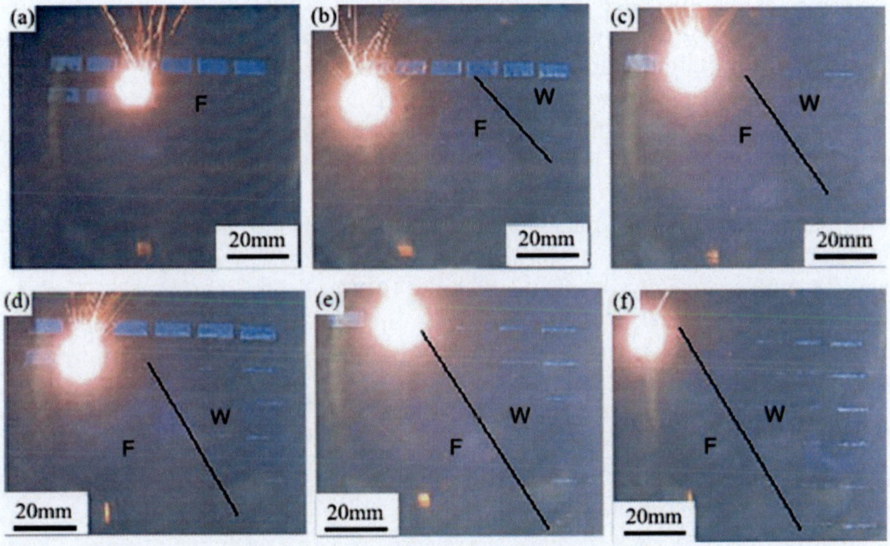

Fig. 4.20 Surface states of the overhanging structure formed by LPBF at different times (W represents warped and F represents flat): **a** the 8th layer; **b** the 15th layer; **c** the 26th layer; **d** the 39th layer; **e** the 45th layer; **f** the 53rd layer

begins to occur at about 15th layer. With the accumulation of processing layers, the degree and scope of warping become larger and larger. It can be found that the warping deformation of the whole building parameter area starts from the upper right corner (the 15th and 26th layer), and slowly migrates and expands to the lower left corner (the 39th layer—the 45th layer—the 53th layer), that is, the overhanging surface with a small scanning speed and small inclined angle starts to warp. It is also found that the same overhanging surface will warp only when it accumulates to a certain number of layers under certain processing parameters, indicating that under the same building conditions, the cumulative stress will increase with the increase of number of layers, which will cause the overhanging surface to warp. As shown in Fig. 4.17, the stress accumulation effect has a great impact on the warping. When the 55th layer is processed, the warping of the overhanging surface with a scanning speed of 200 mm/s becomes very serious, and the processing must be stopped. Considering that the internal stress of the built part cannot be real time measured, the time of deformation of the overhanging surface cannot be determined by quantitative method, but by observing the effect of powder spreading at different times, the database of heat input, the inclined angle of the overhanging surface and the degree of thermal deformation of the overhanging surface is established to facilitate the early data processing and process control of the technologists and designers.

Figure 4.21 shows the powder spreading effect at different times, where can be found that with the superposition and accumulation of layers, the overhanging structure area with a low inclination angle and high energy input has severe warping deformation. The height of the formed surface with warping greatly exceeds the height of the powder bed surface after powder spreading, which not only discards the built parts with the overhanging structure, but also seriously affects the powder spreading safety of the powder spreading device.

Table 4.1 lists the forming defect forms of the overhanging structure during powder spreading and of the final built parts. The defect degree of these forms varies according to the different inclined angle, scanning speed and the number of processing layers. It can be found that the main defects in the built overhanging structures in Table 4.1 include subsidence of the formed surface, warping and fracture, and the existence of these defects affect the LPBF process significantly. The root cause of warping and fracture is thermal deformation. It is very important to ensure that the overhanging structure does not deform during the LPBF process, because the unmelted powder cannot fix the overhanging structure that will warp.

6. Influence of scanning vector length on the manufacturing of overhanging surface

It can be seen from Fig. 4.22 that when the scanning vector length is 80 mm, both ends of the overhanging structure have already completely separated from the supports (100 layers formed), and both ends are severely warped; when the scanning vector length is 20 mm, the overhanging surface is in good shape (250 layers formed), and no deformation can be seen in the overhanging structure in contact with the supports.

Figure 4.22 shows the sample that is under condition of the inclined angle $\theta = 0°$ of the overhanging structure in LPBF. Two ends of the overhanging surface are seriously warped and deformed when the scanning vector length is 80 mm and pulled off the supports below. The experimental results show that the long scanning line accumulates more internal stress than the short scanning line. The analysis shows that when the direction of the scanning line is parallel to the long side of the section, the in-layer shrinkage mainly depends on the longitudinal shrinkage of the scanning

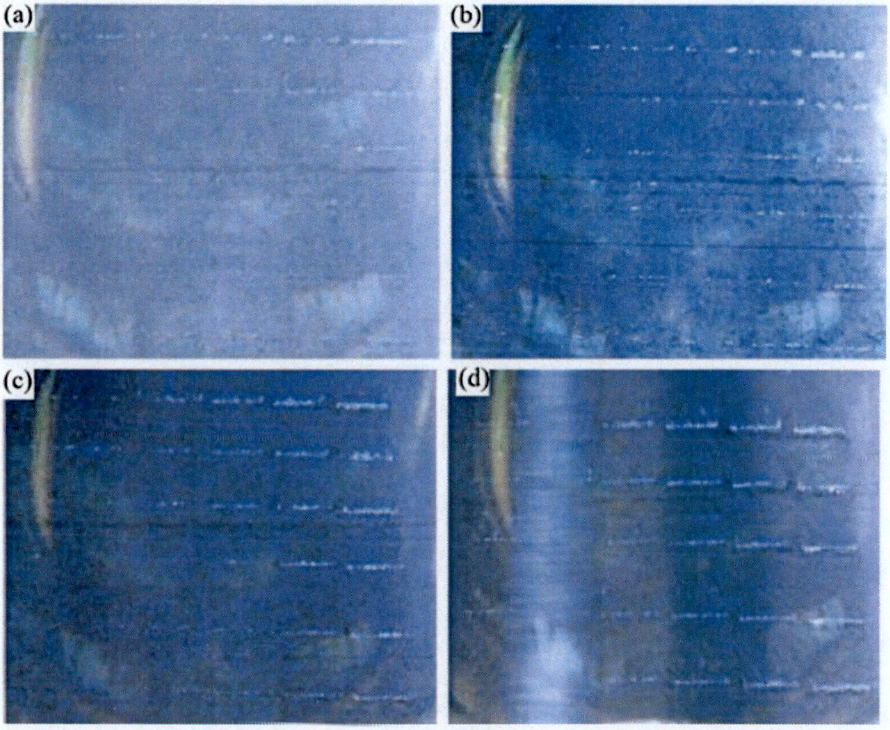

Fig. 4.21 Powder spreading effect shot in different time periods: **a** the 10th layer; **b** the 25th layer; **c** the 45th layer; **d** the 70th layer

Table 4.1 Various defect morphology of overhanging structure during powder spreading and of the final built parts in LPBF

Defects during powder spreading				
Forming defects of overhanging surface				

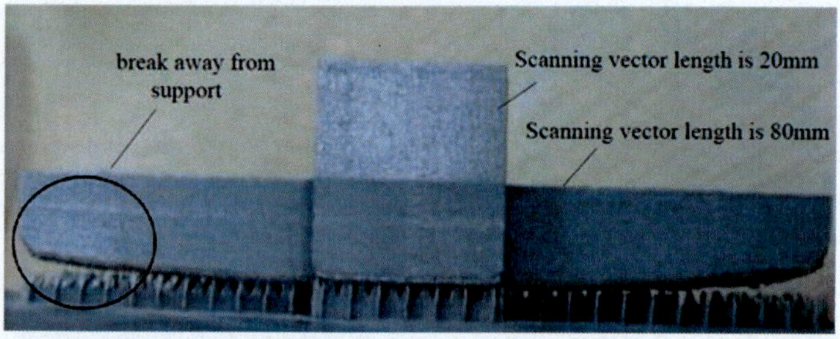

Fig. 4.22 Manufacturing of overhanging structure with different scanning vector lengths (support added on the bottom surface)

line, which makes the shrinkage compensation process inadequate and the stress in the layer is large.

4.3.2 Overhanging Structure of Curved Surface

When the overhanging structure of curved surface is formed by LPBF, the defects such as warping and dross usually appear, which leads to the instability of the processing process. For horizontal or nearly horizontal overhanging structures, only adding supports can ensure forming; for the inclined overhanging surface of the curved surface, it is generally considered that when it is lower than a certain inclined angle, it is a must to add supports to ensure forming. However, the damage to the surface of the parts caused by the support structure must be removed by means of polishing, which increases the time and difficulty of post-treatment. In order to study the forming law of surface overhanging structure, 316L stainless steel powder was used to design a model with circular arc curved surface in the Z-axis direction.

As shown in Fig. 4.23, a 1/4 arc with a radius of 20 mm in the Z-axis direction is designed in the experiment [2]. The width of the part is 10 mm, the thickness is 3 mm, and the scanning area of each layer is 10 mm × 3 mm. In the previous overhanging surface experiments, the ordinary inclined plane overhanging surface model was used, and one model could only verify one angle. When conducting experiments on multiple parameters, the number of experiments was large, and the experimental data processing was also very troublesome. This model overcomes the above shortcomings and can ensure that the inclined angle between the current forming layer and the previous layer changes from 90° at the bottom to 0° at the top. Therefore, the model can simulate the overhanging structure with a inclined angle of 0°–90°, and the lap amount between the upper and lower layers gradually reduces from 100% at the beginning to 0 at the end. As shown in Fig. 4.23a, the overhanging structure is divided into upper and downward surfaces respectively. Points on the surface with inclined angles θ of 80°, 60°, 45°, 30° and 10° are selected for surface topography analysis and roughness measurement. In order to better observe the deformation of the

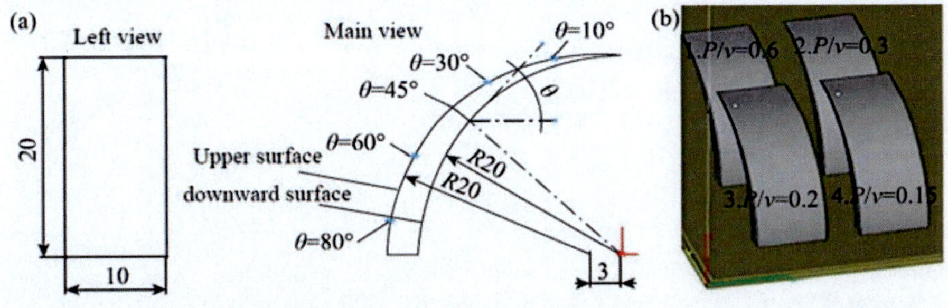

Fig. 4.23 Curved surface overhanging structure model design: **a** curved surface size model; **b** positioning of the model in Magics software

curved overhanging structure and ensure the smooth processing, the stainless-steel comb flexible scraper was used as the powder spreader in the experiment.

Figure 4.24 shows the morphologies of curved overhanging structure under the condition of laser energy input $\psi = 0.15$–0.6 J/mm, where can be seen that the surface quality of the curved surface structure becomes worse when it is far away from the substrate, and serious warping and collapse occur when it reaches the top of the curved surface. As the height of all the warped built parts exceeds the design value by 20 mm, the flexible powder spreading device is used to ensure the complete processing of the four curved parts. With the increase of the height, the surface warping increases, which causes the laser to scan the warping position repeatedly, making the deformation of the top of the surface more serious. In the second half of the experiment, the overhanging area between the upper and lower layers became larger, and began to incline upward due to the internal stress of the built part. The extent of the warping increased with the increase of laser energy input, indicating that the greater the energy input, the greater the internal stress. In addition, when forming the overhanging layer, the overhanging part will not fall due to lack of support but will rise.

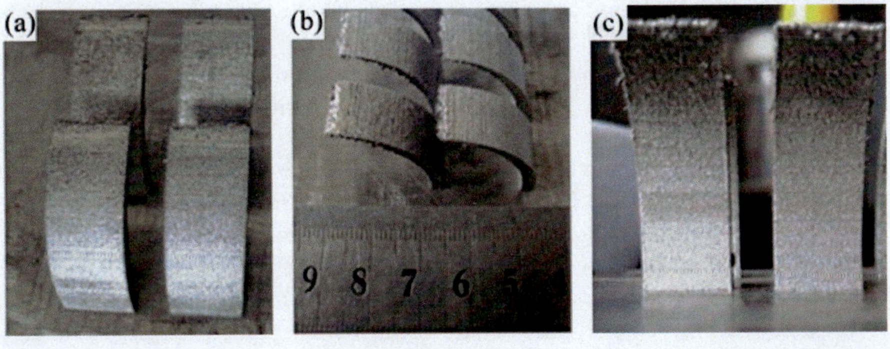

Fig. 4.24 Forming effect of curved overhanging structure: **a** main view; **b** side view; **c** rear view

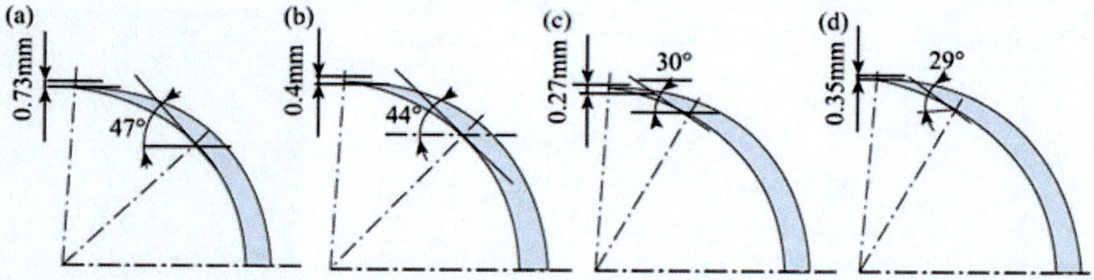

Fig. 4.25 Warping deformation of curved overhanging structure under different laser energy inputs: **a** $\psi = 0.6$ J/mm; **b** $\psi = 0.3$ J/mm; **c** $\psi = 0.2$ J/mm; **d** $\psi = 0.15$ J/mm

The deep melt layer appears on the downward surface of the built part due to the penetration of the laser beam, and the spherical slag particles of different sizes hang on the downward surface. The starting height of slag is also different on the built parts with different energy input. Among them, No. 1 built part with energy input of 0.6 J/mm has the lowest hanging height and the most hanging slag amount. No. 3 and No. 4 built parts with energy input of 0.2 and 0.15 J/mm have similar hanging slag situation less than No. 1 and No. 2. When the curved overhanging structure is formed to the top, the staircase effect on the upper surface is very obvious, and the lap effect of interlayer frits can be seen, resulting in rough surface. It can also be seen from Fig. 4.24 that the quality of the upper and downward surfaces change with the inclination of the surface. With the decrease of the slope of the surface, the quality of the downward surface is much worse than that of the upper surface.

By measuring and analyzing the formed curved surface overhanging structure, it is found that the built parts have warping deformation and overhanging objects on the overhanging surface at the same time. By comparing with the design size, measure the cumulative height of the specimen warping, as shown in Fig. 4.25. It can be found that the higher the energy input is, the greater the warping deformation of the top of the surface will be, and the warping amount will decrease with the decrease of the energy input. When the energy input ψ is 0.2 J/mm, the minimum building height error is 0.27 mm. The forming quality of the specimens with energy input of 0.2 and 0.15 J/mm is almost the similar. It shows that the forming quality of the inclined overhanging surface can be improved by increasing the scanning speed or reducing the scanning energy input, but when the energy input is reduced to 0.15–0.2 J/mm, the improvement of the surface quality by reducing the energy input is already limited. Further reducing the energy input can no longer effectively reduce the deformation and avoid the dross formation, so the limit inclined angle of the overhanging surface that can be well formed is determined as 30°, as shown in Fig. 4.25c. The above results show that the size of laser energy input significantly affects the deformation and building limit angle of the part overhanging structure. Because the excessive laser energy input increases the amount of molten metal and prolongs the cooling and solidification time of the molten pool, the higher the energy input, the greater the warping deformation. In addition, continuously reducing the energy input cannot be used for reducing the degree of warping deformation, because the lower the energy input, the smaller the scanning track penetration will be, which will affect the combination of upper and lower layers, resulting in reduced density or cracking of upper and lower layers.

After measuring the curved surface overhanging structure shown in Fig. 4.25, it is found that the built part scanned with energy input of 0.6 J/mm starts to appear dross drops at the position with an inclination of 47°, and the surface starts to become rough, as shown in Fig. 4.25a. When the energy input is reduced to 0.4 J/mm, the angle of the built part with dross drops to 44°, as shown in Fig. 4.25b. When the energy input is reduced to 0.2 J/mm, the dross will not appear until the building angle is 30°, which is lower than the inclined angle, as shown in Fig. 4.25c, d. For the dross caused by laser penetration, it is found that all of the four built parts under different energy inputs have four areas as shown in Fig. 4.26a: no dross area, dross partial coverage area, dross full coverage area and non building area. The corresponding angles of each area in the overhanging structure formed under different laser energy parameters are also different. Mark the junction angles between different areas of the four curved overhanging structures to determine the angle range of each area, and then study the angle change rule of the dross under different energy input parameters. Figure 4.26b shows the curves obtained by measurement and analysis. The three curves show respectively the inclined angle of the overhanging surface under the energy input parameters where dross starts appear, dross is fully coveraged and it is unable to be lapped and formed.

It can be seen from Fig. 4.27 that the roughness value *Ra* of the upper surface of built part with energy input of 0.15 and 0.2 J/mm has a relatively large error at the point where the inclined angle is 10°. However, with the increase of the inclined angle, the surface roughness *Ra* of the built parts with 0.2 and 0.15 J/mm energy input has little difference at the points at 30°, 45°, 60° and 80° inclined angles. Compare the downward surface roughness values of two parts with energy input of 0.15 and 0.2 J/mm. At the point at inclined angle of 10°, the surface quality of the built part with energy input of 0.1 J/mm is significantly worse than that of the built part with energy input of 0.2 J/mm; however, at the inclined angles of 30°, 45°, 60° and 80°, the downward surface quality of the built part with the energy input of 0.15 J/mm

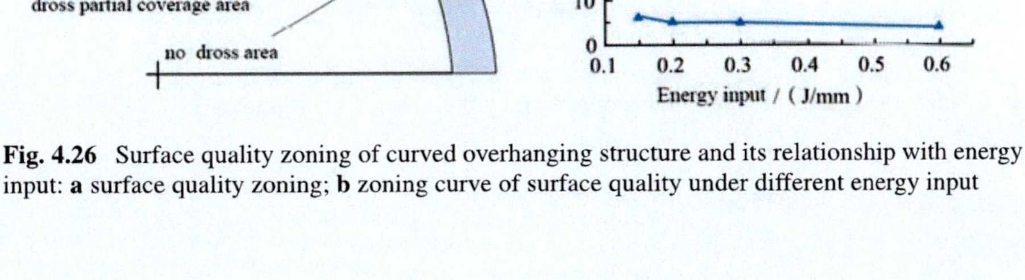

Fig. 4.26 Surface quality zoning of curved overhanging structure and its relationship with energy input: **a** surface quality zoning; **b** zoning curve of surface quality under different energy input

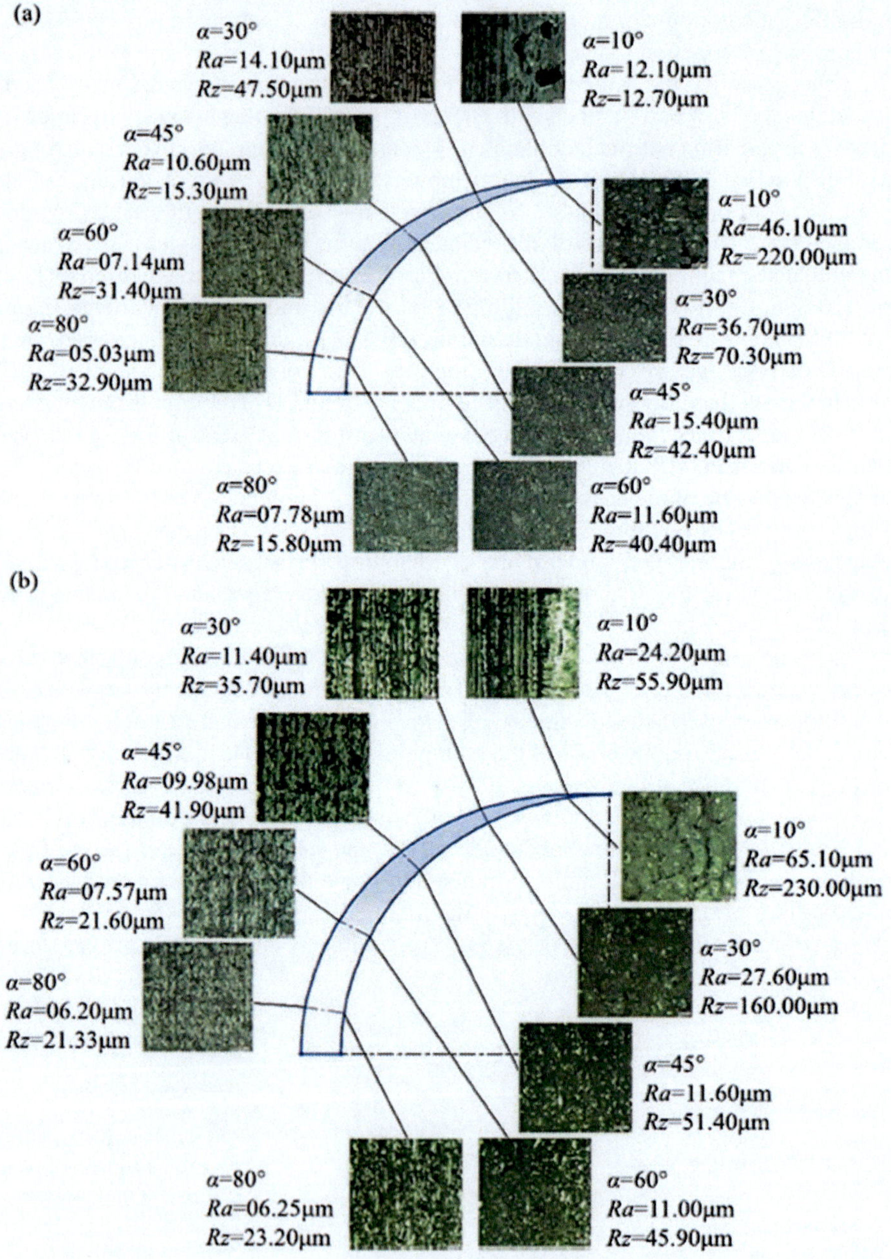

Fig. 4.27 Surface topography and surface roughness at different positions of curved surface overhanging structure with different energy input: **a** 0.2 J/mm; **b** 0.15 J/mm

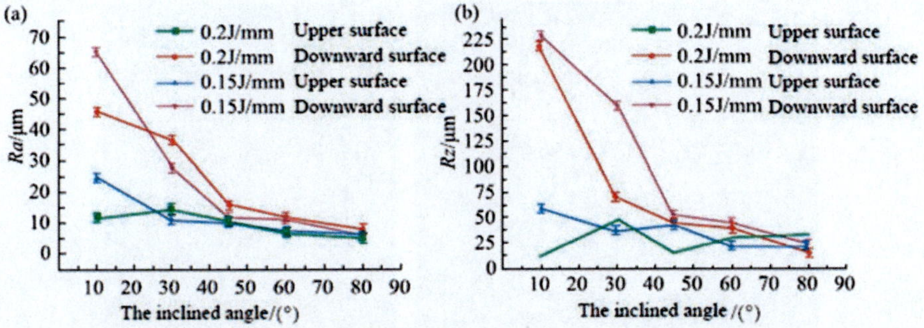

Fig. 4.28 Effect of different energy input on the upper and downward surface roughness of the overhanging surface. **a** Roughness value *Ra*; **b** roughness value *Rz*

is slightly better than that of the built part with energy input of 0.2 J/mm. When the inclined angle is between 10° and 45°, the downward surface roughness *Ra* of the two built parts drops sharply. With the increase of inclined angle, the downward surface roughness decreases gradually. When the inclined angle increases to 80° or even 90°, the upper and downward surface roughness of the two curved surfaces tends to be the same. Through the analysis of Fig. 4.27, the roughness values *Ra* and *Rz* of the upper and downward surfaces of the built parts with energy input of 0.15 and 0.2 J/mm at different inclined angles are tested on the JB—8 rough design, as shown in Fig. 4.28. The variation trend of *Rz* is basically the same as that of *Ra*, but the variation range of *Rz* is much larger than that of *Ra*, and the variation range of *Ra* on the upper surface is from 6 to 20 μm. The upper surface *Rz* varies from 20 to 50 μm; the variation range of *Ra* on the downward surface is from 6 to 65 μm, and the variation range of *Rz* on the upper surface is from 20 to 220 μm. It can also be seen from Fig. 4.28 that the deviation of upper and downward surface roughness values *Ra* and *Rz* becomes larger after the inclined angle is less than 40°.

The surface roughness value at each point on the curved overhanging surface changes with the changing inclined angle, the roughness value of the downward surface of the curve part is obviously higher than that of the upper surface, because the melting track of the inclined surface are built on different bases. The upper surface melting tracks are stacked and solidified on the solid, while the downward surface melting tracks are stacked and solidified on the loose powder. The inclined plane is easy to absorb powder during cooling, resulting in more unmelted powder sticking on the inclined plane. The adhesion of powder particles and the "step" effect significantly increase the surface roughness of curve overhanging parts. Figure 4.29 shows the step effect on the upper and downward surfaces of the curved overhanging structure when the energy input is 0.2 J/mm and $\theta < 45°$. It can be seen from Fig. 4.29a that the lapping effect of the within and between layers melting tracks on the upper surface. It can be seen from Fig. 4.29b that the powder adhered to the downward surface also increases with the increasing stacking height of the curved surface model.

Through the analysis of the above results, it is found that during the manufacturing of the curved overhanging structure, in addition to the lifting of the overhanging part

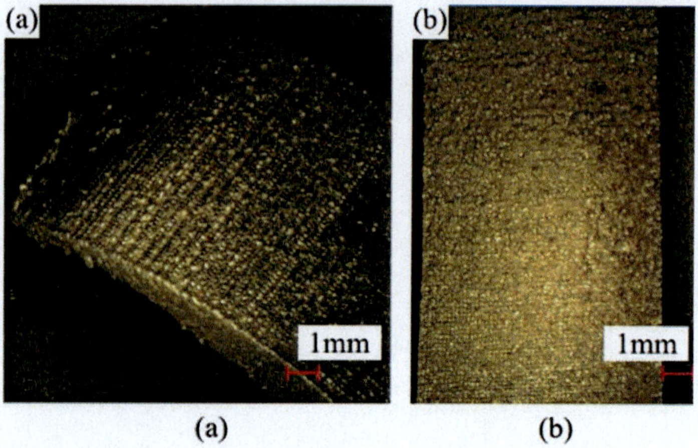

Fig. 4.29 Step effect on the upper and downward surfaces of a curved overhanging structure when the inclined angle $\theta < 45°$ ($\psi = 0.2$ J/mm). **a** Upper surface; **b** downward surface

due to lack of supports, there will also be slag hanging, resulting in a large surface roughness value of the overhanging part of the curved surface. The lifting degree of the overhanging part is proportional to the internal stress caused by the high or low energy input. The amount of laser energy input obviously affects the deformation and building limit angle of the overhanging surface. Under the same layer thickness and spot diameter, too high laser energy input increases the amount of molten metal, prolongs the cooling and solidification time of the molten pool, and increases the deformation time of the overhanging part caused by internal stress. In addition, the laser with higher energy density will melt the powder beyond the design layer thickness, manufacturing the dross and destroying the formation of the overhanging surface.

Supports must be added to the overhanging structure (such as nonlinear surface and inevitable horizontal plane) that still exists beyond the building limit angle after part placement and part modification. Excellent support must be firm, effective and easy to remove from the part. Although Magics software can automatically generate supports according to the parts, the automatically generated supports cannot meet the building requirements of some precision parts' complex curved surface overhanging surface, and the supports of precision parts in special cases need to be added manually. When a nonlinear surface needs to be supported, the support line should be designed according to the slope of the surface slice. When adding support to the surface shown in Fig. 4.30, the lowest point O in the Z direction must be first found. The layer where point O is located will be the first scanned area during part manufacturing, so the support line must start from point O and contact it to ensure the successful modeling of point O and the stability of the part. The denser the contour lines are, the greater the inclined angle of the section is, and no support is needed; the thinner the contour line is, the smaller the inclined angle of the section is, and it needs to be supported.

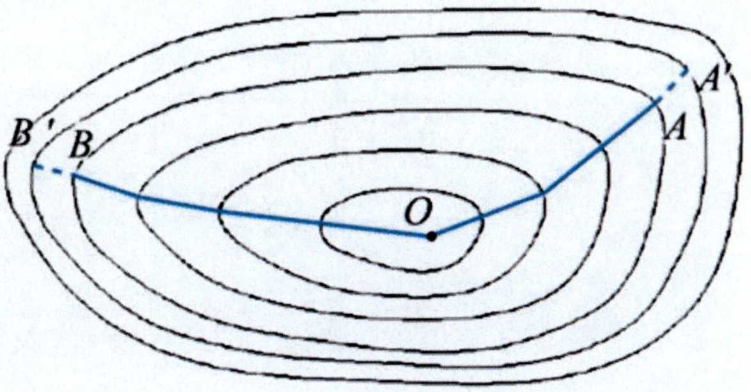

Fig. 4.30 Slice pattern and support line of overhanging surface

In theory, the support line must start from point O and cover all curved areas where the inclined angle of the section is less than the building angle of the process limit.

Figure 4.31 shows a typical circular overhanging surface. In the ($h_1 + h_2$) area, good manufacturing can be achieved through typical process parameters, while in the h_3 area, it is necessary to reduce the energy input (usually by increasing the scanning speed) to reduce warping and scum. How to determine the boundary between h_2 and h_3 regions is the key to laser melt manufacturing of annular cantilever surface in LPBF. The approximate boundary can be basically determined through angle experiment. When manufacturing metal parts with overhanging surfaces with very low inclined angles, it is important to produce the first layer without deformation on the powder bed, because the bottom layer of powder will not limit the deformation, and the energy input is the key to control the deformation. Figure 4.32 shows the comparison between the manufacturing results of circular overhanging surface with or without local energy input control, where can be seen that the surface forming quality of overhanging parts has been significantly improved by controlling local energy input.

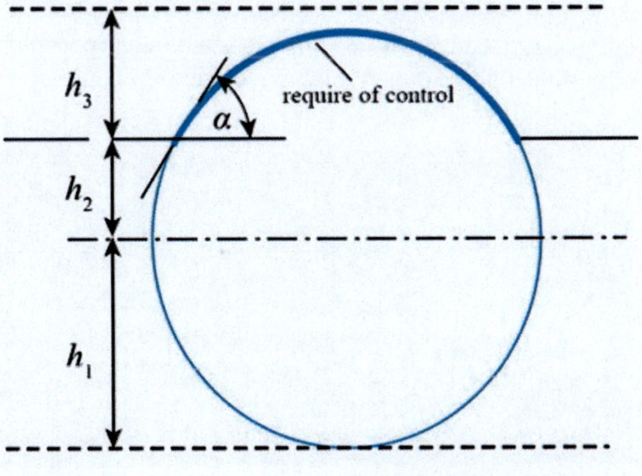

Fig. 4.31 Schematic diagram of circular overhanging surface

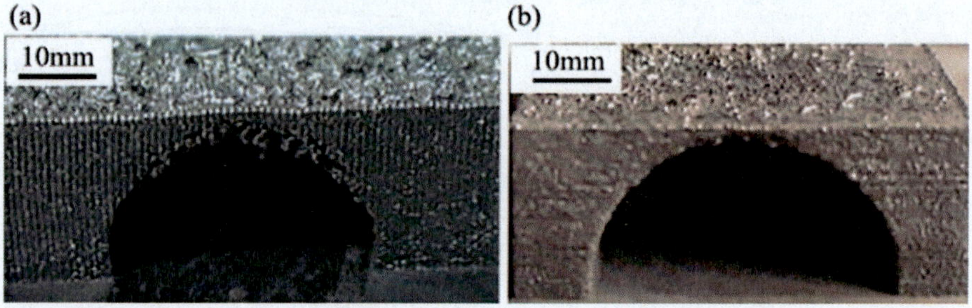

Fig. 4.32 The forming effect of circular overhanging surface. **a** The scanning speed of h3 area is 200 mm/s; **b** the scanning speed of h3 area is 600 mm/s

4.4 Powder Contamination and Its Effect on Defects

4.4.1 Powder Contamination and Its Effects

In the process of L-PBF manufacturing, the metallic powders often have defects such as spattering, oxidation and spheroidization when they are melted by laser irradiation, which causes the contamination of metal powders Fig. 4.33a shows the new powder of 316L stainless steel, and Fig. 4.33b shows the remaining impurities after sieving the powder after a long processing time, which mainly consists of some large size particles, metal spheroidization products oxidized on the surface of the powder, slag falling during the manufacture of the suspended structure, etc.

If the powder is used repeatedly for a long time without sieving, the surface of the manufactured part will be embedded with many inclusions. The composition of these inclusions is the same as that of the slag produced in the steel making process. After analysis, the inclusions are mainly composed of SiO_2, CaO, MnO, etc., as shown in Fig. 4.34. The presence of inclusions in the internal of manufactured parts can seriously affect the mechanical properties and may threaten the stable operation of the powder laying device, because the particle size of the spherical

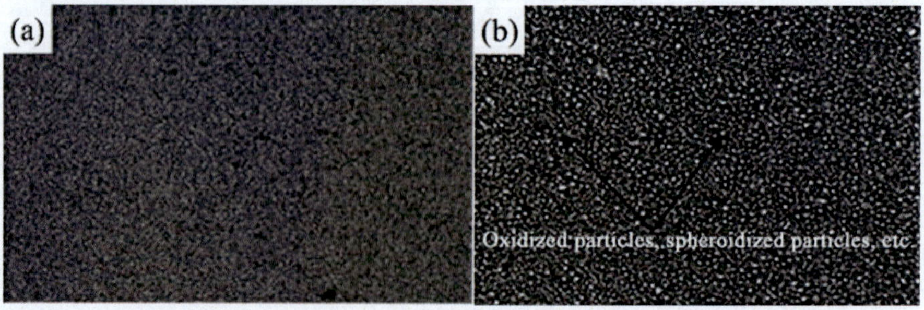

Fig. 4.33 New powder and impurities after sieving of powders used for a long time: **a** 316L new powder before experiment; **b** impurities remaining after sieving of long-processed powders

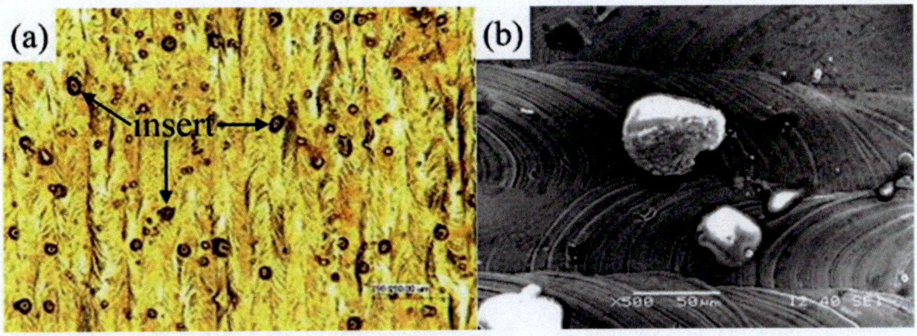

Fig. 4.34 Inclusions embedded in the surface of the manufactured part: **a** surface inserts; **b** microscopic magnification of surface inserts

particles, spatters are generally several times larger than the diameter of the raw powder particles, while the height between the powder laying device and the surface of the manufactured part is generally only a few tens of microns (the size of the set processing layer thickness). When the powder laying device encounters the above-mentioned inclusions, the powder laying device is prone to failure such as jamming. Therefore, the elimination of inserts and the reduction of oxide particles in the powder are important for manufacturing high-quality metal parts. It can be achieved by the following ways: (1) Use single-phase powder as much as possible; (2) The oxygen content of the manufacturing chamber should be kept as low as possible (preferably below 0.1%); (3) Periodically sieve and dry the powder material [1].

4.4.2 Powder Contamination Due to Spatter

The 316L powder used five times was sieved with an aperture of 200 mesh (particles with a particle size less than 75 μm were allowed to pass, and the remaining material was collected) and observed using an ultra-depth 3D microscope as shown in Fig. 4.35a. The SEM image of the spatter particles shown in Fig. 4.35b shows that the spatter particles also show a spherical shape (the larger particle size in the figure), which is due to the free shrinkage of the molten metal into a spherical shape under the action of surface tension after it is separated from the molten pool. A large number of unmelted powder particles are attached to the surface of the spatter particles, which is because the droplets still appear molten or even melted when they fall to the surface of the powder bed, and adhere to the surrounding unmelted powder particles. Figure 4.35c shows the particle size distribution of the spatter particles (quality percentage): D10 < 46.2 μm (10%), D50 < 108.1 μm (50%), D80 < 174.48 μm. The average particle size is 119.7 μm, which is almost three times that of powder particles, and the concentration of particle size distribution is low.

Figure 4.36a, c and e show the SEM images of clean 316L powder, spatter particles and LPBF manufactured parts, respectively. Figure 4.36a shows that the clean powder

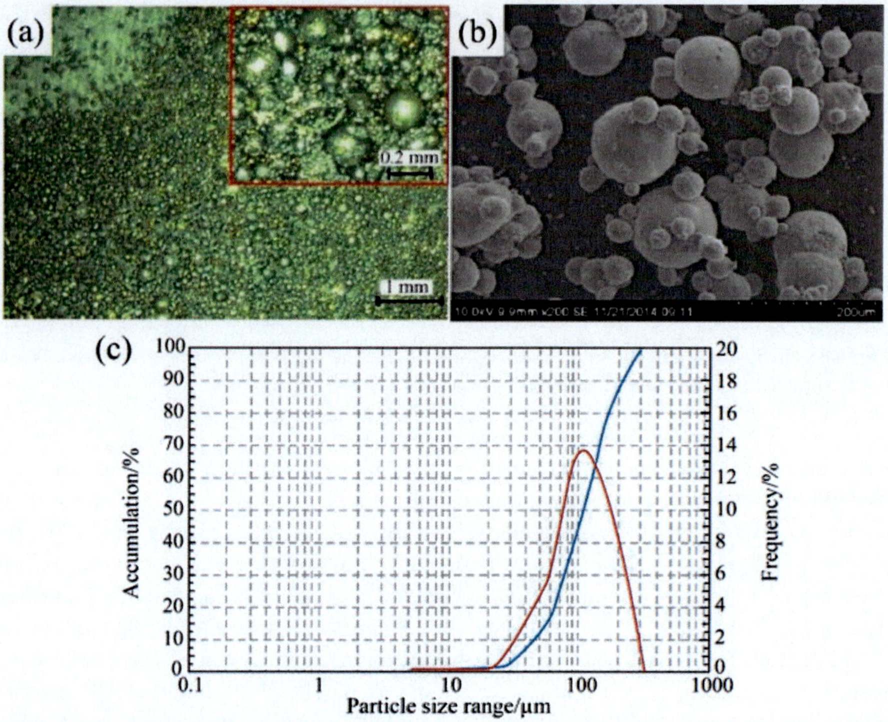

Fig. 4.35 Spatter particles from L-PBF manufacturing: **a** spatter particle morphology; **b** SEM image of spatter particles; **c** spatter particle size distribution

particles have smooth surface and high sphericity with only very few small satellite particles adhering to the surface. Figure 4.36c shows that the diameter of the spatter particles is about 400 μm, and the red dotted lines on the surface are marked as bulges, which are caused by the rapid expansion of the spatter particles when they pass through the laser-irradiated area by absorbing heat. Figure 4.36a, e shows a large number of particles embedded on the surface of the L-PBF part. This is due to the fact that the spatter particles remain at a high temperature when they are scattered on the part surface or on the powder bed during the manufacturing process, which binds the powder in contact with them to the surface. The EDS analysis curves of clean 316L powder, spatter particles and L-PBF manufactured parts are shown in Fig. 4.36b, d, f. Compared with the clean powder, the oxygen content in the spatter particles and parts increased a lot, while the corresponding iron content decreased. This is because the oxygen content in the manufacturing cavity is relatively high, and the material reacts with oxygen to produce oxides.

The XRD curves of clean 316L powder and spatter particles are shown in Fig. 4.37. There are austenite and ferrite in 316L powder, in which austenite is the main phase. In the spatter particles, the content of austenite and ferrite decreases sharply. This is due to the manufacture of oxides $Fe + 2Fe_2 + 3O_4$, SiO_2, MnO_2, etc. The reaction

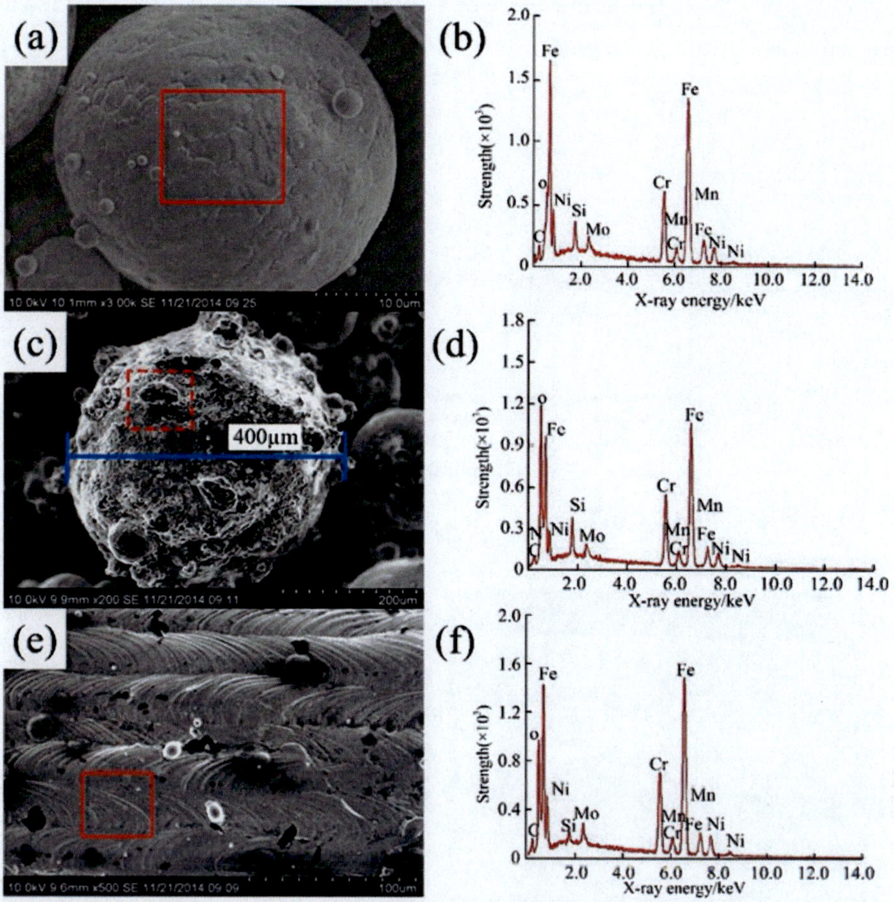

Fig. 4.36 EDS analysis of 316L powder, spatter particles and formed parts: **a** SEM image of clean 316L powder; **b** EDS analysis curve of clean 316L powder; **c** SEM image of spatter particles; **d** EDS analysis curve of spatter particles; **e** SEM image of L-PBFed parts; **f** EDS analysis curve of L-PBFed parts

process is shown in Eq. (4.2). It can be seen that the L-PBF process is very sensitive to oxygen, and the oxygen content in the manufacturing cavity should be reduced as much as possible.

$$\begin{aligned} 3Fe + 2O_2 &\rightarrow Fe_3O_4 + heat \\ Si + O_2 &\rightarrow SiO_2 + heat \\ Mn + O_2 &\rightarrow MnO_2 + heat \end{aligned} \tag{4.2}$$

As shown in Fig. 4.38, some fine spatter particles are embedded in the fabricated part. These spatter particles produce a layer of oxide on the surface when they are

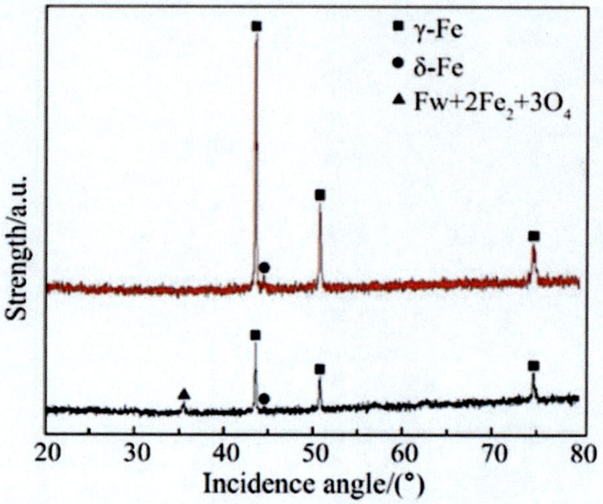

Fig. 4.37 Comparison of XRD patterns of 316L powder and spatter particles

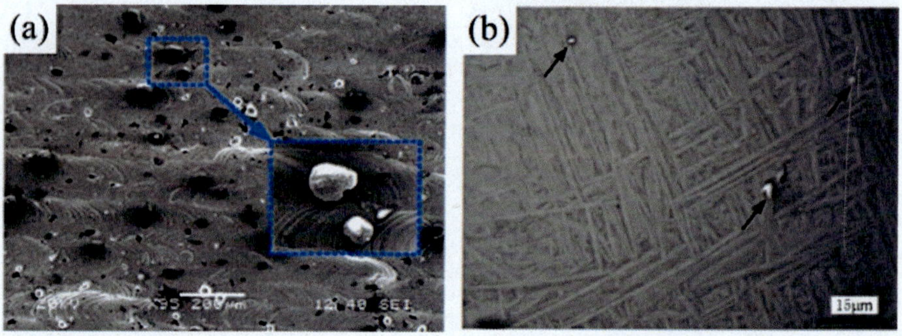

Fig. 4.38 Spatter particles on manufactured parts: **a** spatter particles embedded in the surface of the manufactured part; **b** spatter particles embedded in the inside of the manufactured part

generated. Oxidation is a process of reducing energy, and the wettability of oxide is poor. When the spatter particles are scattered on the surface of the part, the wettability with the substrate or the pre-formed layer is greatly reduced, which prevents the particles from being firmly bonded to the main body of the part. The spatter particles are mixed in the powder and solidified in the matrix of the part, which becomes the fracture source of the part. When subjected to external force, the bonding surface between the spatter particles and the matrix of the part is first destroyed, thereby manufacturing an initial fracture crack, which is further affected by the external force. These cracks gradually expand and eventually lead to the failure of the part [3].

Please refer to Chap. 5 for a study of the manufacture mechanism of spatter particles.

4.4.3 Gas Atmosphere Deaeration and Circulation Purification

Through the manufacturing experiments of stainless steel powder (new powder, old powder) and pure titanium powder under nitrogen and argon protection conditions, it is found that smoke is bound to be generated when the laser interacts with the powder. It was observed that the smoke produced by the new stainless steel powder was significantly reduced, and the black smoke produced by the pure titanium powder was less than that of the new stainless steel powder. The main source of smoke is the combustion and gasification of carbon elements, low melting point alloy elements and impurity elements in the metal powder. The long-term repeated use of the powder has a cumulative effect on the degree of the smoke problem, that is, although the new stainless steel powder produces a small amount of smoke, the black smoke pollutes the powder, and the long-term accumulation causes the smoke to become more and more serious when the laser interacts with the powder. At present, all manufacturers or scientific research institutions have not been able to fundamentally solve the problem of smoke. The main negative effects of smoke include: (1) contamination of the light-transmitting lens; (2) contamination of the powder; (3) contamination of the powder laying guide rail; (4) contamination of the inner wall of the manufacturing cavity.

A very serious consequence of the existence of smoke is that it pollutes the lens. Especially when scanning at low speed, the input of laser energy is large, and the amount of smoke generated is also large. The smoke quickly sticks a layer of black smoke powder on the transparent lens, resulting in serious attenuation of laser power when passing through the lens. Most of the laser energy acts on the lens in the form of thermal energy, and the lens will soon get hoter and hoter until it bursts. When the light-transmitting lens is seriously polluted by the smoke, the laser power incident on the surface of the powder bed is insufficient, and the powder is not fully melted. The manufacturing process must be repeatedly stopped to manually remove the smoke on the lens. It has negative influences on the manufacturing efficiency and the quality of the manufactured parts. Secondly, after the smoke is generated, a small part is blown out of the powder bed by the protective gas, and most of it still falls to the surface of the powder bed that is not used, and is mixed with the powder, which aggravates the pollution degree of the powder. Thirdly, the current L-PBF manufacturing equipment mostly adopts the semi-open powder laying guide installation method, that is, due to the existence of the powder laying arm, there is a slender opening between the manufacturing cavity and the powder laying guide. The dust and black smoke raised during the processing will enter the inside of the guide rail to reduce the lubricity of the guide rail, and even lead to the wear of the guide rail, thereby reducing the accuracy of the guide rail until it cannot be used. Finally, part of the black smoke generated during the process is blown away by the protective gas, and the rest will adhere to the inner wall of the manufacturing cavity, especially the inner wall of the observation window. With the accumulation of black smoke, some black smoke will fall off in the next processing, some will fall on the powder to contaminate the powder, and some will fall on the formed surface to affect the part's quality. The

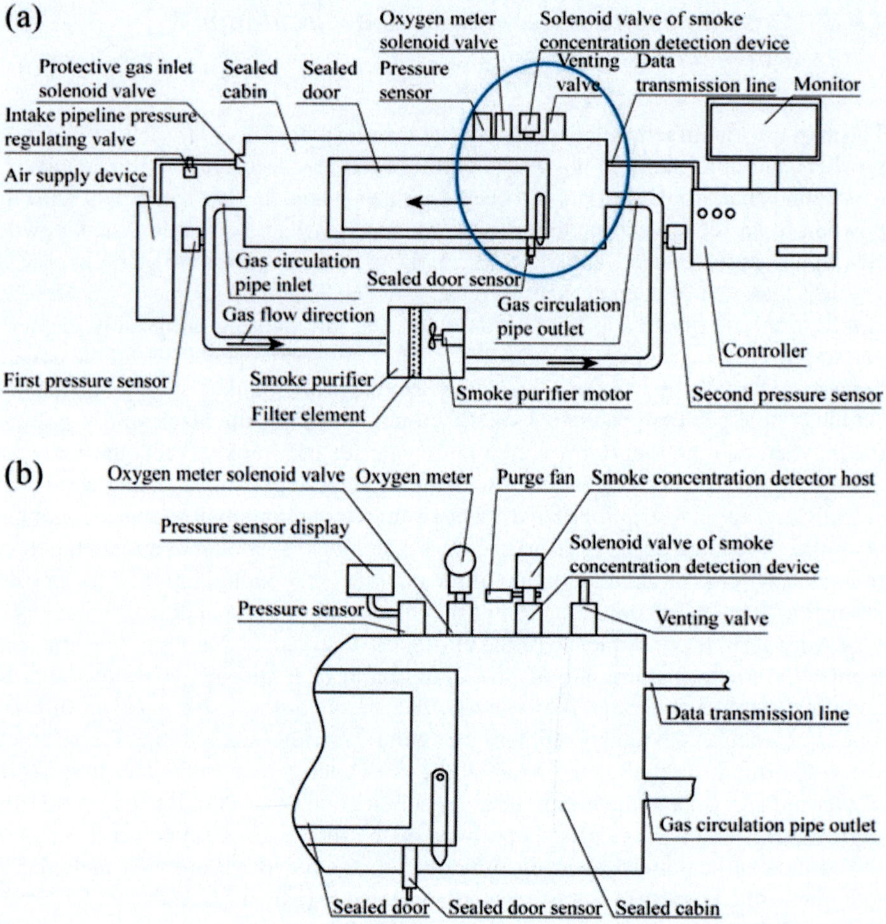

Fig. 4.39 Schematic diagram of the smoke detection and purification system in the sealed cabin: **a** overall view; **b** detail view

black smoke adhering to the side of the observation window will seriously affect the monitoring of the processing process [1].

Aiming at the problem of smoke, a method and equipment for smoke detection and purification in the sealed cabin of a metal AM system are invented here. As shown in Fig. 4.39, the smoke detection and purification equipment in the sealed cabin is mainly composed of the sealed cabin, gas circulation purification system, smoke concentration detection device, pressure detection device, oxygen content detection device, air supply device, exhaust device, display device and control device. The sealed cabin provides an airtight environment for the processing process. The gas circulation purification system is placed at the lower part of the sealed cabin, and is composed of a first pressure sensor of the gas circulation pipe, a second pressure sensor of the gas circulation pipe and a smoke purifier. The control device and the display device are located on the right side of the sealed cabin, including the controller and the display. The smoke concentration detection device is located at the upper right of the sealed cabin, which is used to detect the changes of the smoke concentration in the sealed cabin in real time. The gas supply device provides protective gas for the processing process through the gas inlet. The exhaust device is connected to the sealed cabin through a solenoid valve. The oxygen content detection device and the pressure sensor device are located in the upper right part of the sealed cabin, which are used to detect the oxygen content and pressure in the sealed cabin. The equipment introduces a gas circulation purification system. The smoke concentration in the sealed cabin is detected in real time through the smoke concentration detection device, and the motor speed of the dust collector is continuously adjusted to maintain the gas purity in the sealed cabin. In addition, the controller detects the signal of the smoke concentration detection device and makes corresponding control instructions, and simultaneously transmits the signal of the smoke concentration detection device, the motor speed level of the dust removal device, and the pressure difference signal of the first and second pressure sensors to the display for display. Therefore, the use of smoke detection and purification equipment in the sealed cabin can greatly improve the manufacturing efficiency and quality of L-PBF, reduce powder pollution, and ensure the safety and reliability of the processing process [4].

4.5 Instability Caused by Process Uncertainty

4.5.1 Actual Laser Power

The laser emitted from the fiber laser needs to pass through the optical isolator, beam expander, X/Y axis mirror of scanning galvanometer, and then be focused by the f-θ lens. After multiple absorption or weakening, the laser power that actually irradiates on the powder bed is smaller than the theoretical setting value. Therefore, after the LPBF equipment is assembled, it is usually necessary to debug the equipment and measure the actual output parameters, which can be used only after measurement and calibration. For example, the Leijia DiMetal-280 equipment generally needs to measure the actual incident laser power within a period of 2–3 months. Table 4.2 lists the comparison between the measured value of actual laser power and the laser power set theoretically. When the power is set to 100 W, the measured value is only 86.2W, and the loss is about 14%. When the power loss exceeds a certain value, the tightness of the optical path unit must be checked, and the position of each component on the optical path unit must be corrected until the actual measured value of the laser power after correction reaches 95% of the theoretical value.

Table 4.2 Comparison between laser set value and measured value

Theoretical power	Actual power/W	Power loss	Loss rate (%)
20	20.6	−0.6	−3.20
40	35.8	4.2	10.50
60	52.4	7.6	12.67
80	69.0	11.0	13.75
100	86.2	13.8	13.80
120	101.9	18.1	15.08
140	120.0	20.0	14.29
160	137.0	23.0	14.38
180	153.0	27.0	15.00

4.5.2 Laser Spot

The spot diameter of fiber laser used in LPBF equipment is generally 30–100 μm (generally, the scanning melting track is larger than the spot diameter), as shown in Fig. 4.40. Therefore, the influence of laser spot error on the X/Y axis direction in the LPBF process cannot be ignored, especially when manufacturing small parts, the influence of spot size on the X/Y axis dimension accuracy of built parts is very large. The influence of spot size on the precision of manufacturing size is mainly reflected in the width of the melting track. The forming principle of LPBF technique is to melt the metallic powder by laser beam to form a melting track, then form a surface through the overlap between melting tracks, and then form a solid part by stacking layer by layer. The melting track is as the basic component of the built part, and its width is an important parameter, which is the most basic reason for affecting the dimensional accuracy of the built part. Before printing, the melting track width should be determined according to the actual process, and compensation method should be adopted to eliminate it during the actual data processing.

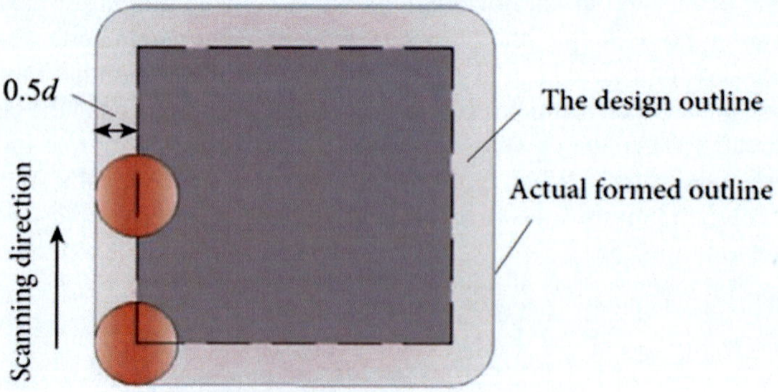

Fig. 4.40 Schematic diagram of influence of spot size on accuracy

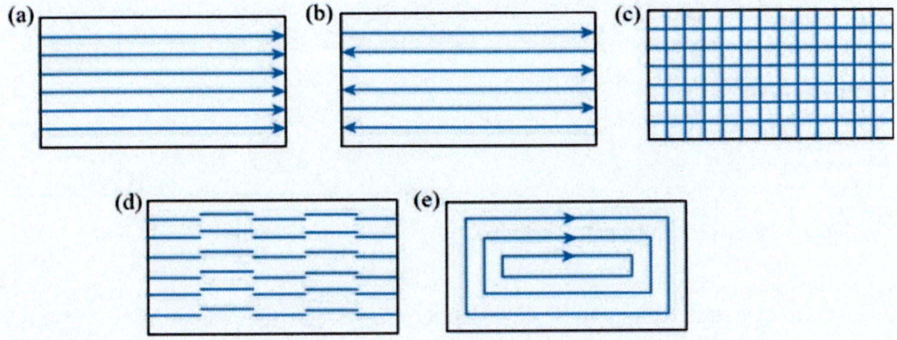

Fig. 4.41 Typical scanning path: **a** one way scanning; **b** "S" scan; **c** cross scan; **d** short linear scanning; **e** isometric progressive scanning

4.5.3 Scanning Strategy

During LPBF process, the change of laser beam moving mode will affect the temperature gradient and temperature field distribution during the forming process, thus generating thermal stress. Proper laser scanning path is helpful to release thermal stress and reduce residual thermal stress during forming. Figure 4.41 shows several typical scanning strategies. One way scanning has the advantage that the control system is easy to handle, but the stress distribution at both ends of the built part is uneven. The "S" scan improves the disadvantage of non-uniform stress distribution in unidirectional scan, but it is easy to warp when the scan line is too long. Cross scanning, namely remelting scanning for each layer, can improve cracks and pores, and reduce stress. Short linear scanning can release part of the stress and reduce the occurrence of warping deformation. Isometric progressive scanning is helpful to reduce the shrinkage deformation due to the continuous change of scanning direction.

4.5.4 Hatch Space

As shown in Fig. 4.42, the hatch space refers to the gap distance between the adjacent tracks. The size of the hatch space is directly related to the laser energy transmission and distribution, and has a great influence on the forming quality of each layer. It is an important process parameter for the LPBF forming.

If the hatch space is too large and the melt tracks are not connected, the metallic powder between the two tracks will absorb less energy, resulting in the metal powder cannot melt, low lap rate between the two melting tracks, uneven surface layer, and poor quality of the built parts. If the hatch space is too small, the scanning area will overlap each other, most of the metal remelting between the areas, easy to cause parts warping deformation, shrinkage, and even material gasification problems, manufacturing efficiency will also be reduced. When the hatch space is appropriate,

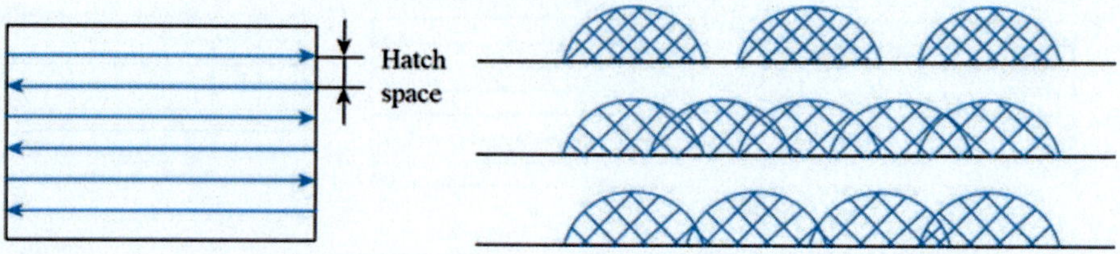

Fig. 4.42 Schematic diagram of hatch space

the scanning area is only partially overlapped, the overlap between the two melting tracks is appropriate, the formed surface is smooth and uniform, and the built parts are of good quality.

4.6 Types of Defects in LPBF-fabricated Parts

4.6.1 Balling Phenomena

Balling phenomena, incomplete melting of metal powder and thermal deformation are easy to occur in the LPBF process. Spheroidizing is a phenomenon that molten metal materials, under the action of interfacial tension between liquid and surrounding medium, try to change the surface shape of liquid metal to a spherical surface with minimum surface area, so that the system composed of liquid and surrounding medium has the minimum free energy. Balling phenomena will produce a large number of pores in the parts, significantly reduce the density and increase the surface roughness and reduce the dimensional accuracy. In the forming process, the spheroidizing effect of the surface of the formed material makes the next layer of powder unable to be laid or the thickness of the powder is uneven, which makes the forming process fail.

Spheroidizing process: when laser irradiates metallic powder, the cross section of the molten pool formed is bowl shaped due to the influence of Gaussian laser energy distribution. As shown in Fig. 4.43a, when the laser cannot penetrate the powder layer and the laser speed is fast, the loose powder at the bottom of the molten pool has no binding force. The shape change of the molten pool is mainly determined by the interfacial tension. During the solidification process, the molten pool quickly rolls into a sphere, causing serious balling phenomena. When the laser can penetrate the molten powder layer and have a certain amount of melting to the solid foundation, the molten pool can be divided into upper and lower parts along the surface of the solid foundation, as shown in Fig. 4.43b. Under the influence of gravity, part of the metal liquid in the upper molten pool will first contact the surface of the solid foundation, manufacturing a new liquid–solid interface (called the second type liquid–solid interface). For the remaining gas–liquid interface, it is possible to make the force of solid particles on liquid particles smaller than the force between liquid

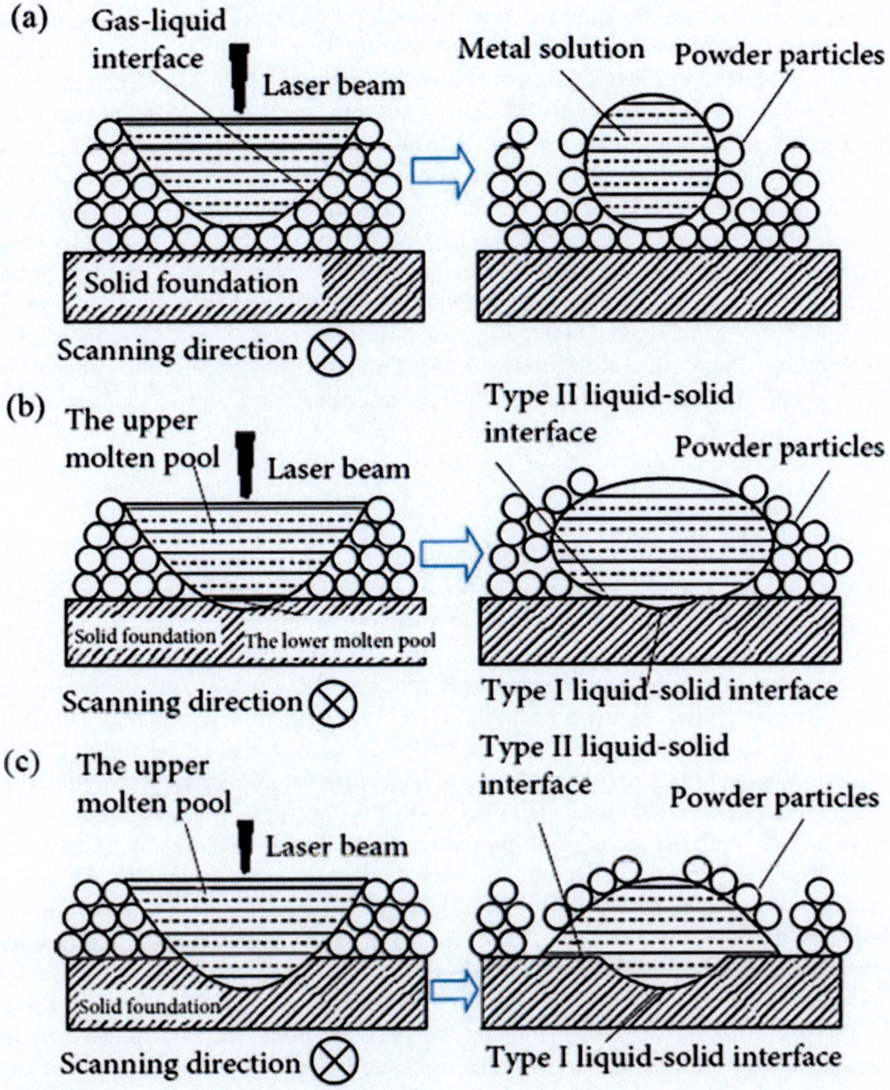

Fig. 4.43 Schematic diagram of spheroidizing evolution of single melting track: **a** laser cannot penetrate the molten powder layer; **b** the laser penetrates the melted powder layer and partially melts the solid foundation; **c** the laser penetrates the melted powder layer and has more melting amount to the solid foundation (*Note* ⊗ means the vertical paper facing inward)

particles. As a result, the total resultant force of particles at the intersection of gas, liquid and solid phases points to the interior of the liquid, so that the liquid surface shrinks inwards into a sphere under the action of interfacial tension. When the laser can penetrate the molten powder layer and has a large amount of melting to the solid foundation, as shown in Fig. 4.43c, the molten liquid in the lower molten pool and the solid foundation are the same material, and the temperature on both sides of the liquid–solid interface (which can be called the first type liquid–solid interface) between the lower molten pool and the solid foundation is very close, the liquid–solid interface is vague and relatively wide, and the adhesive traction of the molten liquid makes this part of the molten pool not spheroidize. Reducing the scanning speed, thickness of powder layer, or increasing the laser power are conducive to increasing the melting amount of solid foundation [5]. Therefore, the balling phenomena can be weakened by setting reasonable LPBF parameters.

4.6.2 Powder Adhesion

Figure 4.44 shows the phenomenon of powder adhesion in the molten pool of single melting track at different scanning speeds. It can be seen from the figure that the metal powder is melted into a molten pool under the action of laser, and the section of the liquid molten pool is rolled in a semicircle arc under the action of surface tension. In the process of rolling in, the powder near the heat affected zone is adhered to form powder adhesion. The powder near the molten pool will also be strongly affected by heat. The gas in the powder particle gap expands rapidly, driving away the fine powder particles around the molten pool, so that the powder particles are not easy to be adsorbed. With the increase of laser energy absorbed by the powder, the driving force increases. Powder adhesion is mainly divided into two categories [6]: one is mosaic powder adhesion (① in Fig. 4.44), which is partially inlaid into the molten pool, and it is difficult to remove; The other is loose powder adhesion (② in Fig. 4.44) with weak adhesion, which can be removed by wiping. This kind of adhesion particles account for the majority of the total powder adhesion.

From the aspect of part shape, the surfaces of parts built by LPBF can be divided into top surface, side wall surface and overhanging surface (including bottom surface). According to the different normal directions, side walls can be divided into three types: normal downward side walls, normal upward side walls and vertical side walls. These sidewalls are formed by the mutual accumulation of molten pool shapes outside a layer of figure outline. Because the sidewall is in direct contact with the powder outside the selected area, the powder adhesion is very serious.

For the vertical sidewall and the normal upward sidewall, the powder adhesion is mainly affected by the powder contacted in the horizontal direction. With the commonly used X–Y orthogonal scanning strategy, after each layer is scanned, a micro plane formed by overlapping the start or end of multiple molten pools is a dentate plane, while the plane formed by a single molten pool is a straight plane, and the dents on the dentate plane are easy to hide powder. Since the layers are orthogonal

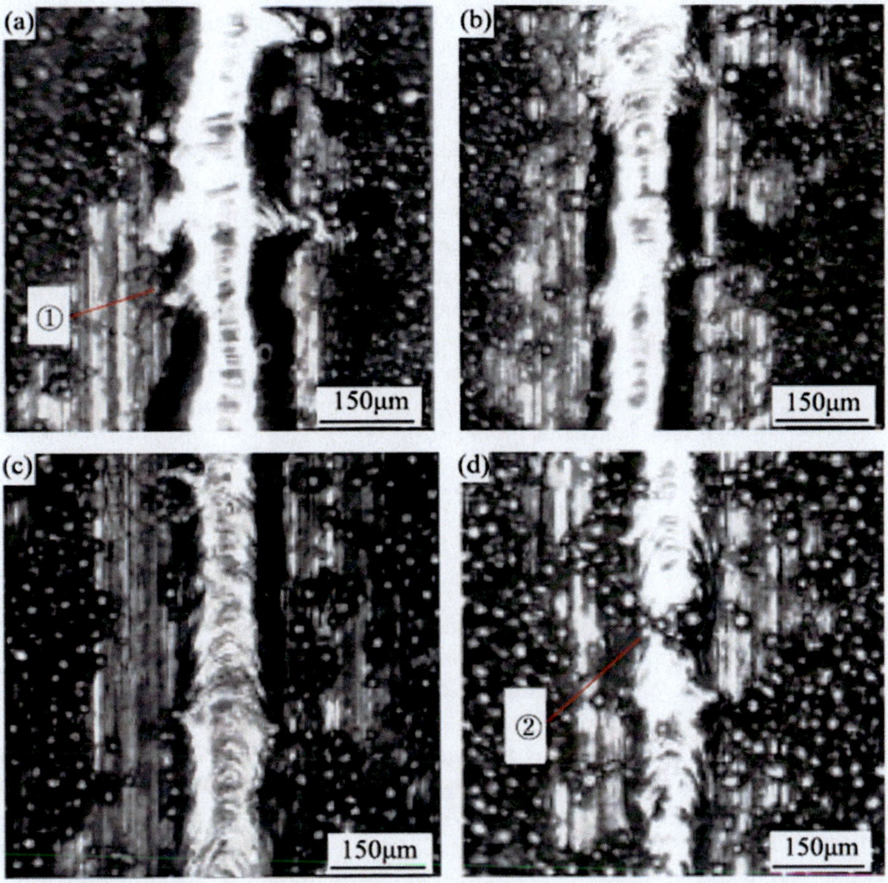

Fig. 4.44 Single line molten pool powder adhesion phenomenon at different scanning speeds (laser power is140 W and layer thickness 40 μm): **a** 50 mm/s; **b** 100 mm/s; **c** 150 mm/s; **d** 200 mm/s. ① Mosaic adhesive particles ② Loose adhesive particles

to each other, the dentate surface of each layer is sandwiched by the linear bread of the previous layer and the subsequent layer. In the process of cladding, the powder particles are firmly adhered to the tooth surface with partial remelting or sintering of the cladding powder, which is difficult to remove later.

For the side wall with the normal direction downward, the powder adhesion is not only the powder adhesion when the single molten pool is formed on the substrate, but also the sintered powder adhesion at the bottom. During manufacturing, the powder below the manufacturing surface is not a solid substrate but a much deeper powder than the layer thickness, and the surface powder is melted under the action of laser. With the increase of the depth of the powder layer, the powder area near the surface of the powder layer is a heat affected zone, and the morphology changes from sintered to free powder. The melted surface molten pool will adhere to the powder in the sintering area to form powder adhesion. Since the adhesion area is at the lower part, the liquid metal will infiltrate under gravity and adhere to the particles in the sintering area. The more liquid metal is, the more obvious the adhesion is.

For the overhanging surface (including the bottom surface), the powder adhesion mechanism is bottom sintered powder adhesion. In the LPBF process, the overhanging surface is easy to cause serious powder adhesion, so the overhanging surface should be reduced as much as possible, generally by adding support. The top surface belongs to a special molten pool lap surface, and its bottom substrate is a hard solid. In the molten pool lap surface in the same layer, except for the figure contour, the powder adhesion on both sides of the molten pool in the filling area will be removed by subsequent lap remelting, so the powder adhesion in the filling area can generally be ignored. The influence of powder adhesion on the manufacturing accuracy can be effectively reduced by using the scanning strategy of multiple hemming and shrinking filling combined with a higher scanning speed.

4.6.3 The Bulge of the Outer Edge

During the LPBF experiments, it was found that the balling phenomena was the most serious at the beginning of the interaction between the laser beam and the powder, which was shown by the serious bulge of track at the edge of the melting layer, which can be defined as "first line scanning balling phenomena". Figure 4.45 shows a single scan track formed at different laser scanning speeds, and it can be found that all four scan tracks have the phenomenon of starting point bulging (specification). When the laser starts to scan, the speed increases from zero to the set value due to the scanning delay, and the energy absorption at the end of the scanning line is too large and the molten pool is too large. At the same time, the beginning point and end point of the scan line are more exposed to metal powder than other points, and the molten pool's

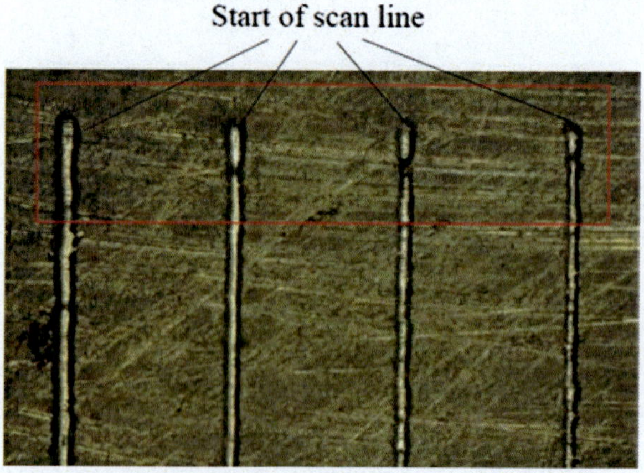

Fig. 4.45 Bulge of the outer edge of starting point of single scan line

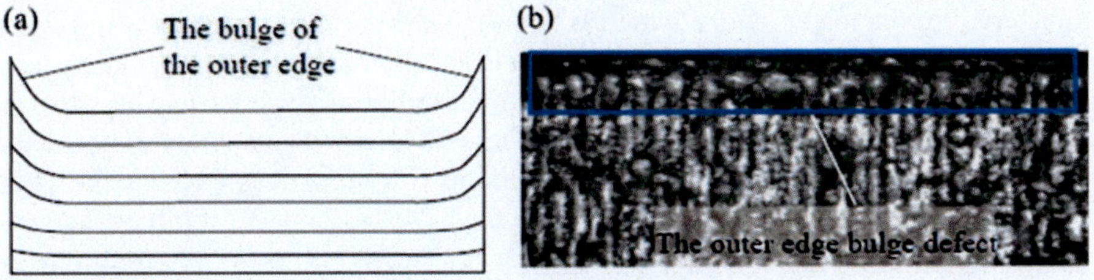

Fig. 4.46 The outer edge bulge defect in LPBF: **a** the bulge of the outer edge and the gradual increase of the layer thickness; **b** actual rendering of the outer edge bulge

suction effect on the metal powder further increases the melting width and height at the beginning point of the scan line. Accordingly, the scan lines start scanning, powder absorbs energy from the laser, there is a sudden increase significantly with the surrounding is not laser scanning large temperature gradient between powder. In the area of the role of laser, the powder absorbs energy and melts rapidly when it reaches the melting point and the formed liquid phase can be rapidly cooled and solidified without spreading around.

The first line scanning balling phenomena is extremely unfavorable to the LPBF process. It not only forms a bulge frame on the outside, affects the powder laying of the next layer, but also causes warping deformation. The problem of first line scanning balling phenomena can be solved by gradually increasing the laser power of the starting edge to the set value. The problem of first line scanning balling phenomena can also be solved by using a higher scanning speed on the premise of ensuring complete melting of materials.

The bulge of the outer edge will accumulate in the process of multi-layer manufacturing, which increases the thickness of the inner powder layer of the parts built by LPBF, and the powder melting shrinkage causes this defect to be more serious. As shown in Fig. 4.46, the melting width and melting height of the starting point and the end point of the scanning line become larger, like a fence enclosing other areas of the part. The border part is fully contacted with the powder, and the internal area of the part is seriously shrunk after being filled with powder. Figure 4.46a shows the schematic diagram of the bulge of the outer frame and the gradual thickening of the layer thickness, and Fig. 4.46b shows the actual effect of the bulge of the outer frame in the orthogonal test sample. In the case of edge hook, although the surface finish of the sample can be improved, it aggravated the outer edge bulge defect.

4.6.4 Warping and Cracking

During the LPBF process, layers of metallic powder are melted by laser beam and stacked on the formed layer to form a three-dimensional entity. When a new layer of powder is added on the previous formed layer, the heat brought will be quickly transferred to the lower layers, which is bound to change the temperature and stress distribution of the formed layers. This process is called subsequent thermal cycling (STC). The subsequent added materials will cause the increase of tensile stress in the following materials during the cooling stage, which will gradually increase the residual stress in the parts. When the maximum residual stress exceeds the yield strength of the material, the part cracks in the layer or between the part and the substrate. With the continuous accumulation of stress, the cracking continues to expand, and finally threatens the safety of the manufacturing process and is forced to stop, as shown in Fig. 4.47.

In the LPBF parts, the thermal stress at both ends of the part is the largest, so the possibility of cracking at both ends of the built part is the largest. When parts tend to warp upwards, because the tooth top of the support structure is directly connected with the lower part of the built part, it has a downward "pull" effect on the built part, so it can prevent the built part from separating from the support structure and reduce the risk of warping deformation of the built part. However, if the strength of supports at the two ends is not enough to "pull" the warping of the built part, the built part will crack at both ends [3]. Under the combined action of the stress along the height direction and scanning direction, the built part will bend to the middle to form a certain angle of warping, as shown in Fig. 4.48.

The greater the material shrinkage, the more serious its warping deformation. Shrinkage is mainly caused by state of material, temperature change and laser energy. First, the change of state of material causes shrinkage. When the powder particles are stacked in solid state and not compacted, the density is generally only about 50% of the full density, while the density of the parts formed by laser melting in the powder bed is generally more than 95%. Therefore, the change of density during the LPBF process will inevitably cause the shrinkage of the parts. The second is the shrinkage caused by temperature change. Because the working temperature is much higher than the room temperature, when the work piece is cooled to the room temperature, the overall shrinkage will occur. The shrinkage is mainly determined by the geometric shape of the material and the work piece. Its shrinkage characteristics

Fig. 4.47 Cracking and warping defects in LPBF process

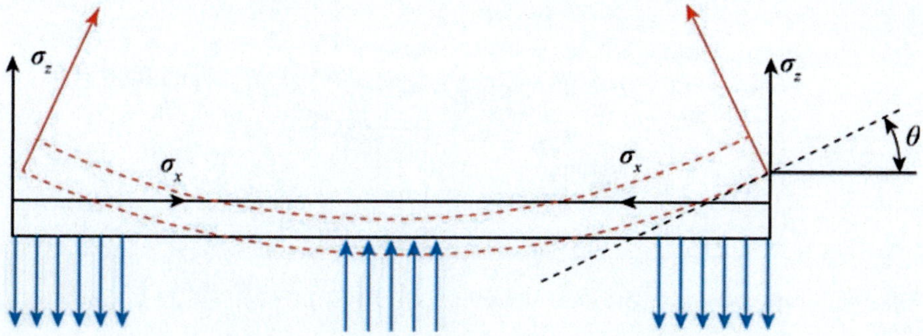

Fig. 4.48 Schematic diagram of warping deformation

are fundamentally different from that of the state of material. The state of material shrinkage is the superposition of local shrinkage of the melting track or surface, while the temperature change shrinkage is a whole shrinkage process, and the shrinkage rates in three directions are nearly equal. Finally, there is the shrinkage caused by the uneven energy of the laser spot. Assuming that the laser beam is a Gaussian beam, the energy in the center of the laser spot that is perpendicular to the incidence is the highest, and the point gradually weakens outwards. Therefore, the laser energy received by the powder at different positions within the range of the laser spot will be different, and due to the large porosity in the powder layer, the heat conduction will be greatly reduced, so that the energy obtained by the lower part of the powder layer is much less than that obtained by the upper part. The uneven energy obtained by the upper and lower parts will cause uneven temperature rise above and below the powder layer. The upper part of the powder layer has high energy, fast temperature rise, fast heat dissipation, and large volume shrinkage, while the lower part has less energy, slow temperature rise, slow heat dissipation, and small volume shrinkage. As a result, uneven spot energy in the formed layer will also lead to warping.

Warping deformation has a great impact on the precision of built parts, causing large size and shape errors, and even leading to processing failure or scrapping of metal parts, as shown in Fig. 4.49. Figure 4.49a shows the warping deformation during the LPBF process of the overhanging structure, and Fig. 4.49b shows that the supports are seriously pulled off due to warping deformation during the LPBF process of the overhanging structure with supports.

4.6.5 Pores

With the rapid development of LPBF technique, the LPBF equipment is becoming more mature, and the process is also gradually stable. The density of the built parts can be nearly 100%, but the LPBF process still inevitably produces porosity defects ranging from a few microns to dozens of microns, as shown in Fig. 4.50. Pores in

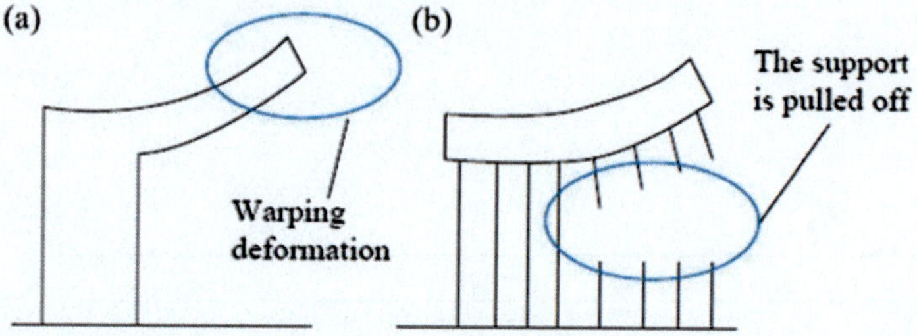

Fig. 4.49 Shape and size errors caused by warping deformation: **a** warping; **b** the support is pulled off

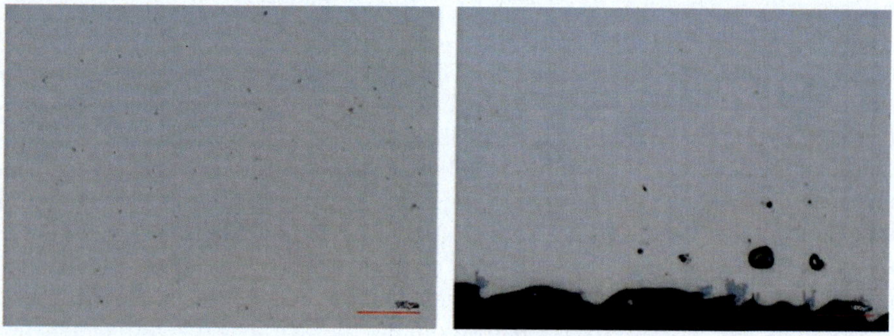

Fig. 4.50 SEM diagram of porosity defects

the built parts will seriously affect the mechanical properties, especially the fatigue life and reduce the density. The effect of larger pores on the built parts is greater than that of small pores, and the stress concentration caused by irregular holes is more significant than that of spherical pores.

Pores defects are mainly divided into two types: one is irregular metallurgical defects formed by poor lapping of molten pool, the other is spherical porosity defect. Metallurgical defects are mostly caused by unreasonable process plan or insufficient laser energy input, insufficient lapping between the molten pool and the matrix during the forming process, resulting in poor metallurgical bonding, or pores caused by incomplete melting of metal powder. Pores defects can be divided into two categories: one is "keyhole" similar to laser deep penetration welding process; The other is the pore in the powder with a diameter of several microns [7]. The metallic powder prepared by atomization method has pores of different sizes in the interior, which is mainly caused by the formation of bubbles when inert atomized gas enters the metal liquid during powder preparation. Keyhole is a kind of high energy laser that heats molten metallic powder and vaporizes metal molten pool during LPBF process. The gas suddenly generated increases the local pressure rapidly, and exerts pressure on the free metal surface to "punch" out of the small pores. The gas and plasma in the pore expand violently under the action of high temperature and pressure, and then erupt. As the gas in the pore decreases, the gas in the pore cannot maintain the existence of the pore. The hole begins to close gradually and involves some metal vapor and shielding gas to form a pore defect. With the movement of the laser, the LPBF process of the next pore begins to appear.

In order to reduce the pore defect, several measures must be taken. Appropriate forming process parameters shall be selected, so that each layer of metallic powder can be fully melted. The metallic powder with good sphericity, uniformity and small particle size shall be selected as far as possible to reduce the filling of shielding gas between the powders. The thickness of powder spreading shall be reduce, and preheat the substrate and powder before manufacturing to reduce the temperature gradient, allowing more time for the gas to exit the molten pool. These measures can effectively inhibit the occurrence of pore defects.

4.6.6 Microstructural Nonuniformity

Figure 4.51 shows the internal structure of 316L stainless steel parts fabricated by LPBF. It can be seen that there is the distribution of irregular molten pool in the internal structure of the built part, and it has the following defects: ① the height of two adjacent scanning melting tracks in the same layer is different; ② the same scanning tracks is wavy rather than horizontal; ③ unequal thickness between adjacent layers.

The reason why the adjacent scanning melting tracks described in the defect ① have different melting tracks is that the energy input is too high, and the powder around the molten pool is blown away due to the material vaporization. When the hatch space of the LPBF is small, the next molten pool will be scanned in the powder free zone. In addition, the scanning strategy of interlaminar stagger and orthogonalization is adopted during forming, that is, the scanning line of the next layer is

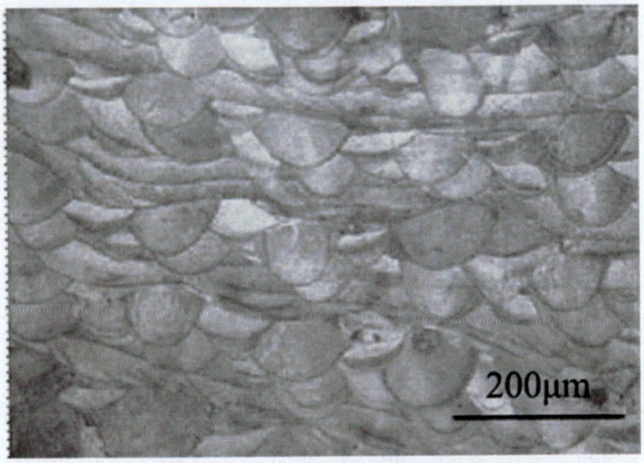

Fig. 4.51 Microstructure defects of LPBF-fabricated 316L stainless steel

formed between the two scanning lines of the previous layer, which is also an important reason for the difference in the height of the adjacent scanning lines. The above two reasons have different effects on the forming quality of adjacent melting tracks, but it can be seen from Fig. 4.52 that there is a regular combination of high and low order between adjacent melting tracks.

The same scanning line described in defect ② is wavy. The first reason is that the thickness of the processing layer in LPBF forming is very thin, generally only 20–50 μm. In addition, there are occasional bulges in the forming process, resulting in different processing layers of the same layer. The second reason is that the irradiation direction of the laser is not perpendicular to the surface of the powder bed, but in the form of a light cone. The third reason is that when the powder spreading scraper pushes the powder forward, it shakes due to the mechanical assembly or the inequality of the formed surface. When the flexible tooth elastic powder spreading device pushes the powder forward, it will also cause powder leakage, elastic deformation, etc., resulting in a certain degree of unevenness of the powder spreading surface.

The reason for the unequal thickness between adjacent layers described in defect ③ is that the actual layer thickness is not equal to the set layer thickness due to the solidification shrinkage of the material. The second reason is that the rising and falling amount of the building cylinder is unstable. Figure 4.52 shows the comparison between the measured value and the theoretical value of the falling amount of the building cylinder (when measuring, the layer thickness is set to 25 μm). It is found from the measurement results in Fig. 4.52 that the actual drop of the building cylinder changes at the theoretical value attachment, and the cumulative errors calculated by multiple measurements are respectively −15 μm/39 times, 23.4 μm/38 times.

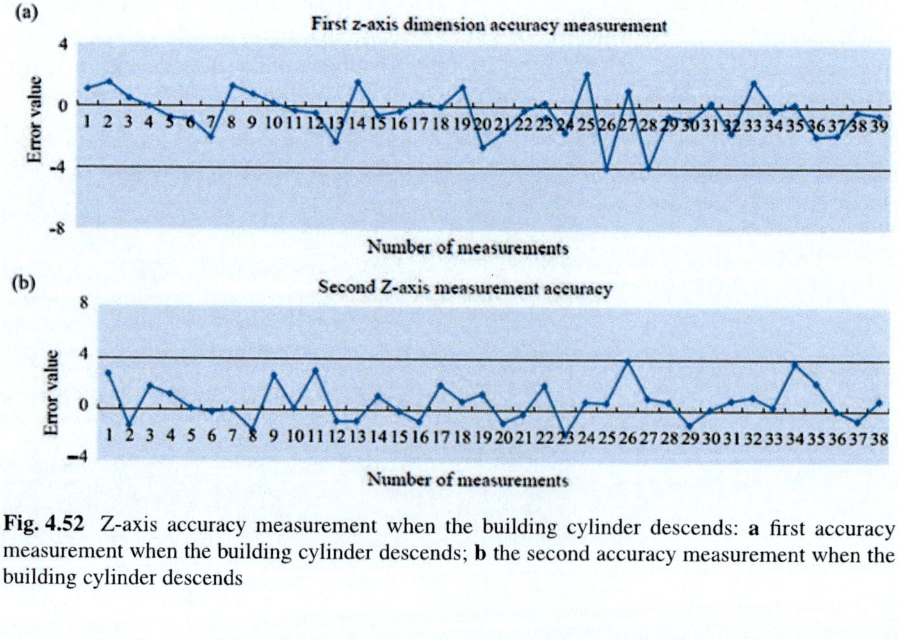

Fig. 4.52 Z-axis accuracy measurement when the building cylinder descends: **a** first accuracy measurement when the building cylinder descends; **b** the second accuracy measurement when the building cylinder descends

From the measurement results, it can be seen that there are positive and negative accumulations of the rising and falling errors of the building cylinder, which may be related to the rubber pad in the building cylinder. It can also be seen from the measurement curve that the single error value of building cylinder is generally within 5 μm.

4.7 Element Evaporation During LPBF

The LPBF technique uses a tiny focused laser spot to quickly melt the preset powder layer by layer. Under the action of high energy density laser, the metal powder melts instantaneously, and the instantaneous temperature in the molten pool can reach more than 3000 °C. Some alloy elements may vaporize, become metal vapor and escape from the molten pool, causing alloy element loss. Some researchers found that the loss of Al element in the AM process of Ti–Al alloy reached more than 15%, and the large element loss caused the composition and structure of the formed alloy to deviate from the design value, resulting in such defects as alloy composition segregation, element depletion and abnormal structure [8, 9]. When the loss of alloy elements reaches a certain level due to the disturbance of materials, processes or processing environment, the element evaporation becomes a problem that cannot be ignored in the LPBF. For example, for producing Ti-6Al-4V alloy using LPBF, the evaporation rate of Al element is hundreds or even tens of thousands of times than that of Ti and V element. Selective element evaporation occurs during the forming process, so the content of Al element in the formed sample is relatively reduced, while the content of Ti and V element is relatively increased. Because the Al element is α phase stable element, which leads to less α/α′ phase content and more β phase content. This brings great challenges to the application of LPBF technique in the aerospace field, where high material composition and performance is required.

Element evaporation is a typical thermophysical phenomenon in LPBF. As shown in Fig. 4.53, the temperature and kinetic energy of alloy elements increase after absorbing laser energy, and they quickly migrate to the surface of the molten pool, vaporizing at the liquid/gas interface of the molten pool. When they have enough speed, they will escape from the liquid surface and diffuse to the environment through the gas boundary layer. In this process, the combined force of vapor recoil force, surface tension and ambient air pressure in the molten pool determines the angle, speed and difficulty of alloy elements escaping from the liquid surface, and then determines the intensity of element evaporation.

Figure 4.54 shows the XRD patterns of smoke and dust generated during magnesium alloy AZ91D (Mg-Al-Zn) using LPBF. It can be seen from the figure that all diffraction peaks correspond to α-Mg. It can be inferred that Mg burning loss mainly occurred in AZ91D alloy during LPBF processing. This is mainly because the activity coefficient of magnesium element is much lower than that of Al and Zn elements. Although all three elements will evaporate during the forming process, the evaporation of Mg element is much higher than that of the other two elements, resulting in a

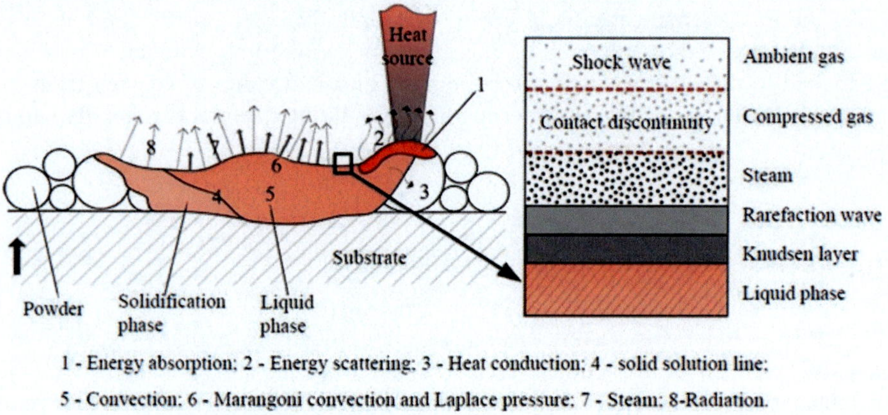

Fig. 4.53 Schematic diagram of metal vapor generation in the molten pool of AM. Reprinted with permission from a work by Xie [10]

relatively lower Mg content in the final formed sample, while the content of Al and Zn elements increases [11].

The element evaporation process in LPBF follows the traditional thermodynamic theory of evaporation, that is, the generation of metal vapor in the liquid molten pool mainly includes four stages, namely, the migration of alloy elements from the liquid phase to the melt surface, the liquid/gas phase transition of the melt surface, the evaporation of the liquid/gas interface, and the diffusion of vapor in the building cavity. These processes are closely related to the temperature in the molten pool. On the one hand, higher molten pool temperature is bound to increase the saturated vapor pressure and activity coefficient of alloy elements (the main basis for judging the evaporation trend of alloy elements), reducing the difficulty of element evaporation. On the other hand, it will increase the counter flow mass transfer in the molten pool and the pressure gradient and concentration gradient of the liquid surface, driving the alloy elements to migrate to the surface of the molten pool by convection diffusion,

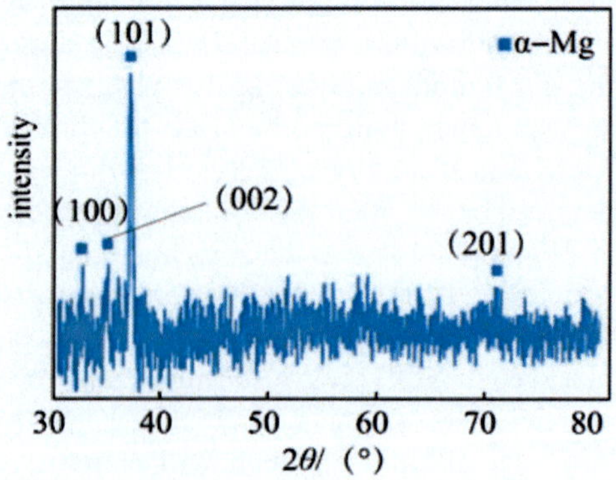

Fig. 4.54 XRD spectrum of smoke and dust generated during manufacturing magnesium alloy AZ91D using LPBF. Reprinted with permission from a work by Wei et al. [11]

therefore, the alloy element molecules are more likely to break free from the binding of the liquid surface of the molten pool and enter the gas phase driven by the surface tension and vapor recoil force in the molten pool. Obviously, the evaporation of elements is closely related to the process parameters of LPBF forming. In theory, the greater the laser energy input, the more energy absorbed by the powder, the greater the burning loss of alloy elements. However, the smoke and dust of the evaporated alloy elements block the laser beam during the rising process, and the energy actually absorbed by the alloy powder will still be reduced (despite the existence of protective gas flow). When the laser energy density is high (high laser power or low scanning speed), the burning loss of alloy elements will not change too much. When the laser energy absorbed by the powder is less, the evaporated alloy element soot will also be reduced. As the laser beam is shielded less, more laser energy will shine on the powder bed, and the alloy powder will absorb more energy. With the increase of laser energy input, the evaporation of alloy elements will also increase.

As mentioned above, the evaporation process of alloy elements includes migration in the liquid phase, volatilization reaction at the interface and diffusion in the gas phase. After absorbing energy, the molecular kinetic energy of alloy element molecules increases and vaporizes at the liquid/gas interface of the molten pool. When it has enough speed, it can escape from the melt surface and enter the vacuum chamber smoothly through the gas boundary layer.

In the process of LPBF manufacturing, the building chamber is generally filled with inert nitrogen gas or argon gas. On the one hand, it can be used as a protective gas to prevent material oxidation. On the other hand, when the protective gas pressure is large enough, alloy element molecules escaping from the liquid/gas interface will be re-scattered back to the molten body after collision with the external gas, thus preventing the evaporation of alloy elements. Therefore, the coupling between the metal vapor recoil force in the molten pool and the external pressure on the surface of the molten pool is the main mechanical effect of element evaporation. It can be believed that when the surface tension and vapor recoil force caused by high temperature in the molten pool reach equilibrium with the atmospheric pressure, the evaporation of alloy elements will also reach a dynamic equilibrium. For the LPBF process, the evaporation of alloy elements is inevitable. However, the dynamic balance can be achieved by adjusting the process parameters and the evaporation of alloy elements under atmospheric pressure, which can minimize the element loss and ensure the homogeneity of the alloy composition, structure and properties of the built parts.

References

1. Wang D (2011) Study on the fabrication properties and process of stainless steel parts by selective laser melting. South China University of Technology, Guangzhou
2. Wang D, Mai S, Xiao D et al (2016) Surface quality of the curved overhanging structure manufactured from 316-L stainless steel by SLM. Int J Adv Manuf Technol 86(1–4):781–792

3. Liu Y (2015) Research on the mechanism of selective laser melting and direct manufacturing of structural features. South China university of technology, Guangzhou
4. Wang D, Bai Y, Yangy. A metal 3D printer sealed cabin atmosphere deoxygenation and circulation purification method and equipment: China, CN104353832A. 2015-02-18
5. Wu W, Yang Y, Wang D (2010) Balling phenomena in selective laser melting progress. J South China Univ Technol (Nat Sci Ed) 038(005):110–115
6. Wu W, Xiao D, Yang Y et al (2016) Analysis on powder adhesion problems in selective laser melting forming process. Hot Work Technol 24:43–47
7. An CH, Yang Y, Zhang J et al (2018) Experimental study on density and pore defects of cobalt-chromium aiioy fabricated by selective laser melting. Appl Laser 38(05):30–37
8. Powell A, Pal U, Van Den Avyle J et al (1997) Analysis of multicomponent evaporation in electron beam melting and refining of titanium alloys. Metall and Mater Trans B 28(6):1227–1239
9. Klassen A, Scharowsky T, Körner C (2014) Evaporation model for beam based additive manufacturing using free surface lattice Boltzmann methods. J Phys D Appl Phys 47(27):1–10
10. Xie ZH (2013) Research on processing and mechanism of AZ91D magnesium alloy by selective laser melting. Huazhong University of Science and Technology, Wuhan
11. Wei K, Wang Z, Zeng X (2016) Element loss of AZ91D magnesium alloy during selective laser melting process. Acta Metall Sin 52:184–190

Chapter 5
Formation Mechanism of Spatter and Its Influence on Mechanical Properties in Process of Laser Powder Bed Fusion

Due to the application of laser powder bed fusion (L-PBF) technique in the high-tech field of large complex parts such as aero-engine impeller, aerospace lightweight parts, injection molds, its processing stability has become more and more important. The stability of large size complex components in long time processing has been recognized as one of the key and difficult problems to solve. The main reasons are:

(1) The thickness of L-PBF accumulation layer is 20–50 μm, large-size components generally need thousands of processing repetition of powder laying-powder melting, which takes up to hundreds of hours.
(2) L-PBF each layer and by a number of melting tracks interwoven;
(3) The influence factors of the L-PBF process are complex. During the process, the forming environment such as element vaporization and spatter influence, and the surface disturbance factors of parts cause undetermined disturbance.

Compared with some conventional machining process, the L-PBF process is a multi-factor interaction process, which leads to the complex, unstable and more sensitive multi-layer stacking process some factors of small perturbations are likely to trigger a chain reaction, leading to the development of unstable stacking accumulation in the subsequent accumulation, and even make the accumulation layer growth process instability.

Spark and spatter are common phenomena in laser processing. In laser welding, cutting, drilling, surface strengthening and other processes, the power density of laser beam acting on the processed material is up to 10^6–10^7 W/cm^2. Under the action of continuous laser beam radiation, the surface of the material not only melts, and even vaporize to form a plasma to help the material absorb the laser, the molten material is squeezed under vaporization recoil pressure, Part of the liquid material fly away out of the molten pool at a certain speed and form spatter. The higher the laser energy input, the stronger the metal vapor generated, leading to intensification of liquid metal flow in the molten pool, while sputtering more droplets under the recoil pressure, and the spatter speed and the height reached also increase. Therefore, as a subsidiary product of LPBF process, the behavior characteristics and formation mechanism of

© National Defense Industry Press 2024

D. Wang et al., *Laser Powder Bed Fusion of Additive Manufacturing Technology*, Additive Manufacturing Technology, https://doi.org/10.1007/978-981-99-5513-8_5

spatter are directly determined by the interaction between laser and powder bed, the negative effect of spatter particles on LPBFed parts was also gradually found in actual production. This chapter will comprehensively introduce and discuss the spatter behavior in the forming process of LPBF combined with the research team.

5.1 Factors Influencing Spatter Behavior

Dr. Liu Yang, one of members of the research team, took the LPBF process of 316L stainless steel metal powder as an example, and conducted detailed discussion and analysis on the influence of various LPBF process parameters on the spatter behavior.

5.1.1 Influence of Scan Lines

Figure 5.1a shows the spatter behavior at different positions of the scan line, at the beginning of the scan line, the laser has just irradiated onto the powder, due to the high speed of laser scanning, the time of laser spot on metal powder particle is only 0.5–2 ms, in such a short time, heat has not enough time to conduct, molten pool has not formed (At this time the molten pool size is very small), The spot produced by the molten pool is also weak, the spatter has not been formed yet. At the midpoint of the scan line, the molten pool has maintained dynamic stability, and the spatter has also maintained dynamic stability. At a higher energy input, the molten pool produces a dazzling flash, and the liquid metal column can reach a height of 5 cm. The unmelted powder particles are impacted and raised at a certain angle, and the spatter is in a divergent state. At the end of the scan line, the spatter is more intense. Due to the reduced speed of the laser beam, too much laser energy is concentrated in a small area, causing the material to absorb more energy and the temperature to rise sharply, the size of the molten pool becomes larger, and a dazzling flash is emitted. At the same time, violent turbulence occurs in the molten pool, and the powder around the molten pool is sucked into the molten pool due to capillary action, which makes the melting track at the end wider and higher, forming a bulge, which is the "First line scan balling" effect.

5.1.2 Influence of Scanning Speed

Figure 5.2 shows the dynamic process of the spatter when the laser power is fixed at 200 W and the scanning speed is 200 mm/s and 800 mm/s, respectively, it can be found that the scanning speed has a great influence on the spatter. The number of spatter in Fig. 5.2a is large with a maximum jetting height of about 5 cm, and the degree of spatter aggregation is low. Figure 5.2b shows that the spatter is very small

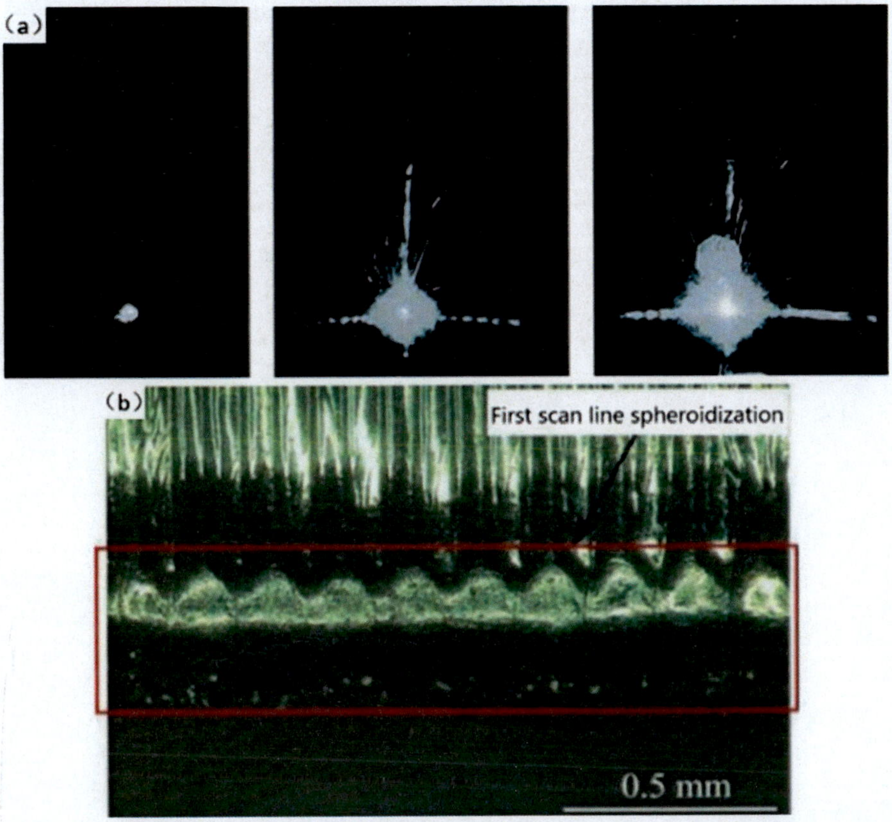

Fig. 5.1 Spatter behavior at different positions of scanning line: **a** spatter morphology at each point of scanning line; **b** first line scanning spheroidization

and exhibits hysteresis. Figure 5.2c shows a sharp increase in the number of spatter and decrease in size, and a jetting height of 11 cm. The reason is that under a certain laser power, the higher the scanning speed, the smaller the energy input, as a result, the energy in the molten pool is small, the intensity of the generated metal vapor is small, and only a small amount of molten metal can be driven out of the molten pool to form molten droplets, and the number of spatter is small. At higher scanning speed, the inclination angle θ of the front wall of the small hole in the molten pool increases, the laser spot is directly irradiated on the front wall of the small hole, and the metal droplets formed under the action of the recoil force of the metal vapor are sprayed along the back wall of the small hole. In addition, Fig. 5.2c shows that the spatter intensity exhibits a certain periodic fluctuation, this is because at lower scan speeds, the molten pool undergoes violent evaporation and vaporization, a cloud of metal vapor, powder, and liquid metal forms directly above the molten pool, shielding part of the laser energy, the energy obtained in the molten pool decreases, the intensity

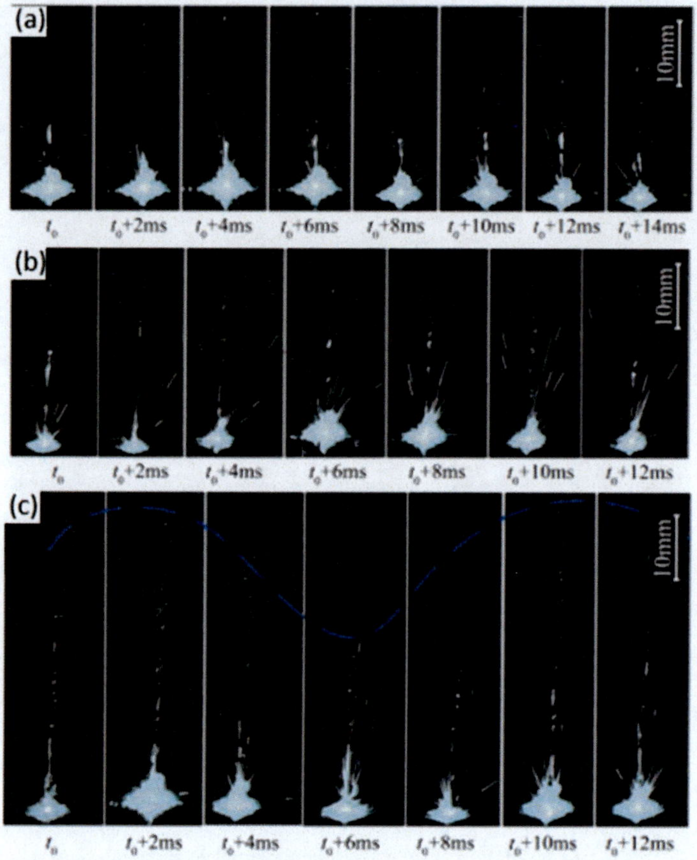

Fig. 5.2 Effect of scanning speed on spatter behavior: **a** scanning speed 800 mm/s; **b** scanning speed: 200 mm/s; **c** scanning speed: 50 mm/s

of the cloud decreases, and the energy irradiated to the powder bed increases again, as shown in Fig. 5.2c at times $t_0 + 6$ ms and $t_0 + 8$ ms.

5.1.3 Influence of Laser Power

Figure 5.3 shows the dynamic process of the spatter at speed of 50 mm/s and laser power of 100 W and 150 W, respectively. Figure 5.3a shows that the intensity of spatter is weak, the size is large, and the jetting height is about 5 cm. In Fig. 5.3b, the intensity of spatter increases, but the direction is disordered, and the jetting height is about 7 cm. The reason is that the greater the energy input, the stronger the impact of the metal vapor and plasma inside and outside the molten pool, and the greater the impact on the liquid jet. The smaller the droplet is pulverized, the greater the initial

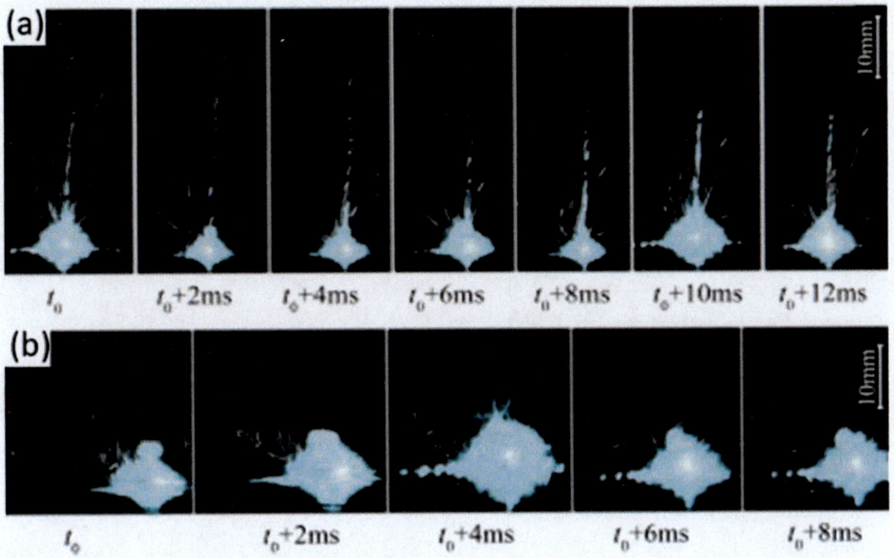

Fig. 5.3 Influence of laser power on spatter behavior: **a** the laser power: 150 W; **b** the laser power: 100 W. Reprinted with permission from a work by Wang et al. [1]

momentum gained, and the greater the jet velocity and the jetting height reached. In addition, it can be found that the larger the energy input, the higher the concentration of the spatter, as shown in Fig. 5.2c. This is because when the energy input is large, the molten pool of a larger size will be formed, and the strength of the metal vapor in the molten pool is large, leading to violent turbulence in the molten pool, and the powder around the molten pool is sucked into the molten pool due to capillary action. Therefore, it can be determined that when the energy input is small, the powder spatter is more, when the energy input is high, the molten metal will spatter more.

5.1.4 Influence of Oxygen Content in Building Chamber

Before the LPBF process, the oxygen in the building chamber needs to be pumped out, it can be formed when the oxygen concentration in the building chamber is less than about 0.1%. However, when the sealing type of the equipment is poor, the oxygen content in the building chamber is high, and the metal material will oxidize with oxygen, or even burn (such as pure titanium powder). Fe, C, Si, Mn, Ti, Ca contained in the material will react with oxygen to form corresponding oxides, such as SiO_2, MnO, CaO, etc. Some researchers have shown that oxygen in the protective atmosphere is an important cause of metal globularization. The formation of a layer of oxide on the surface of the liquid metal will reduce the wettability of the liquid metal,

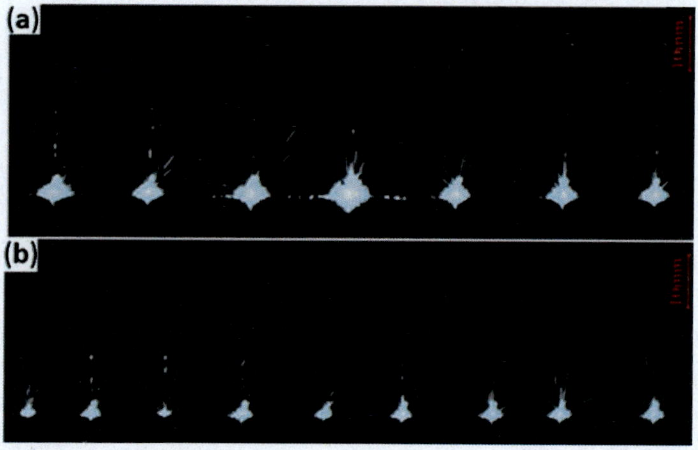

Fig. 5.4 Effect of oxygen content on spatter behavior (laser power 150 W, scanning speed 50 mm/s): **a** spatter without shielding gas; **b** spatter with shielding gas

resulting in greatly reduced wettability between the liquid metal and the substrate or the front forming layer, which is easy to cause interlayer cracking. As shown in Fig. 5.4a, b, when high-purity nitrogen N2 (≥99.99%, ensuring oxygen content below 0.2%) is used to protect the building chamber, the spatter is weak and the jetting height is about 7 cm. In the absence of nitrogen, however, the spatter was very strong, reaching a maximum of 11 cm. This is because the oxygen content is high, and the molten pool in the material produces severe oxidation reaction, oxidation reaction process produces lots of heat, cause the molten pool size bigger, also contributed to the "boil" in the molten pool and have a strong metallic vapor, leading to more liquid metal escape from the molten pool and more powder particles around the molten pool under the impact force, thus forming spatters.

As shown in Fig. 5.5a, b, under nitrogen protection, the sparks are relatively concentrated and the spatters are concentrated at an angle of about 75°, while under no nitrogen protection, the sparks diverge and the angle is about 115°. The powder bed after laser scanning is also very different. When there is gas protection, because there is gas flow in the building chamber, the smoke and dust generated can be blown away in time, so the powder bed is relatively clean with few dust and spatter particles. When there is no gas protection, there is no gas flow in the building chamber, and the generated gas is scattered around the forming area, resulting in an obvious layer of smoke and spatter particles accumulated on the powder bed, as shown in Fig. 5.5c, d. These dust and spatter will be mixed in the powder during the next forming cycle, and will be scanned by the laser and mixed in the part, seriously affecting the mechanical and microscopic properties of the LPBFed parts. The dust deposited on the transparent lens during the forming process leads to serious attenuation of laser energy input, and the material can't absorb enough energy and not melt enough. Whereas, the lens absorbs too much laser energy and heats up, which further causes the reduction of laser transmittance. The diameter of spatter particles is generally

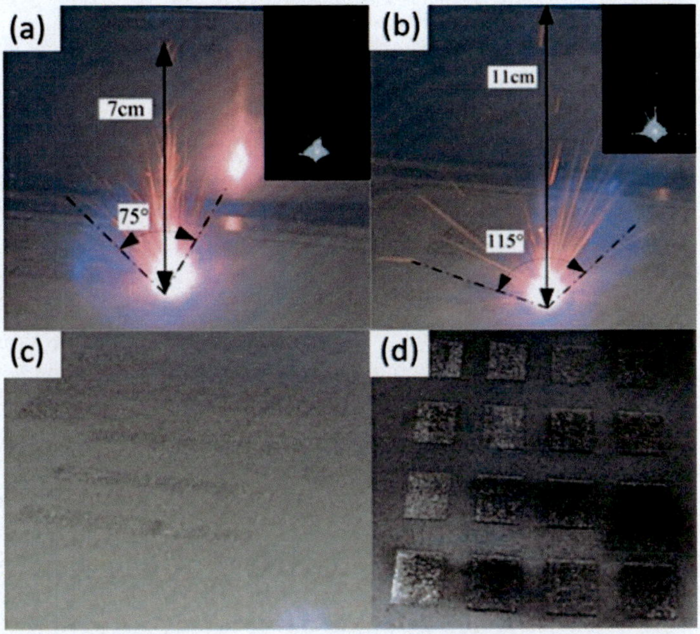

Fig. 5.5 Effect of oxygen content on spark behavior: **a** sparks with gas protection; **b** sparks without gas protection; **c** powder bed surface with protection; **d** powder bed surface without gas protection

100–300 μm, the largest particles can even reach 500 μm, several times larger than the powder particle, and the thickness of each layer of powder is only 20–50 μm. If it is a recoating device, when the spraying powder tool encounters the large size particles on the surface of the parts, it may cause the powder laying device to be stuck or impacted, thus affecting the accuracy of the device. If it is a flexible powder laying device, it may destroy the rack of the powder laying device, resulting in poor quality of subsequent powder laying.

5.2 Types and Mechanism of Spatter Formation

5.2.1 Spatter During Traditional Welding Process

The molten pool and spatter mechanism in laser welding (LW) process have been studied in depth. The spatter in LW process is considered to be the result of the escape of molten liquid caused by the recoil pressure generated by the surface energy transfer. Under the continuous radiation action of laser beam with power density up to 10^6–10^7 W/cm^2, metal materials not only melt, but even vaporize to produce metal vapor and plasma, metal vapor and plasma hoarded inside the molten pool contribute to the keyhole structure. The escape of vapor and plasma forms a plume jet, and the escape behavior of molten liquid in molten pool under the recoil pressure of metal vapor constitutes the spatter in LW process.

Vladimir and Akira [2] published an article in 1997 that calculated and analyzed the energy balance, melt surface temperature, keyhole/cutting front propagation velocity and melt ejection velocity in laser welding and cutting processes based on a priori physical model, concluded that in the laser welding process, the metal vapor force is the main reason of molten pool motion and molten metal loss, and the recoil force induced by evaporation increases with the increase of molten pool surface temperature, resulting in melt ejection from the interaction zone.

The high-speed imaging technology of welding process under special sandwich sample is widely used in the research of deep fusion welding pool and spatter, and the characteristics of keyhole, vapor plume and spatter generated by welding process can be displayed dynamically. As shown in Fig. 5.6, Zhang et al. [3] filmed the high-power low-speed fiber laser welding process. During the low-speed incomplete penetration welding process, they found that there were typical fold structures on the front wall of the keyhole with small angle inclination due to surface tension. Under the action of the evaporation pressure of the laser on the material, the fold grows rapidly and moves to the keyhole. The tip part may even break away from the front wall of the keyhole to form micro-droplets and be accelerated to about 11 m/s to fly away from the keyhole to become spatters. In addition, the acceleration process of the plume pair through the spatter droplets was recorded. The spatter droplets entering the plume at 0.17 m/s were accelerated to 1.2 m/s and flew out of the plume.

In order to discuss the fundamental mechanism of spatter during laser welding, Li et al. [4] processed the plate with a large angle tilt laser beam by changing the defocus amount of laser and the angle of laser incidence to simulate the state of laser beam oblique irradiation on the front wall of keyhole. As shown in Fig. 5.7, the

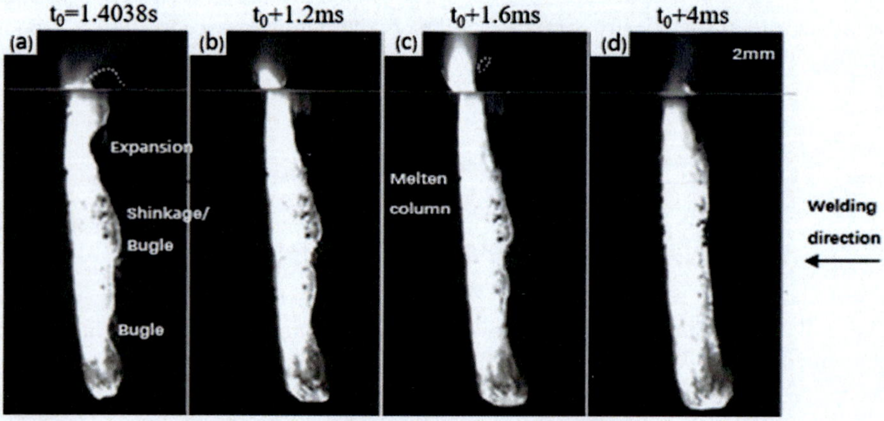

Fig. 5.6 Spatter generation under the action of keyhole structure and metal vapor during laser welding. Reprinted with permission from a work by Zhang et al. [3]

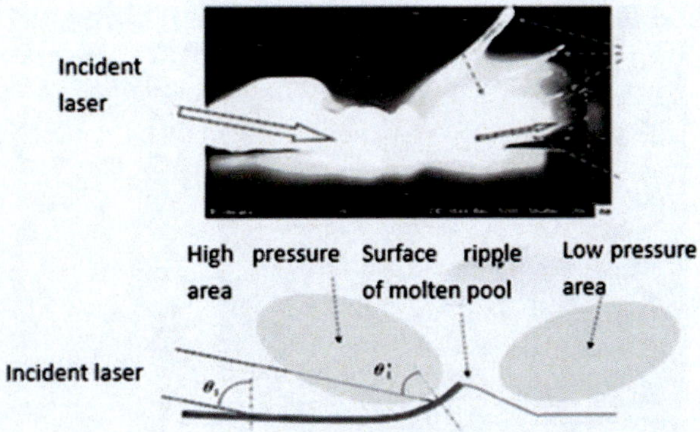

Fig. 5.7 Large angle laser beam processing simulates the laser radiation process of the front wall of the keyhole. Reprinted with permission from a work by Li et al. [4]

molten pool surface under laser tilt irradiation fluctuates frequently to form liquid ripples due to unstable vaporization, and the existence of ripples changes the laser absorption efficiency, and the intensified local vaporization process and the increased steam recoil effect cause tearing and spatter on the molten pool surface.

The simulation process also proves that the laser reflection of the front wall of the keyhole with small inclination angle is not the main reason for the bulge formation of the back wall of the keyhole. On this basis, Cheng et al. [5] found that the molten pool fluctuation on the front wall of the keyhole and the heterogeneous high-pressure plasma generated by the evaporation of the keyhole wall under uneven laser energy absorption caused the instability and rapid fluctuation of the vapor flow field inside the keyhole. Then the initial fluctuation of the back wall of the keyhole is generated. The upward molten liquid steam promotes the growth of the bulge of the back wall of the keyhole, providing the main kinetic energy for it. The mechanics of the process is the essence of steam and the friction between the melt and flow fluctuation pressure destroyed the ideal state of equilibrium.

The LPBF process has some similarities with the laser welding process: both use the thermal effect of laser to process metal materials, in the process of processing through the protective gas, metal materials after absorbing energy will produce metal vapor, molten pool, there are rapid solidification of molten pool and other phenomena. However, due to the different original material forms, that is, the laser welding material is metallic bulks, and the LPBF material is metal powder particles, resulting in the difference between the two phenomena. Compared with the laser welding, the laser action spot used in the LPBF process is much smaller, the laser scanning speed is much higher, and the molten pool size and depth are smaller, as shown in Fig. 5.8. In addition, the material around the molten pool is more likely to be taken away during the LPBF process, forming a source of spatter or being carried into the molten pool. Due to the LPBF material is powder, the heat conduction inside the powder and the heat and mass transfer inside the molten pool are more complex than the heat and mass transfer in the corresponding bulk material. The process involves the rapid melting and partial evaporation of powder, the flow of molten metal, the spatter and redistribution of powder particles or liquid, the rapid solidification mechanism and

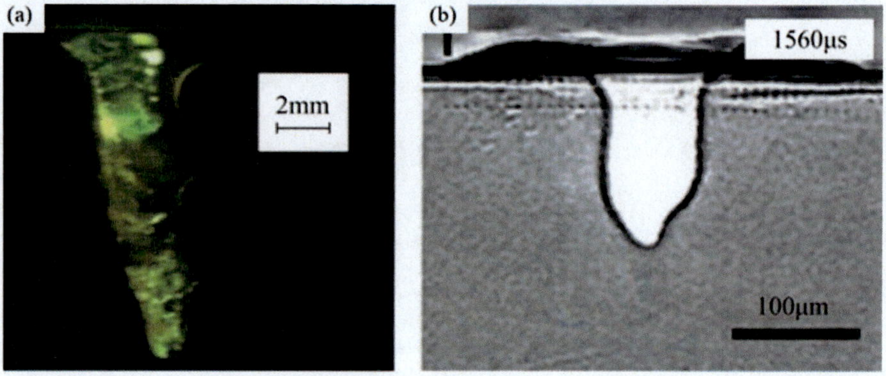

Fig. 5.8 Typical weld pool image: **a** welding process; **b** laser melting process in powder bed. Reprinted with permission from a work by Andani et al. [6]

other transient and dynamic physical phenomena, this makes the spatter mechanism during LPBF process more complicated.

5.2.2 Mechanism of Spatter Formation in LPBF process

The spatter during the LPBF process is the product of the interaction between laser and powder material. In experiments or other occasions, many researchers have observed the morphology of the processed spatter and found that the sputtered particles are mainly spherical and the size is larger than the initial powder particles. According to the spatter formation mechanism in the process of manufacturing 316L stainless steel, Yang Liu et al. (South China University of Technology) divided the spatter into droplet spatter and powder spatter. According to the different forms and sources of spatters, Wang et al. [1] divided spatter into three types: (1) From the original metal powder, only need the laser power to reach a certain threshold, enough metal vapor or ambient gas disturbance, so that the metal powder to produce the phenomenon of flying; (2) Due to the shrinkage of the molten pool and its surface tension during the laser action, the metal powder around the molten pool enters the molten pool and becomes spatters under the action of metal vapor; (3) The molten pool melt escapes from the molten pool and becomes spatter after undergoing the process of bulging and necking, as shown in Fig. 5.9.

Gunenthiram et al. [7] used volume energy density (VED) and powder layer thickness as variables to study the influence of process parameters on spattering links under two kinds of materials. Through quantitative analysis, it is found that under the action of powder bed heat conduction, the metallic powder in contact with the front and side of the molten pool melts and easily agglomerates to form larger droplets under the action of heat radiation and evaporation. When the mass density is lower than the threshold (10 J/mm^3), the droplets will melt into the molten pool.

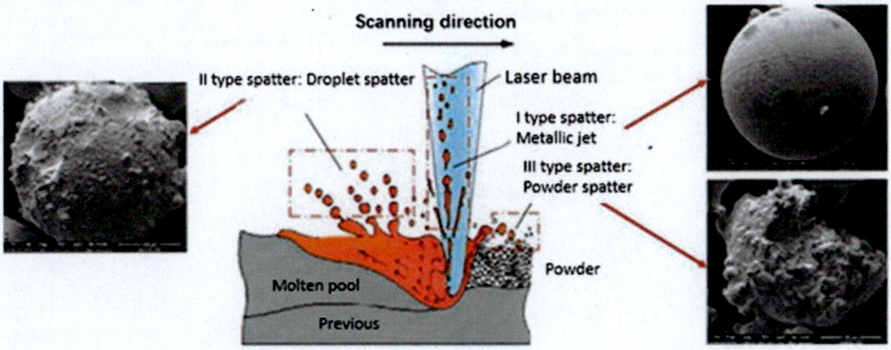

Fig. 5.9 Three classifications of spatter particles [6]

The backward acceleration of the fluid and the upward metal vapor create a steep molten pool boundary, which is considered to be a sign of the beginning of the spatter. Combined with the influence of molten track cross section on energy density discussed in this paper, and combined with the spatter mechanism, the possible processing states of the whole LPBF process are divided into four categories: At low energy density input, especially at high scanning speed, an unstable process that is prone to a large number of globularization and a large amount of spatter. A normal machining with low penetration obtained at low energy density input and relatively suitable scanning speed, in which the number of spatterer is greatly reduced. A high penetration steady-state machining with keyhole structure with slightly increased spatter (65 J/mm^3, 316L) at suitable energy input and scanning speed. A kind of unstable processing of the energy input "hump" melt, as shown in Fig. 5.10 [7]. The effect of input energy on the state of molten pool directly determines the state of spatter, denudation and keyhole structure. According to the fluctuation of spattering state, the evaporation of molten pool is considered as the driving force for spattering.

Bidare et al. [8] used high-speed camera and schlieren imaging technology to capture spatter denudation and plume phenomena under single fuse and multi-layer forming. When laser power and scanning speed were used as control variables, the spatter types were divided into vertical direction, acute angle and obtuse angle according to the angle between sputtering direction and scanning direction, and the degree of denudation was the lightest in the vertical scanning direction, the reason is explained by plume movement under schlieren imaging (including iron vapor, plasma gas and protection gas Ar). As shown in Fig. 5.11, the performance of the plume is verified from the perspective of thermal distribution and ambient gas flow in the form of numerical models. In this paper, the causes of spatter and denudation mechanism are explained by combining the motion characteristics of spatter particles and visualized plumes with numerical models, and the important roles of protective gas and laser plume near the laser operating point in the laser interaction process are emphasized.

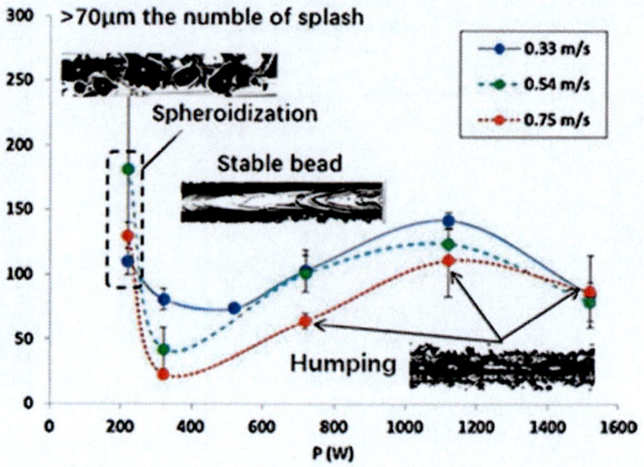

Fig. 5.10 Effect of different laser energy on large size spatter. Reprinted with permission from a work by Gunenthiram et al. [7]

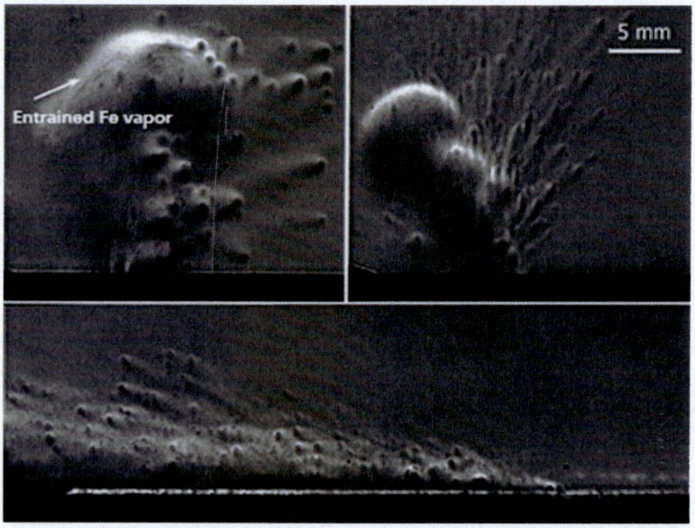

Fig. 5.11 Plume process schlieren imaging at three sputtering angles. Reprinted with permission from a work by Bidare et al. [8]

Ly et al. [9] analyzed the spatter process of LPBF process through high-speed camera and simulation analysis, and believed that the recoil effect produced by the laser action process was not the main reason for the formation of spatter. Ly et al. [9] analyzed the spatter motion process in the LPBF process through high-speed camera and simulation analysis, and believed that the recoil effect generated by the laser action process was not the main cause of spatter formation, but the entraining effect caused by the violent flow of protective gas around the laser action point,

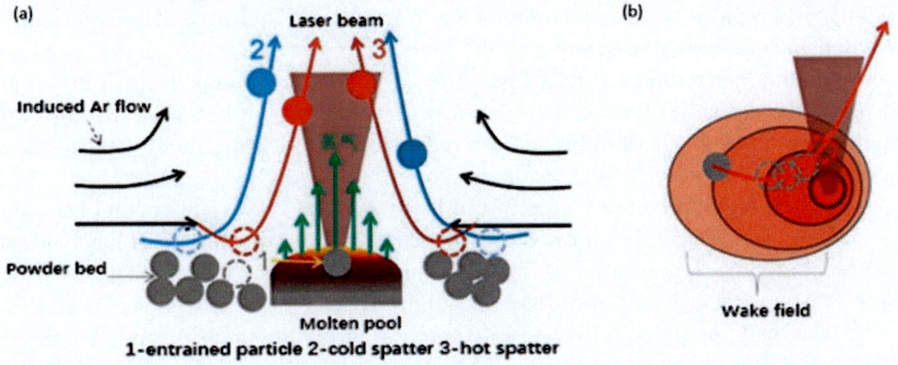

Fig. 5.12 Formation process of three spatter processes: **a** three kinds of spatter formation processes; **b** the excitation process of powder spatter. Reprinted with permission from a work by Ly et al. [9]

which was similar to the study of P. Bidare. However, different from the former study by P. Bidare, Sonny Ly et al. explained the generation of two kinds of spatter in a simulation way from the influence of metal vapor pressure on the shape of keyhole. When the evaporation effect of laser power and scanning speed on metal is enough to form a deep and vertical powder bed molten pool, vertical spatter injection will be generated, and on the contrary, obtuse angle spatter will be obtained. Based on the high-speed camera, this paper shows three types of spatter generation mechanisms, as shown in Fig. 5.12. Under the recoil force, the metal melt raises, elongates its neck and finally escapes into a spatter, which accounts for 15% of the total number of spatteres, and the speed is about 3–8 m/s. Some metallic particles on the powder bed are pulled up under the entraining of intense flowing atmospheric gas, or become hot spatter by laser radiation (about 60% of spatter, 6–20 m/s), or become cold spatter without heating (about 25% of spatter, 2–4 m/s).

For the molten pool, the characterization ability of high-speed cameras is general, and the internal contour and depth of the molten pool cannot be clarified. Zhao et al. [10] used X-ray and diffraction techniques to photograph the molten pool under the LPBF process, and directly calculated and analyzed the dynamics of the molten pool, the powder spatter process and the rapid solidification process, especially the development and transformation process of microstructure. In the laser processing with laser power of 520 W, the process lasts for 1 ms. Under the joint action of Marlangoni effect and recoil pressure, heat and mass are rapidly absorbed and diffused to the powder around the working point, and the spatter is formed under the recoil pressure generated by metal evaporation. Although the formation mechanism of spatter is relatively simple in this literature, the movement track and characteristics of spatter particles are more intuitively and clearly expressed under the X-ray technology. More detailed spatter movements include: (1) At the beginning of the action, it is shot out of the molten pool by the reaction force of metal vapor. (2) Under the action of molten surface tension and protective gas flow, it slowly approaches the laser operating point and is rapidly excited and emitted into a spatter. (3) Under the action of surface tension and complex air entrainment, the two spatter particles approach the laser and fuse into a larger spatter particle. In addition, through the contrast tests of different powers, the dynamic flow of the molten pool is caused by the non-uniformity of the surface tension distribution and the recoil force, which eventually leads to the asymmetric distribution of the spatter angle.

Based on high-speed photography and X-ray imaging, Nakamura et al. [11] used tungsten carbide tracer to achieve easier detection of the flow and spatter mechanism of the molten pool, as shown in Fig. 5.13. In this paper, the origin location of the spatter is analyzed based on different scanning speeds with laser power of 10 kW. When the scanning speed is low (17–50 mm/s), 80% of the spatter is generated in the head and side of the keyhole structure. During the process from 50 to 100 mm/s, the spatter generation obviously shifts from the head of the keyhole structure to the rear of the molten pool. When the scanning speed reaches 100 mm/s, the spatter generated at the side of the keyhole structure occupies the main body. When the scanning speed reaches 300 mm/s, the spatter is mainly generated at the rear of the keyhole structure. The addition of tracer particles provides intuitive and reliable measurement data for the description of the flow, velocity and spatter trajectory in the molten pool, and provides a basis for explaining the change of spatter position.

5.3 Influence of Spatter on Mechanical Properties

According to the above content, it can be clear that the spatter particles produced in the LPBF process include two types:

(1) The spatter affects the stability of LPBF process, the flatness of new powder layer and the energy absorption rate of laser irradiated powder bed, which will adversely affect the stability of liquid molten pool.
(2) Spatter particles mixed in clean powder will also affect the microstructure and mechanical properties of parts, resulting in the formation of micropores and inclusion defects in the microstructure, and eventually leading to a serious decrease in the elongation and fatigue strength of parts.

The author's research team has carried out related researches on the influence of usage times of raw powder on the compact and mechanical properties of LPBF, including the influence of metal powder and spatter particles on the LPBF process under different pollution levels, and the compaction and mechanical properties of the forming quality were tested.

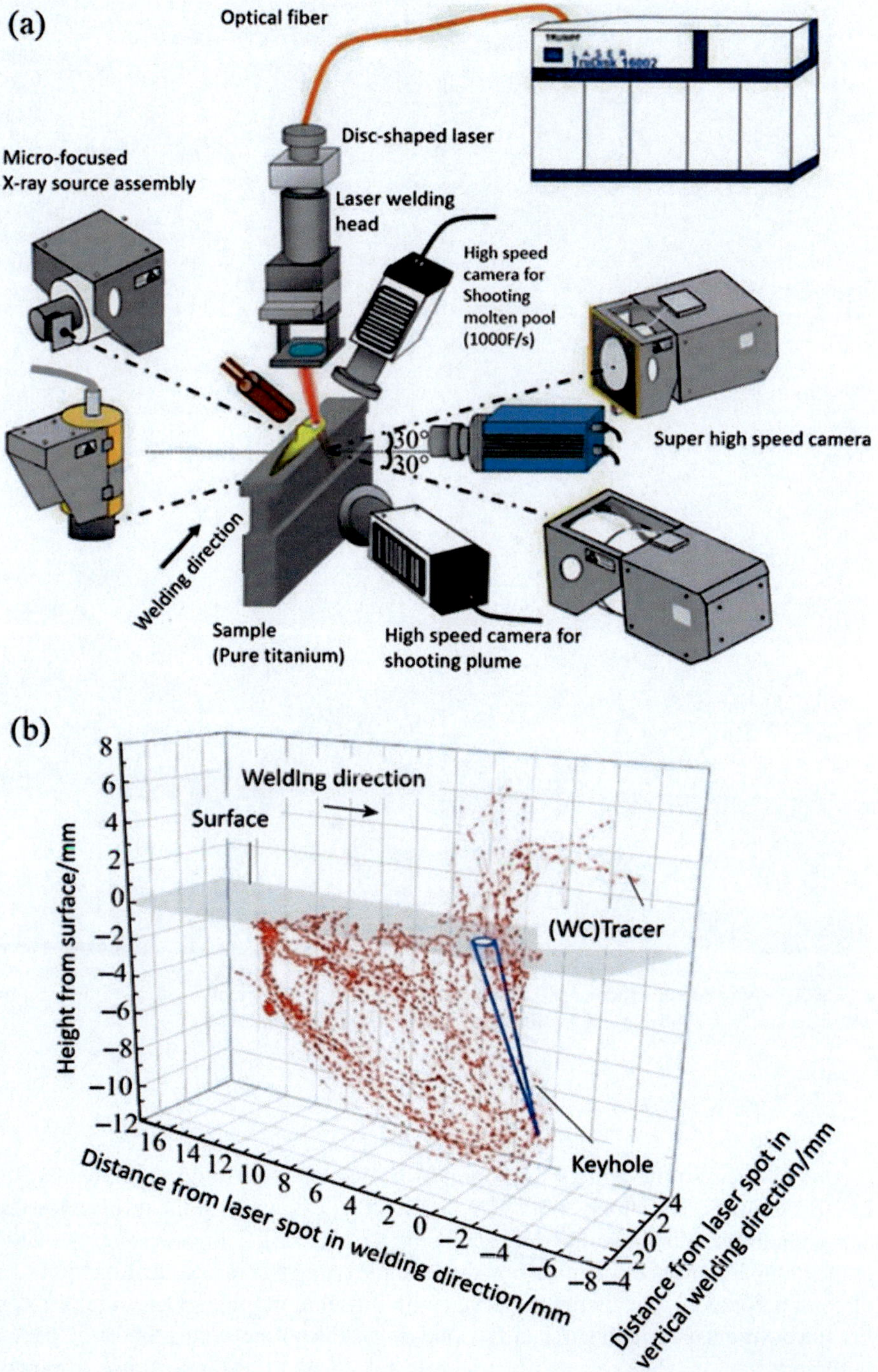

Fig. 5.13 Three dimensional motion track of X-ray imaging system and tracer: **a** schematic diagram of X-ray imaging system; **b** motion track of missing particles. Reprinted with permission from a work by Nakamura et al. [11]

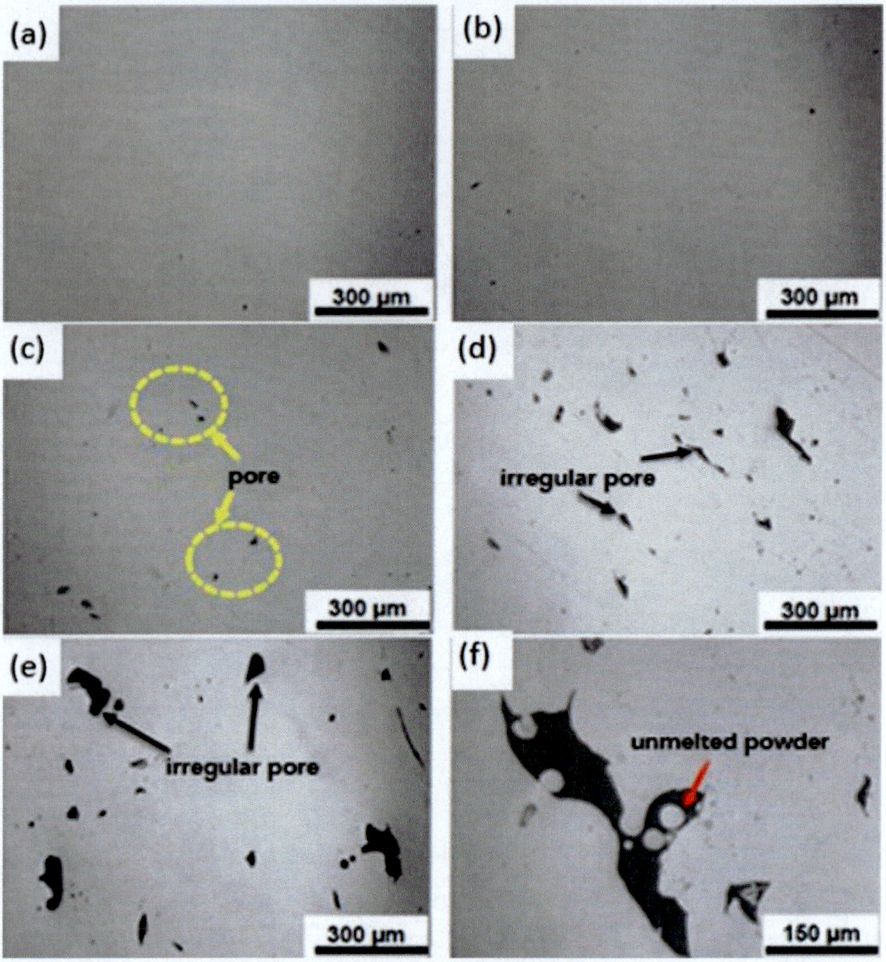

Fig. 5.14 Macro defect structure of LPBFed part section under different powder use times: **a** first use; **b** second use; **c** third use; **d** the fourth use; **e** the fifth use; **f** sixth use

5.3.1 Densification

Figure 5.14 shows the optical micrographs of densification and pore defects of the LPBF-fabricated test blocks obtained with different powder application times. The experimental results show that the compactness of the LPBF process is very sensitive to the number of uses of metallic powder, and the compactness and residual pore size of parts prepared from powder materials with different number of uses are different. For the powders used in the first and second times, the obtained parts had high density with very few micro pores, as illustrated in Fig. 5.14a, b. The third use of powder resulted in significant decrease in the density of the samples due to coagulation in the

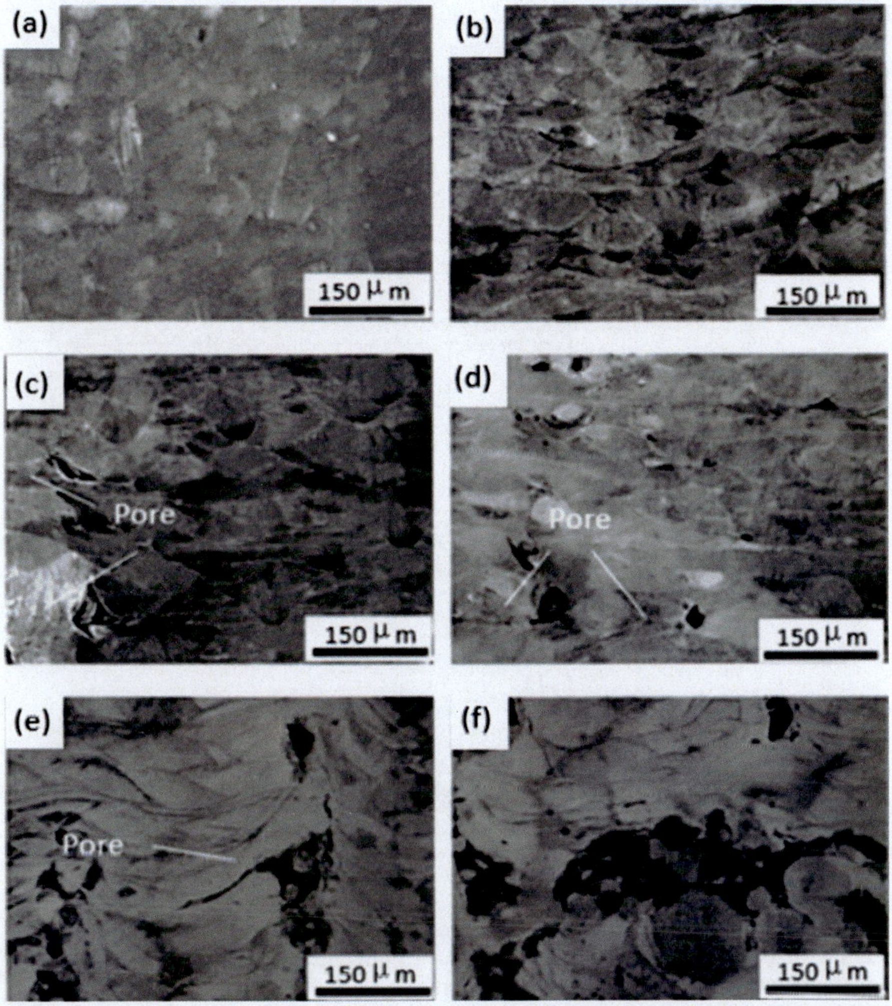

Fig. 5.15 Macro defect structure of LPBFed part section under different powder usage times: **a** first use; **b** second use; **c** third use; **d** the fourth use; **e** the fifth use; **f** sixth use

tiny appearance and smooth pores (Fig. 5.14c). With the fourth use, some irregular pores with sharp corners were obviously generated inside the samples (Fig. 5.14d). When the powder was used for the fifth time, the average size of irregular pores reached 150 μm (Fig. 5.14e). As the powder was applied up to six times, unmelted powder began to appear in narrow pores with an average size of 300 μm (Fig. 5.14f).

Through further corrosion and microstructure observation, the macroscopic melting track defect structure diagram shown in Fig. 5.15 was obtained. Under the influence of powder usage times, the stability of fusion pool in terms of lap and morphology is directly affected. In the state of Fig. 5.15a, b, the melt tracks show

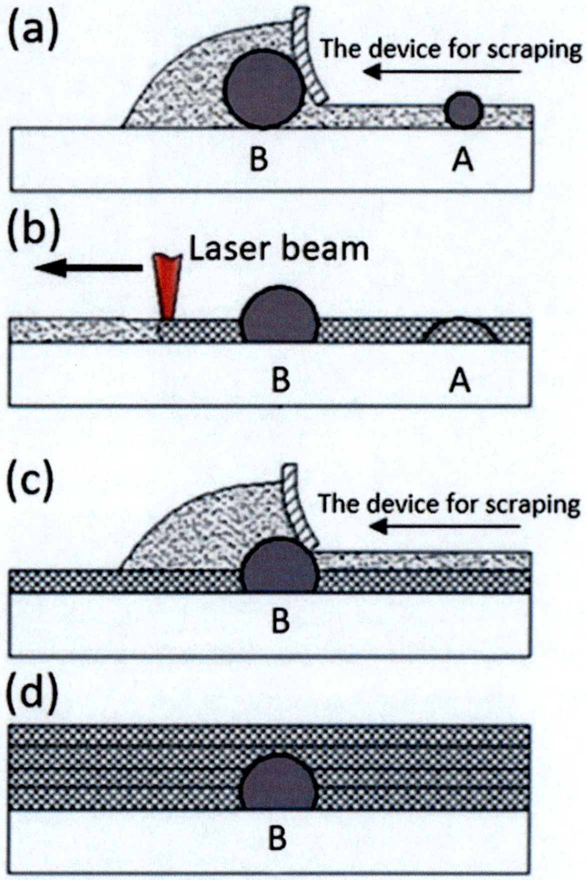

Fig. 5.16 The mechanism of spatter particles with different sizes: **a** powder feeding for the N-th layer; **b** laser scanning for N-th layer; **c** powder feeding for the N + T-th layer; **d** end of N + M-th layer

normal shape and overlapping effect, and a good metallurgical bonding effect is obtained between the melt tracks, and finally shows good compactness. With the use of metallic powder reaching three and four times, micro pores begin to appear along the boundary of the molten pool. When the use number reaches more than five times, the size of the pores increases greatly and the inclusion of unmelted powder particles will be accompanied. At the same time, multiple pores will join along the boundaries of the molten pool to form a larger size hole.

According to the formation mechanism of pore structure under different usage times, the influence mechanism of spatter particles in the LPBF process was summarized. Figure 5.16 shows the influence of spatter particles with different sizes on the LPBFed part. Spatter particles A and B are scattered on the surface of the LPBFed parts. When spreading the next layer of powder, due to the small size, the spatter particle A can be completely melted by the laser beam, thereby merging together with the LPBFed parts, but may also create pores and may themselves become inclusions in the track (Fig. 5.17a). Since the size of B particle is much larger than the thickness

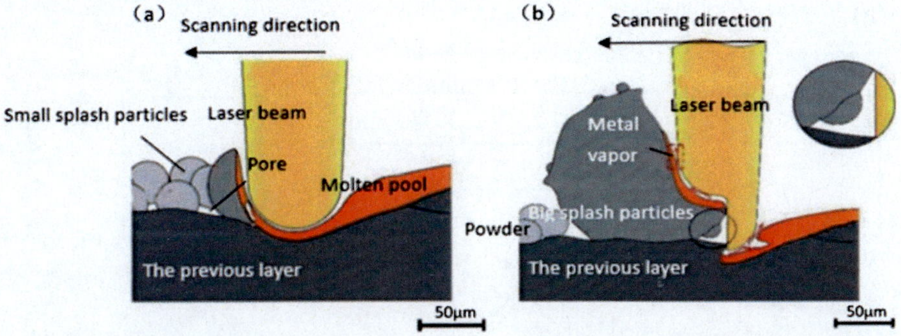

Fig. 5.17 Cladding mechanism of particles: **a** cladding mechanism of particle A; **b** cladding mechanism of B particles

of the powder, it cannot be completely melted by the laser beam (Fig. 5.17b), which prevents the laser from melting the metal powder obscured by B particle, resulting in the unmelted powder phenomenon in Fig. 5.16.

5.3.2 Mechanical Properties

In the initial stage, the author's team designed a group of comparative experiments based on 316L stainless steel powder in order to study the effect of spatter particles on the mechanical properties of LPBF-fabricated parts. Two groups of plate shaped tensile test specimens were formed by using clean powder and polluted powder (powder that had been used for 5 times but not sieved), and the size of the specimens met the national standard GB/T 228-2002. The same process conditions were used. The parameters are shown in Table 5.1.

It can be seen from Fig. 5.18 that the fracture of the clean powder sample is in the middle and has a small necking, while the fracture of the contaminated powder sample has no necking phenomenon. Figure 5.19 shows the stress–strain curve, showing that both groups of samples are ductile fractures. When the fracture occurs, the maximum tensile strength and yield strength of the clean sample is as high as 678 and 516 MPa, and the maximum strain is 31.3%. The maximum tensile strength and yield strength of the contaminated sample is only 517 and 450 MPa, and the maximum strain is

Table 5.1 Process parameters of LPBF-fabricated 316L stainless steel

Parameters	Laser power (W)	Scanning speed (mm/s)	Hatch space (mm)	Layer thickness (mm)	Scanning strategy
Value	150	600	0.08	0.04	Staggered scanning

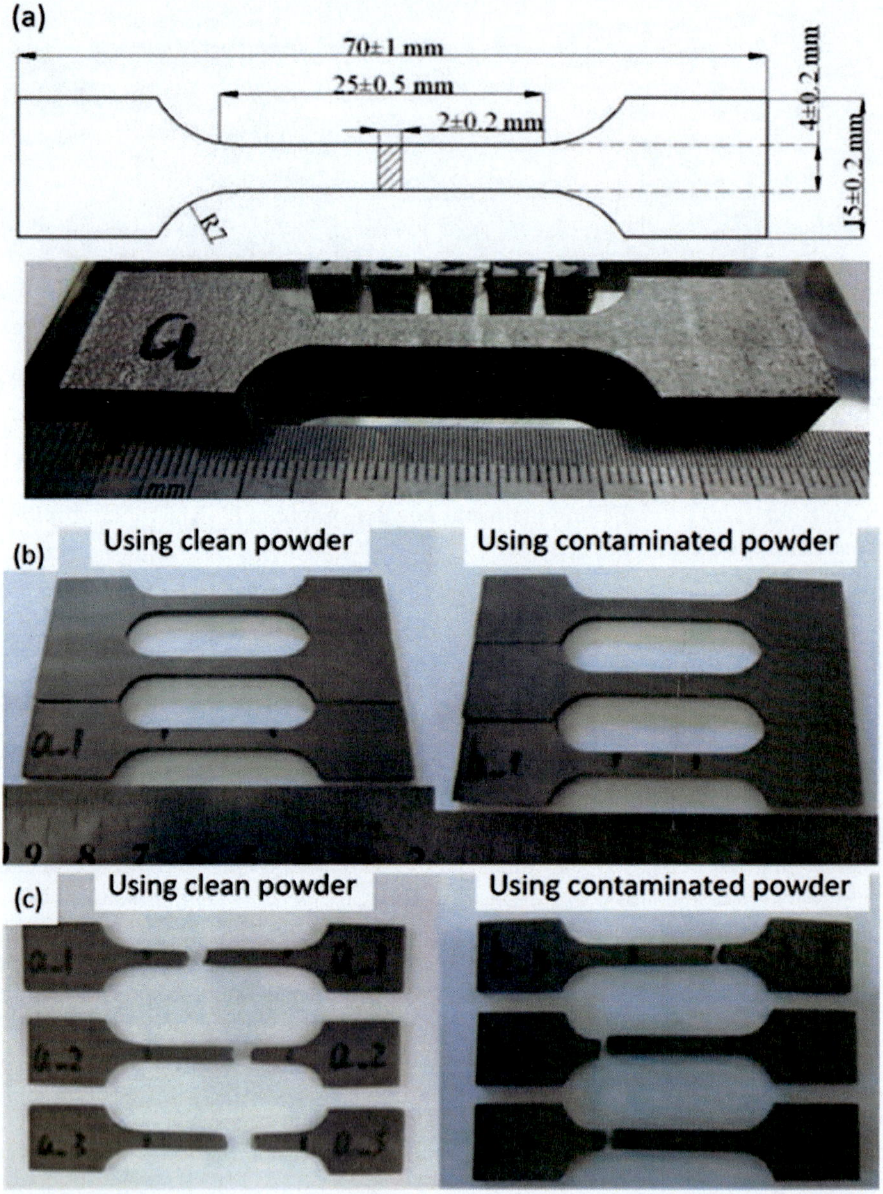

Fig. 5.18 Tensile specimens: **a** tensile specimen size; **b** specimen before tensile test; **c** specimen after tensile test

only 15.7%. The average value of the two groups of samples is shown in Table 5.2. It can be seen that the tensile properties of the contaminated powder are much lower than those of the clean powder.

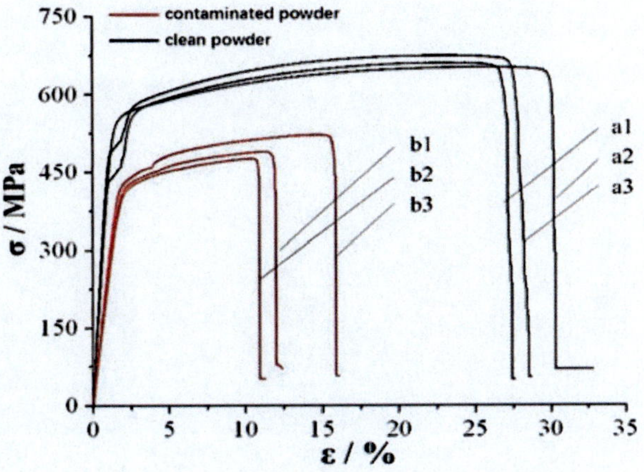

Fig. 5.19 Stress strain curve

Table 5.2 Tensile test result

Mechanical properties	UTS (Mpa)	$\sigma_{0.2}$ (Mpa)	EL (%)
Clear powder	664.7 ± 17.5	488.9 ± 16.7	29.3 ± 2.1
Contaminated powder	499.8 ± 19.9	436.3 ± 13.8	13.7 ± 2.8

To further the analysis of the fracture mechanism of two kinds of samples, using SEM observation of fracture morphology. Figure 5.20 shows the fracture morphology of the clean powder sample. Figure 5.20a shows the low magnification (500× magnification) of the clean sample, from which it can be seen that there are a small number of holes at the fracture, and the surface is uneven. Figure 5.20b shows the high-magnification (10,000 times magnification) of the clean sample, which shows that a large number of irregular small and shallow dimples are distributed at the fracture, and the distribution of dimples is not uniform. There are also some large pits, which indicates that the sample is ductile fracture. Figure 5.20c shows the fracture morphology of the contaminated powder sample. Compared with Fig. 5.20a, there are a lot of holes on the section shown in Fig. 5.20c, and the surface is rough and uneven. Figure 5.20d shows the high magnification (10,000 times magnification) of the contaminated sample. As shown in Fig. 5.20b, a large number of irregular small and shallow dimples are distributed at the fracture, and the distribution of dimples is not uniform, indicating that the sample is also ductile fracture. Considering Fig. 5.20c, the reason for the reduction of tensile strength of the contaminated sample is that there is a large amount of spatter impurities in the sample. In the tensile test, the joint surface between the spatter particles and the main part is first destroyed, thus forming the initial fracture crack. Under the further action of external force, these cracks gradually expand and eventually lead to the failure of the part. Therefore, the fracture mechanism of the contaminated specimen is the ductile fracture caused by defects.

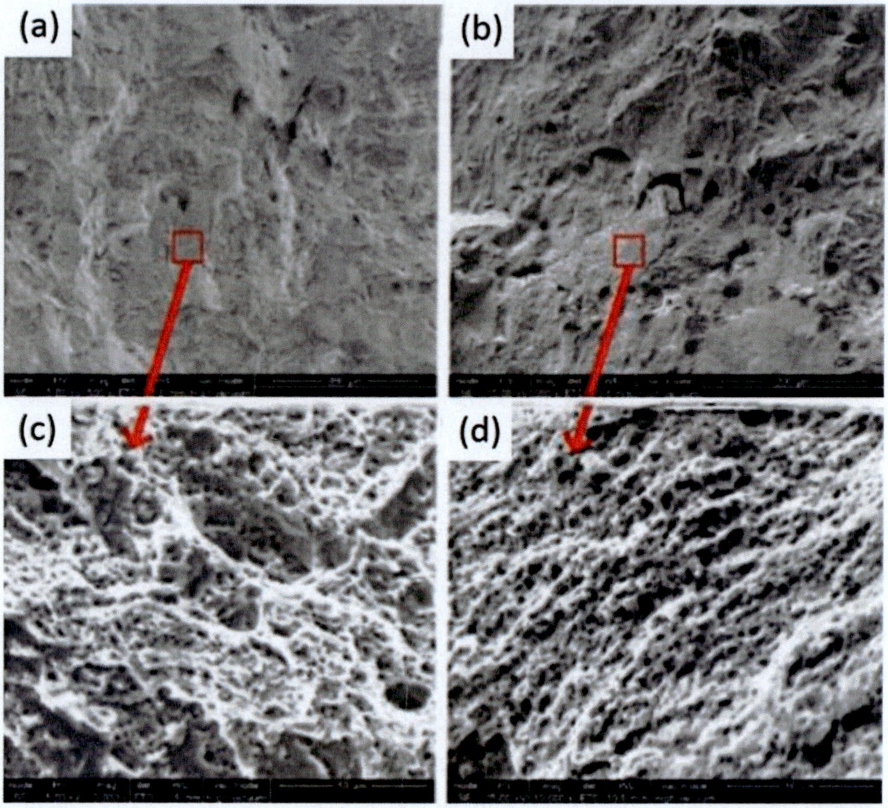

Fig. 5.20 SEM morphology of tensile fractures: **a** low magnification diagram of tensile fracture surface of clean powder; **b** high magnification diagram of tensile fracture surface of clean powder; **c** micrograph of tensile fracture surface of contaminated powder; **d** high magnification diagram of tensile fracture surface of contaminated powder

Based on the above studies, as shown in Fig. 5.21, the team carried out mechanical properties tests of LPBF CoCrW parts under different use times. The size of the tensile bar is shown in Fig. 5.21a, the tensile bar fabricated by LPBF is shown in Fig. 5.21b, and the tensile sample after room temperature tensile experiment is shown in Fig. 5.21c, all fractures occur within the canonical area. In addition, with the increase of powder application times, the fracture position of the tensile part is closer to the clamping end. The stress–strain curve of the tensile sample is shown in Fig. 5.22, and the tensile strength, yield strength and elongation are respectively shown in Fig. 5.22. It can be seen from the fracture morphology and fracture curve that the fracture mode of the specimen is brittle fracture. With the increase of powder usage times, the tensile properties of the samples decreased, and the tensile strength decreased from 1284.12 to 874.57 MPa. At the same time, the yield strength decreased from 875.85 to 689.18 MPa, which also showed a similar trend. In the tensile test, the elongation rate of the new powder formed part decreased

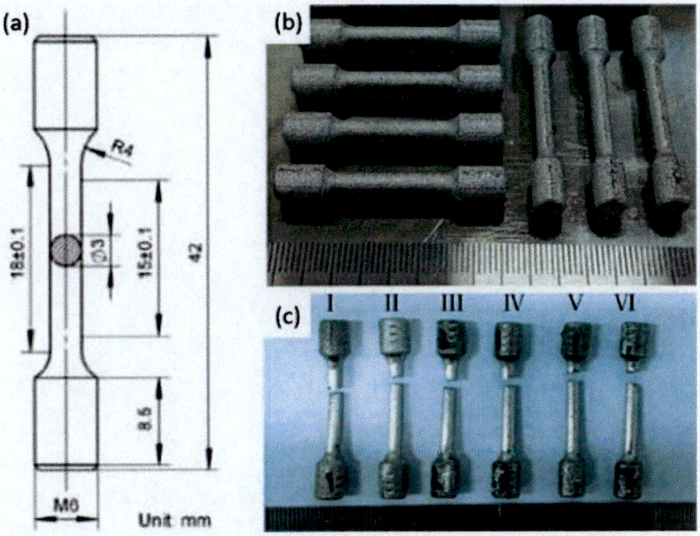

Fig. 5.21 LPBF-fabricated CoCrW alloy tensile parts obtained from contaminated powder (six groups of I-VI are obtained according to the number of times the powder is used): **a** schematic diagram of standard stretch piece; **b** stretch piece molded piece; **c** macro morphology of tensile pieces in each batch

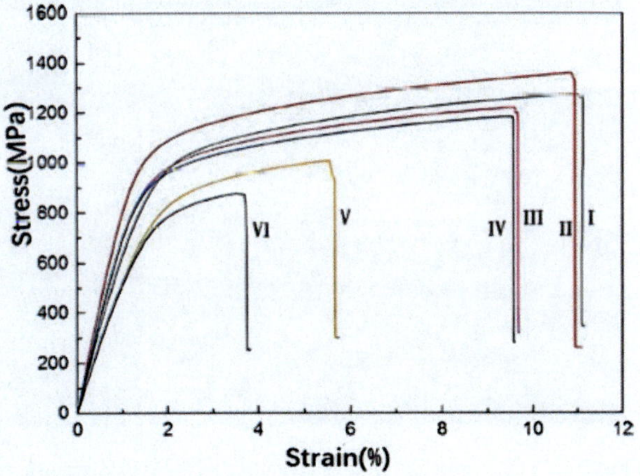

Fig. 5.22 Stress strain curves of six groups of tensile parts

from 11.15 to 9.60% at the fourth use, and decreased sharply to 3.82% at the sixth use. The experimental results of tensile strength and elongation confirm that the powder contamination caused by spatter has a significant effect on the tensile properties of LPBFed parts, especially after four times of repeated use.

Figure 5.23 shows the SEM morphologies of tensile fracture under different powder usage times. The results show that the fracture morphology has obvious

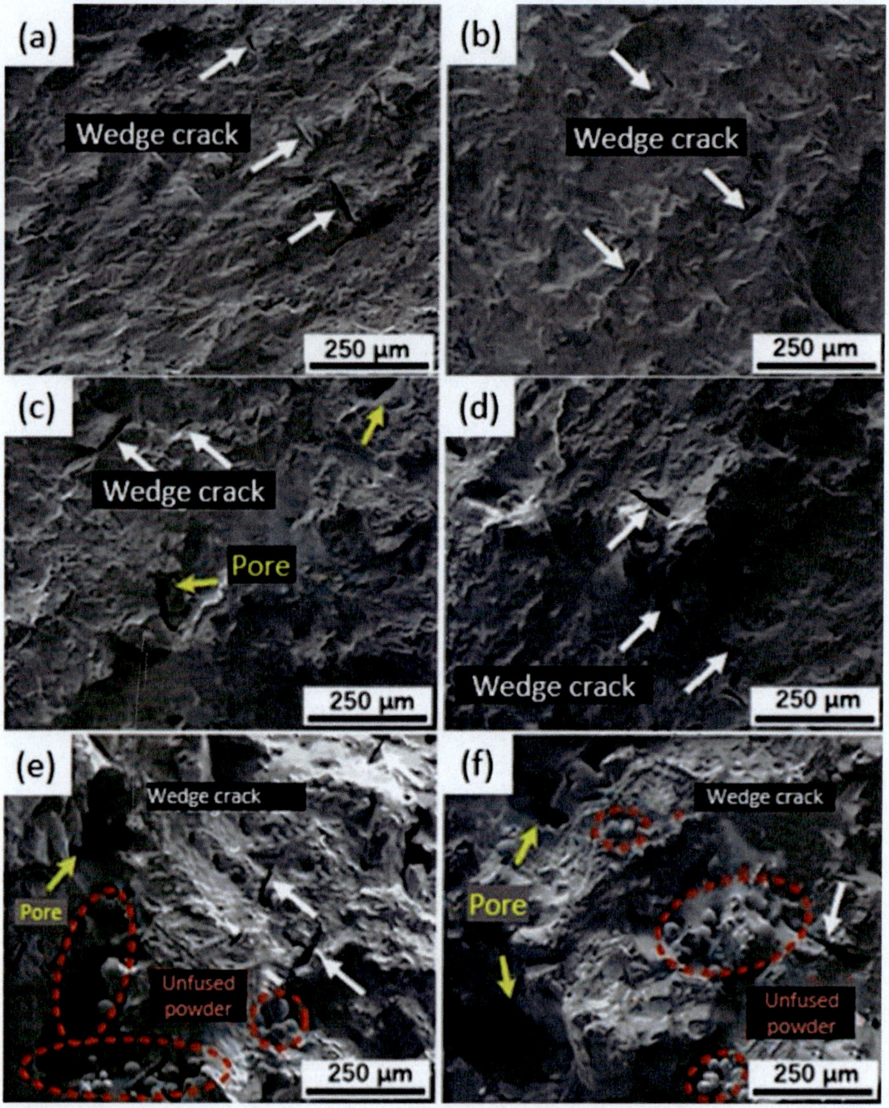

Fig. 5.23 SEM morphologies of fracture surfaces of six groups of tensile pieces: **a** first use; **b** second use; **c** third use; **d** the fourth use; **e** the fifth use; **f** sixth use

cleavage plane and wedge crack. According to Fig. 5.23c, the macroscopic fracture angle is perpendicular to the tensile direction, and there is no obvious neck shrinkage phenomenon, so it can be confirmed that brittle cleavage fracture occurs in the CoCrW tensile samples. When the powder was used once or twice, the macroscopic fracture of the tensile sample was plane fracture, and there were a few parallel wedge cracks at the fracture (Fig. 5.23a, b). When the CoCrW powder was applied for the third and fourth times, the fracture was obviously uneven, the wedge crack increased, and a small amount of unmelted powder particles were embedded in the macropores (Fig. 5.23c, d). When the powder material was used for the fifth and sixth times, the macroscopic fracture of the tensile sample was more uneven, with a large number of wedge cracks, pores and unfused powder particles (Fig. 5.23c, e, f).

Through the research team's research on the uses number of the raw powder, from the new powder to the sixth use of the powder, it can be clearly observed that the formed part begins to appear defects from the microstructure level, and the pores and inclusions defects are significantly enhanced and directly lead to the sharp attenuation of the mechanical properties of the LPBFed parts. Therefore, special treatment is recommended for reused powder materials, especially those used more than four times. Although in this experiment, in order to obtain obvious comparison, the gas circulation power of the forming system is only set to 50% of the normal power, but in the actual industrial LPBF equipment, the efficiency of spatter purification is obviously better than the experiments. In addition, Ardila et al. [12] studied the effect of repeated use of Inconel 718 recycled powder on the performance of LPBFed parts. Using more optimized cycling conditions, they found that after 14 reuses of the powder, the powder properties showed only minor changes. The reason for this may be that the high-efficiency circulation device is mounted on one side of the forming area and filters the powder after each printing task.

5.4 Droplet Spatter Behavior and State of Processing

Compared with powder spatter, the formation process of droplet spatter and the process of inducing macroscopic defects have more prominent effects on the quality of LPBF, the study of droplet spatter behavior is very important to understand the LPBF process. In order to have a deeper understanding of the processing principle of droplet spatter behavior, this book attempts to deepen the understanding of the interaction mechanism of laser and powder and the production mechanism of droplet spatter by processing and analyzing high-speed images of droplet spatter behavior.

5.4.1 Droplet Spatter and Its Image

In order to discuss the motion characteristics of droplet spatter behavior, the processing parameters proposed in Table 5.3 are used to conduct comparative experiments of droplet spatter behavior under three typical processing states. Under the laser power of 160 W and the laser scanning speed of 500, 750 and 1500 mm/s, the single melt track morphology is shown in Fig. 5.24. The single melt track shows three conditions: overmelting state, normal forming state and unmelting state, respectively. It can be seen from Fig. 5.24a that the melt track width in the state of high laser energy density is large, and the melt track is continuous and shiny. However,

Table 5.3 Experimental parameters

The serial number	Layer thickness *h* (mm)	Hatch space *d* (mm)	Actual laser power *P* (W)	Scanning speed *v* (mm/s)	Laser energy density Ψ (J/mm)	The forming condition
1	0.03	0.06	160	500	0.320	Overmelting forming
2				750	0.213	Normal forming
3				1500	0.106	Unmelting forming

due to the high input energy, the metal molten liquid is redistributed under the action of surface tension before full solidification and forms bulges or globularization. In addition, the intense spatter process will also lead to the reduction of molten fluid in the molten pool, and eventually lead to the instability of the width and height of the single-track. Under the normal energy density (Fig. 5.24b), the morphology of the track is continuous and consistent, the surface is smooth and rich in metallic luster, and there is no globularization and convex structure. Under the action of low energy density (Fig. 5.24c), the absorption of laser energy by metal powder and matrix is insufficient, and the depth of molten pool is small, the wetting time of molten liquid and substrate is shortened, and the surface tension of molten liquid cannot be overcome to form a more serious globularization structure, and the forming continuity of melt track is poor.

The droplet spatter behavior corresponding to the single-track process under three typical processing states is shown in Fig. 5.25. In the superfusion processing state, the droplet spatter process mainly presents a violent droplet spraying process perpendicular to the scanning direction. With the increase of scanning speed and the decrease of line energy density, the angle between the spray direction and the scanning direction of droplet and plume gradually increases, while the number of droplet particles gradually decreases. In different forming processes (Fig. 5.25a, b), the distribution of droplet spatter shows a significant fan-shaped divergence characteristic, and the intense laser machining process gives the droplet spatter process unique complexity and randomness. In addition, with the increase of the scanning speed, the trailing distance of the molten pool gradually increases at the same cooling rate. In the unmelting process shown in Fig. 5.25c, the increase of the pool length makes it easier for the molten liquid to escape at the pool tail far from the keyhole and form droplets spatter. The direction of plume ejection is highly consistent with the distribution direction of droplet spatter. The abrupt change of plume ejection angle will also lead to the change of droplet particle formation and ejection angle near the keyhole, which is determined by the formation mechanism of droplet spatter.

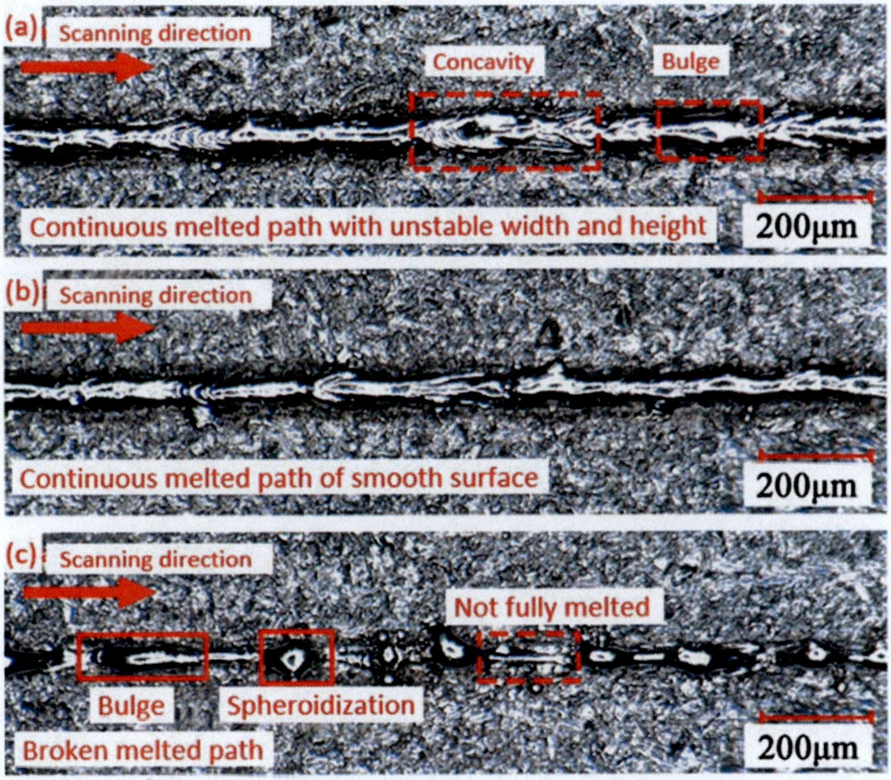

Fig. 5.24 Morphology of single-track formed under three typical processing conditions: **a** single-track of overmelting (the linear energy density is 0.3 J/mm); **b** single-track of normal forming (linear energy density of 0.2 J/mm); **c** single-track of unmelting forming (linear energy density is 0.1 J/mm)

5.4.2 Image Processing of Droplet Spatter

In order to improve the ability of spatter image information to reflect the action process of laser and metal powder, this book takes the original image as an example to perform related image processing operations to achieve the purpose of image enhancement. The test noise introduced by the camera itself in the hardware transformation output gray image is removed by median filtering. In addition, according to the gray value characteristics of molten pool, plume, spatter particles and defocus information in the image, K-means clustering algorithm is used to screen and compress the image content to obtain the spatter behavior information in the two-dimensional depth of field plane. According to the clustering results, the image segmentation result based on the gray value information can achieved and most defocus information or halo spots can be removed, then the three-dimensional spatter behavior is compressed into the spatter characteristics in the two-dimensional vertical plane,

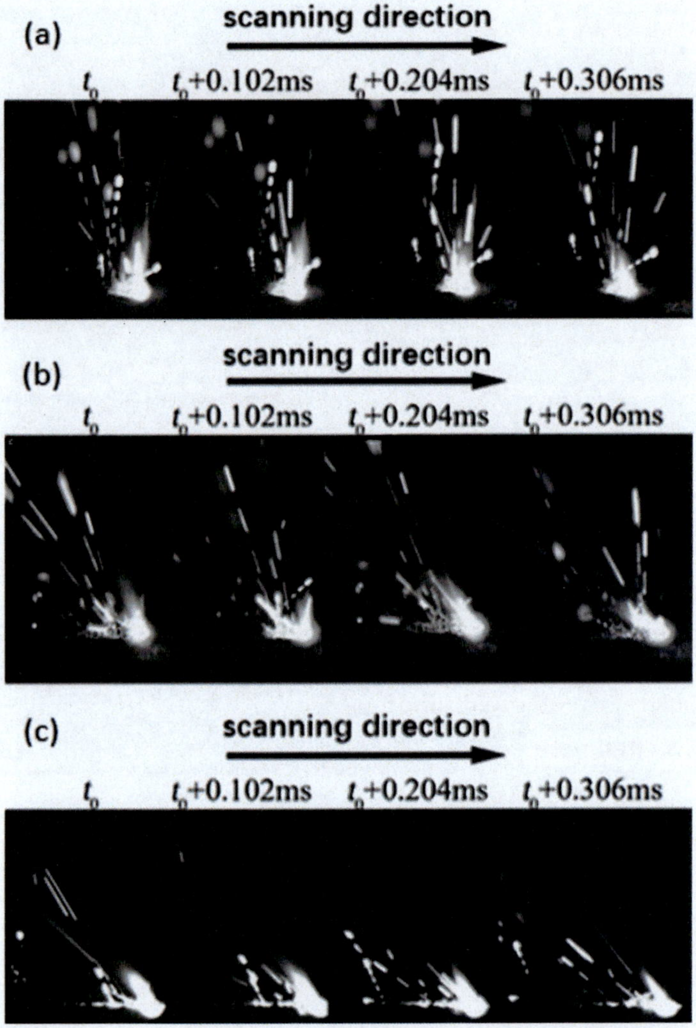

Fig. 5.25 Forming process of droplet spatter under three typical processing conditions: **a** liquid droplet spatters of overmelting forming (the linear energy density is 0.3 J/mm); **b** liquid droplet spatters of normal forming (linear energy density of 0.2 J/mm); **c** liquid droplet spatters of unmelting forming (linear energy density is 0.1 J/mm)

which facilitates the discussion and analysis of the mechanism and characteristics of the laser molten pool system in this chapter.

Figure 5.26 shows the image interception frame of droplet spatter behavior, set the cluster centroid k = 4 and perform K-means clustering operation on the original image. Among them, the determination of K value is determined by the contents contained in the image, including background, molten pool or droplet particles, plume, spot or halo, which have significant differences in gray value during imaging.

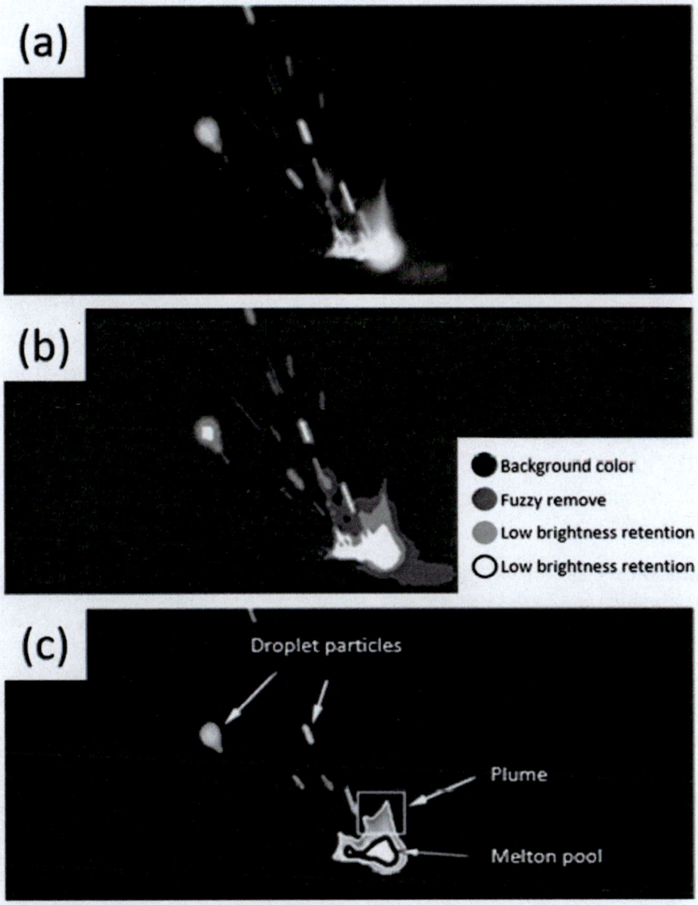

Fig. 5.26 K-means clustering segmentation: **a** the original image; **b** K-means clustering results; **c** image segmentation and enhancement effect

As a mixture of plasma and metal vapor, the plume is translucent and brighter than halo and diffuse light. Therefore, either too large or too small K value will cause excessive expression of image content, and finally obtain the clustering result based on pixel gray value as shown in Fig. 5.26b. The high brightness retained part is the projection of high temperature strong light molten pool or spatter particles in the plane of depth of focus, the low-brightness remnant consists mostly of erupting plasma (plume structures) and some spatter particles, the blur removal part extracts most defocus information and halo spots in the original image. The target image is obtained by calculating the dot product between the high-brightness and low-brightness preserved parts and the original image, as shown in Fig. 5.26c.

According to the imaging principle of the high-speed camera, the photosensitive chip captures the light radiation information of the droplet particles during the camera exposure time, each frame of the high-speed imaging system contains information about the movement of the droplet particles during the exposure time, even if the imaging system has shorter exposure time and higher acquisition frame rate, the images of droplet particles obtained are the movement information of droplet spatter during the exposure time rather than the actual size of the particles. Based on this imaging feature, after positioning the spatter particles in each image, use the residual image of the droplet particle image for calculate the sputtering angle of the droplet spatter, as shown in Fig. 5.27. After locating the spatter particles in the target image, calculated the minimum envelope ellipse O of the target particle image, and the angle between the long axis of the ellipse and the horizontal reference line in the opposite direction of scanning is defined as the sputtering angle α of the spatter particles. At the same time, the pixel size and number information of the local droplet spatter image are calculated, in which the pixel area of the droplet spatter is expressed by the number of pixels it occupies. In order to reduce the interference of abnormal spatter particles, the characteristic information of droplet spatter behavior is counted in the form of the average information of all droplet particles in the frame.

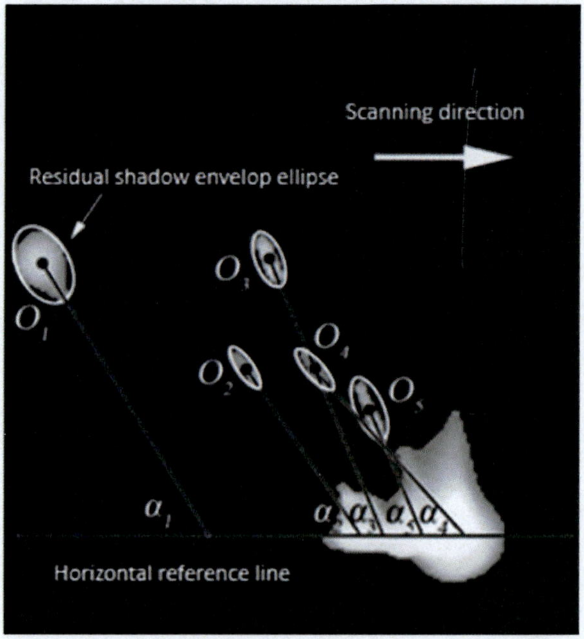

Fig. 5.27 Statistical chart of spatter characteristics and angle calculation method

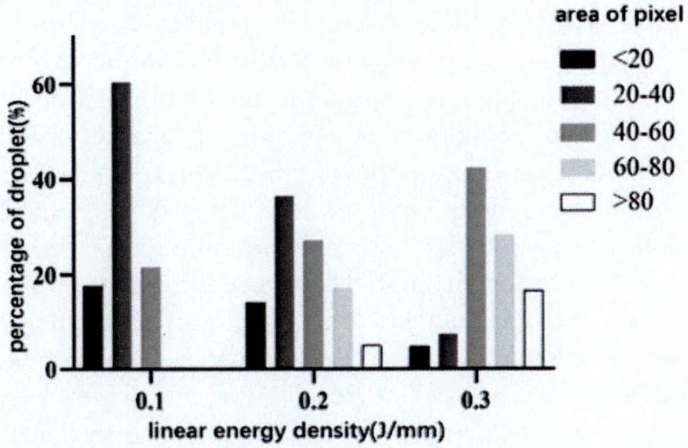

Fig. 5.28 Number proportion of droplet spatter particle area under different linear energy density

5.4.3 Characteristics of Droplet Spatter Behavior

In order to further explore the generation, evolution mechanism and behavior characteristics of droplet spatter contained in the statistical data of characteristic information, Fig. 5.28 shows the number proportion of each spatter particle area in the processing of three linear energy densities. In the process of single-track manufacturing with low energy density input, the small spatter particles with pixel area between 20 and 40 pixels occupy an absolute dominant position in the particle area distribution. This phenomenon indicates that the process of migration and excitation of the original powder is the main part of the droplet spatter behavior in the unmelting state. In addition, at a higher scanning speed, the interaction time between the laser beam and the metallic powder is short, and a large amount of metal vapor cannot be generated. The weak steam pressure can't continuously penetrate into the melt or matrix, thus weakening the positive role of the keyhole structure in the heat transfer process. The low temperature gradient can't make the molten liquid obtain strong fluidity relying on the Marangoni convection process. At the same time, the steam recoil force and Marangoni effect can't promote the overflow of the molten liquid, which makes the micro molten pool system unable to supplement the metal molten liquid in the form of capturing the surrounding powder by surface tension. Although the molten liquid has the tendency of escaping to form droplet spatter under the recoil force, it is unable to form large droplet particles due to the limitation of the molten liquid supply. With the decrease of laser scanning speed and the increase of linear energy density, the proportion of droplet particles whose pixel area is greater than 40 pixels gradually increases, indicating that droplet spatter is supplemented by sufficient liquid during the formation process. However, the higher energy input not only increases the depth of keyhole and molten pool, but also leads to stronger metal evaporation effect, larger temperature gradient of molten pool and more obvious surface tension gradient, under the positive Marangoni effect and metal vapor expansion extrusion, the molten liquid has more complex and intense fluidity, which promotes the more frequent and intense escaping process of molten pool. The more violent metal vapor eruption process is a direct promoting factor for the powder spatter formation process under the entrainment effect. At the same time, with the more frequent powder migration and laser excitation process, the probability of micro droplet spatter colliding and merging into larger droplet spatter at the initial stage of formation is greatly improved, although this process is random. It is worth noting that even under the input of high energy density, small spatter particles with a pixel area less than 40 pixels still account for a certain proportion of the total amount. This result shows that in the state of overmelting processing, there is still a process of rapid escape that can't obtain sufficient liquid supplement with high acceleration, but it can't rule out the existence of the process that the powder spatter is transformed into small droplet spatter by thermal radiation. However, droplet spatter with a pixel area greater than 60 pixels only occurs in high energy density, and its proportion increases with the increase of energy density. This data phenomenon shows the generation and intensification of droplet spatter in the process of liquid escape.

The sputtering angle of droplet spatter is considered by researchers as a spatter characteristic that can directly reflect the state of molten pool, the shape of keyhole and the state of metal vapor jet. It can be seen from the distribution of spatter angle in different energy density processing, the Fig. 5.29 shows that with the change of linear energy density, the spatter angle in each processing process shows significant difference in distribution. When the online energy density is 0.1 J/mm, the droplet spatter with the sputtering angle less than 60° (61.95%) becomes the main part of the sputtering angle distribution in the state of unmelting processing. With the decrease of scanning speed, the linear energy density increases, the particles sputtered at angle of 60°–70° become the main body of the spatter, and the number of particles sputtered at an angle greater than 70° also increases significantly. The further increase of linear energy density directly led to the significant increase of the percentage of droplet particles with sputtering angle greater than 80° to 66.77%,

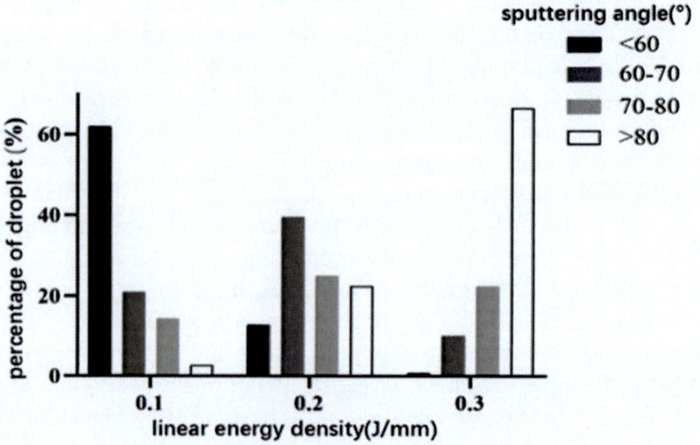

Fig. 5.29 Number proportion of droplet sputtering angles under different linear energy densities

while the droplet distribution with sputtering angle less than 80° decreased to varying degrees. The spatter angle of the spatter particles during the unmelting process and the overmelting process shows extreme distribution characteristics. In contrast, the sputtering angle distribution in the normal forming process shows an obvious transition trend compared with the other two processes. At this time, the proportion of spatter particles with a sputtering angle greater than 60° has reached 87.05%, Even so, the proportion of particles in each sputtering angle shows a small difference, and finally shows an abnormal fluctuation. In conclusion, from the number of droplet particles at different pixel sizes and the number distribution of droplet at different sputtering angles, the change of linear energy density makes the formation mechanism and sputtering behavior of droplet particles change significantly.

In combination with the above droplet spatter behavior characteristics and droplet spatter formation mechanism, the droplet spatter formation mechanism is divided into four categorics (including types 1–4) according to the location of different droplet production, as shown in Fig. 5.30. The formation mechanism of the four types of droplet spatter includes ①–③ type droplet spatteres formed by the liquid in the front, side and rear of the keyhole, which obtain greater power under the Marangoni effect and recoil force and overcome the surface tension escape, the sputtering angle is determined by the fluidity of the molten liquid, and because the flowing ambient gas carries the original powder to the molten pool, the fourth type of droplet spatter,

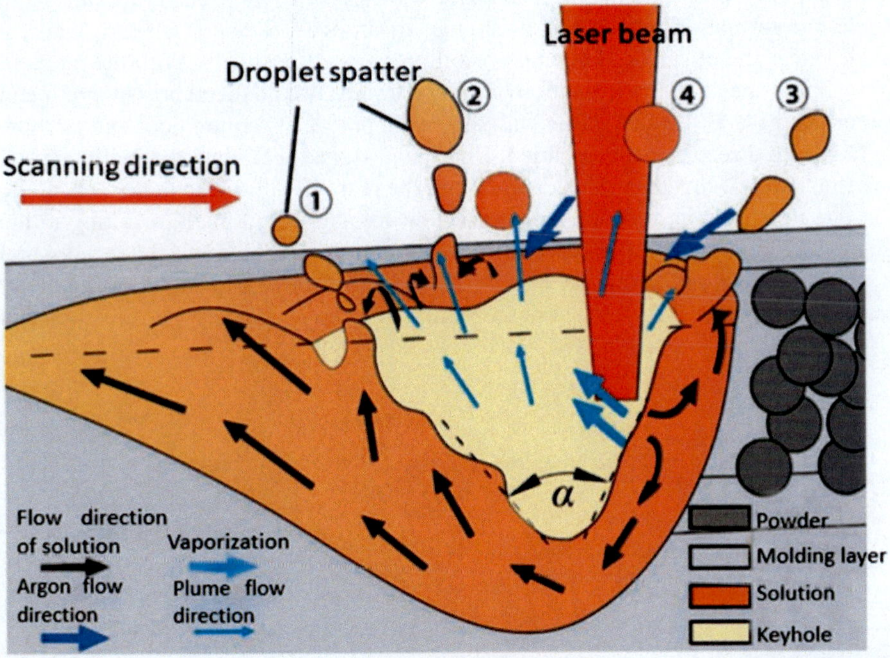

Fig. 5.30 Formation mechanism of droplet spatter and schematic diagram of molten pool and keyhole under different linear energy density

which is formed with metal vapor in the form of droplets after laser radiation, is directly determined by the jet angle of the plume. During the interaction between the laser beam and the metallic powder, the depression formed by the recoil pressure of the metal vapor in the molten pool is the keyhole structure, and its shape, size and other characteristics are determined by the laser power and scanning speed. Under the action of Gaussian distribution light spot, the molten liquid on the front wall of the keyhole continuously vaporizes to produce a large amount of metal vapor and stores it in the keyhole structure, while the angle between the front wall and the back wall of the keyhole α. The escape path of metal vapor is restricted, that is, the jet angle of the plume is determined. The escape angle of the plume not only directly affects the sputtering angle of type ④ droplet spatter in Fig. 5.30, but also directly affects the shape of the molten pool, the movement and acceleration direction of the molten liquid due to the change of the direction of vapor recoil pressure, which leads to the change of the sputtering angle of type ①–③ droplet spatter shown in Fig. 5.30.

When the online energy density is 0.3 J/mm in the low scanning speed, the laser parameters determine the tilt angle of the front wall of the keyhole. Under the intense steam recoil pressure, the keyhole structure gets a greater depth in the vertical direction, so that the molten liquid gets a greater motion component in the vertical direction. Finally, the morphology of the molten pool and keyhole shown in Fig. 5.31a and the lower level of the molten pool pixel length in the single-track experiment are obtained. In terms of droplet spatter behavior mechanism, as the escape process of metal vapor is limited by the front and back walls of the keyhole and erupts at a large angle, the formation process of ①–④ types of droplets escape at a nearly vertical angle under the effect of plume and recoil pressure respectively. With the increase of scanning speed, the vaporization of molten liquid and the recoil pressure of metal vapor decrease at the same time, and the development of molten pool and keyhole in the depth direction is constrained; Under high-speed scanning, the inclination of the front wall of the keyhole decreases, and the rear wall of the keyhole tends to be flat due to the increase of the tangential flow velocity of the melt, resulting in the angle α of the key wall increase, appear the evolution process of molten pool and keyhole morphology shown in Fig. 5.31b, c. The change of keyhole morphology leads to the change of metal vapor escape direction and recoil pressure direction, and

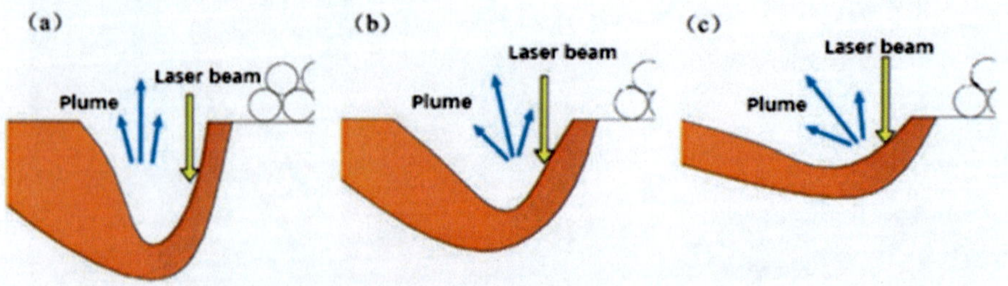

Fig. 5.31 Pixel length of molten pool and shape of molten pool and keyhole in each processing state: **a** molten pool morphology of overmelting forming; **b** molten pool shape of normal forming; **c** molten pool morphology of unmelting forming

then the original powder and molten liquid in molten state change significantly in the sputtering direction. It is worth noting that due to the weakening of the evaporation process and recoil pressure of the metal melt, the fluidity of the molten liquid in the molten pool is weakened, which means that the development process of the ①–③ droplet spatter in Fig. 5.30 will be inhibited, and the original powder migration, excitation and sputtering processes represented by the ④ process become the dominant mechanism for the formation of droplet spatter.

5.5 Building Chamber Gas Circulation System

The above contents introduce and discuss in detail the spatter phenomenon and its influence on the LPBFed parts in terms of defect formation mechanism, behavior characteristics, etc. In the actual production state, the gas circulation purification device equipped with the LPBF equipment is the most effective means to filter the laser processing products such as spatter particles and harmful gases. This section will introduce the relevant research contents of the gas purification system in the experiment.

5.5.1 Air Flow Distribution of DiMetal-100 Building Chamber

Figure 5.32 shows a simplified model of DiMetal-100 building chamber. The model is simplified according to the sealing chamber of the forming equipment, and the basic structural dimensions of the chamber and the original design structure and dimensions of the air inlet/outlet structure are retained. The chamber is 100 mm in

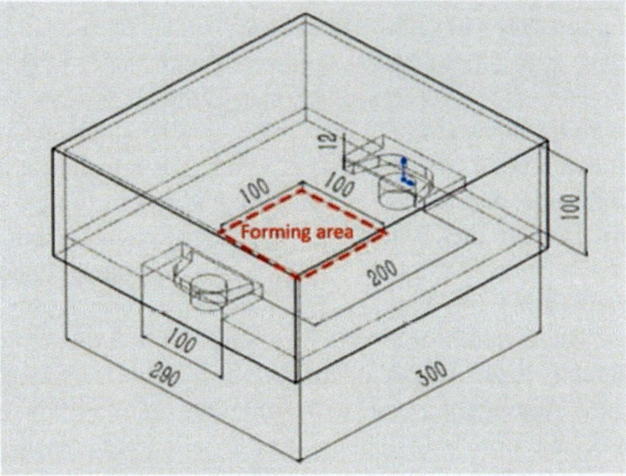

Fig. 5.32 Simplified model of Dimetal-100 building chamber (unit: mm)

height, 290 mm in width, 300 mm in length, and the air inlet/outlet pipe diameter is 32 mm, the square opening of the air inlet/outlet is 12 mm in height and 76 mm in width, the air outlet and air inlet are symmetrically distributed, with a spacing of 200 mm. Forming area is 100 mm × 100 mm, and consider that the powder bed and the surface of the building chamber are in the same plane. Solidworks Flow Simulation module is used to simulate the flow of shielding gas in the building chamber involved in the study. The flow medium is pure argon, and the initial conditions of the building chamber are set as 293.2 K temperature and one standard atmospheric pressure. The model solution accuracy of the simulation process is 4 levels, and the model grid is divided into adaptive Cartesian grid. According to the actual production and processing parameters, the boundary condition at the air inlet is set as 10 m^3/h of uniform input volume flow, and the boundary condition at the air outlet is set as a standard atmospheric pressure.

Figure 5.33 shows the flow field distribution diagram of building chamber obtained under the condition of actual forming equipment. It can be seen from the streamline distribution of the building chamber shown in Fig. 5.33a that when the argon shielding gas is pumped into the air inlet pipe, it enters the building chamber at a horizontal angle after being guided through the air inlet device and forms the horizontal laminar flow shown in Area I. Under the guidance of a fan-shaped beam expander, the protective gas of horizontal laminar flow preferentially diffuses in the horizontal plane at the fan-shaped angle shown in Fig. 5.33c. However, due to the large diffusion angle, most of the shielding gas in Area I can't directly enter the air outlet, but directly hits the wall of the building chamber at the air outlet side and flows vertically upward along the wall. When the air flow reaches the top of the chamber, it continues to turn and forms the air flow vortex (Area II) shown in Fig. 5.33b. The protective gas part constituting the vortex can leave the chamber directly through the air outlet; About 1/3 of the vortex shielding gas meets the horizontal laminar flow in Area I and forms a contact surface bounded by B-C-D in Fig. 5.33a. Taking B-C-D as the dividing line, the protective airflow field in the vertical plane can be divided into the horizontal laminar flow part I of the air inlet pointing to the air outlet and the vortex airflow part II occupying the top of the building chamber. As the horizontal laminar flow has a higher velocity, under the Bernoulli effect, the flow direction of the vortex shielding gas connected with the B-C-D boundary changes sharply and flows to the outlet. However, part of the air flow in Area II fails to converge directly with the horizontal laminar flow (Area III is distinguished by the D-E boundary), so it continues to develop and flow towards the wall of the building chamber at the air inlet side, and forms a symmetrical vortex structure shown in Fig. 5.33d at the side near the air inlet. Due to the significant height advantage of region III, the downward argon can continuously participate in the convergence process of horizontal laminar flow from the C-D boundary.

The different interaction processes of vortex flow and horizontal laminar flow directly lead to the weakening of the flow performance of shielding gas in the forming area. Among them, due to the antagonism of air flow in areas I and II at the air outlet side, the flow field change angle of about 13° (section B-C) is finally formed. Due to the further development of vortex flow, the upper limit of horizontal laminar flow

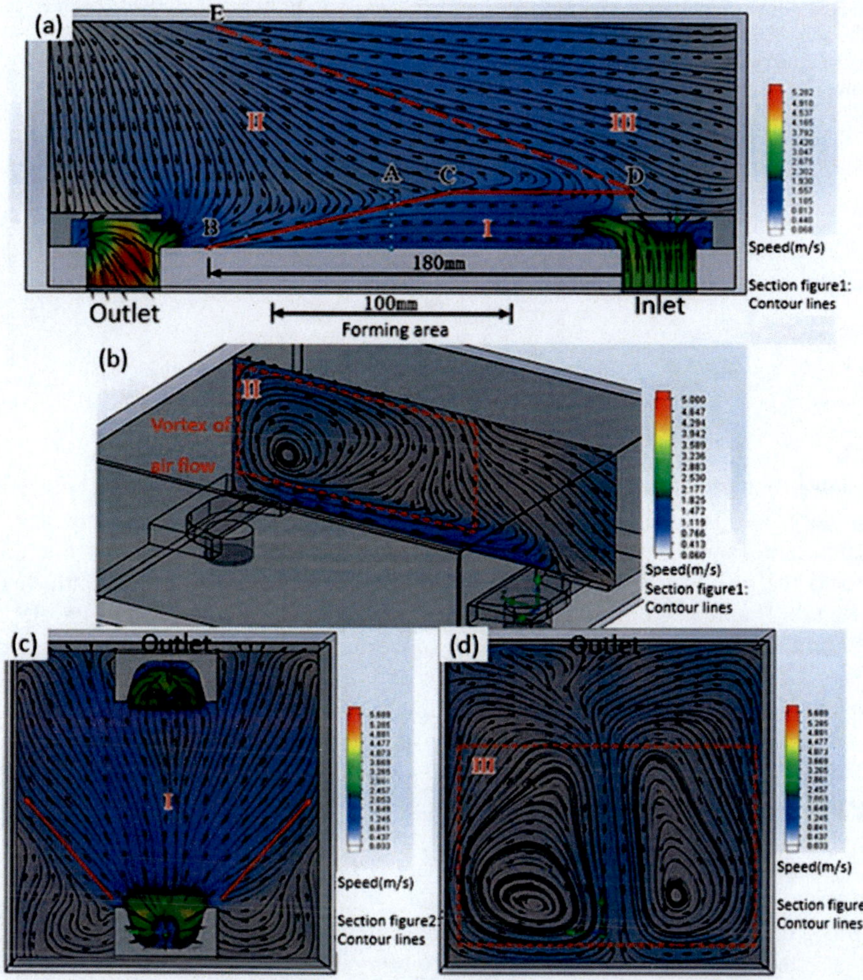

Fig. 5.33 Flow field distribution of building chamber: **a** side view of building chamber flow field; **b** vortex movement of airflows; **c** flow field distribution in the water plane at 0.01 mm height; **d** horizontal flow field distribution at 0.05 mm height

height represented by area I is limited to about 25 mm from the melt layer under the restriction of horizontal vortex flow pressure in area III. Finally, the horizontal laminar flow in the area I with trapezoidal structure becomes an effective protective atmosphere for filtering the spatter, harmful gas and other laser processing products.

In order to further discuss the flow performance of the effective protective atmosphere under the combined action of vortex and horizontal laminar flow, the multipoint flow velocity in the central area of the forming range represented by position A was measured and the curve shown in Fig. 5.34 was obtained. The flow velocity shows that the horizontal flow velocity component of the shielding gas in area I at

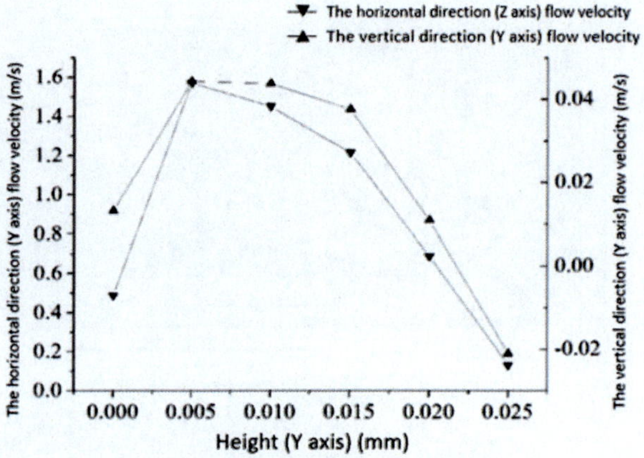

Fig. 5.34 Flow velocity of multi-point measurement points in the center of the forming area

the lower boundary (formed surface side) and the upper boundary (B-C-D divided boundary side) is significantly lower than that in the middle layer, the former is the normal performance of the boundary layer under the friction resistance, while the latter is the result of the weakening of the flow performance after the confluence of Region II and Region I, finally, the horizontal laminar flow in region I develops into gradually weakened gradual flow. It is worth noting that the entrainment and erosion of the air flow on the powder bed is an important factor restricting the flow performance of the shielding gas in the impurities filtering link such as spatteres based on the circulating protection air flow, although the radical circulation strategy can improve the purification effect of spatter, it will cause the loss of powder in the powder bed. In Fig. 5.34, in the height range of 0–25 mm in the height direction, the velocity component in the vertical direction shows a trend of increasing first and then decreasing, with the increase of the height, the positive motion component in the vertical direction is conducive to improving the entrainment effect of air flow to improve the removal effect of spatter particles. In the actual application process, the actual effect of dust accumulation in the building chamber as shown in Fig. 5.35a, b is obtained, which can be verified by the flow field distribution at the top and side of the air outlet in the observation window as shown in Fig. 5.33a, c. The dust accumulation in the building chamber window and on the top of the air outlet is the result of the dust entrainment in the air outlet side by the horizontal laminar flow after experiencing the vortex in region II. The dust accumulation on the side of the air outlet is the result of the entrainment of the fan-shaped diffusing air flow in the horizontal plane. At the same time, due to the entrainment of powder in the forming area of the horizontal laminar flow region at low height, too high circulating fan power will lead to the bare substrate/forming layer as shown in Fig. 5.35c.

The research shows that although the existing technology can't completely remove the spatter particles, improving the uniformity and average flow velocity of the shielding gas flow above the forming area is the basic method to extend the trajectory of the spatter and improve the particle purification effect in the forming area. Through the simulation of the protective gas fluid in the DeMatel-100 building chamber, the

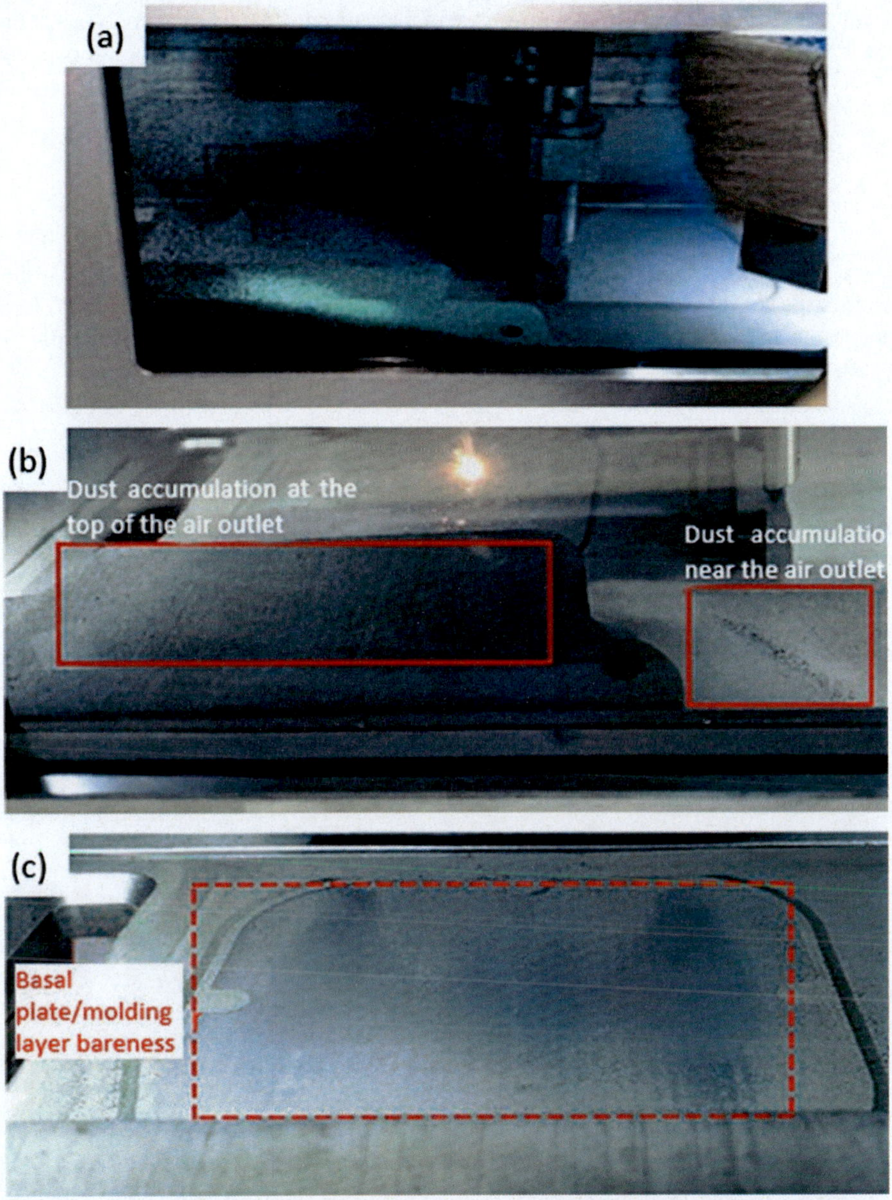

Fig. 5.35 Defects in gas circulation during actual processing: **a** accumulation of black dust in the chamber forming window; **b** dust accumulation at the top and side of the air outlet; **c** bare substrate or formed layer

airflow distribution and flow performance in the building chamber are analyzed. This book analyzes the distribution and development of the effective protective atmosphere in the forming area.

5.5.2 *Influence of Building Chamber Parameters on Air Flow*

Based on the flow field distribution in the DeMatel-100 building chamber, the influence of the size parameters of the building chamber on the flow field distribution characteristics in the chamber was discussed, and the size parameters that can effectively protect the atmosphere and improve the flow performance were obtained.

5.5.2.1 Influence of Building Chamber Length

According to the chamber model shown in Fig. 5.32, the air inlet and outlet spacing is doubled to 400 mm, and the other model parameters and boundary conditions remain unchanged to obtain the flow field distribution diagram of the building chamber shown in Fig. 5.36a. Compared with the flow field distribution in Fig. 5.33a, the extension of the chamber length makes the air flow in Area II appear more gas accumulation in front of the air outlet, which has a blocking effect on the emission process of spatteres and soot. At the same time, the gas accumulation process causes the cyclonic flow to oppose the horizontal laminar flow, and forms a small vortex at the head of area I. Although the increase of the chamber length improves the length

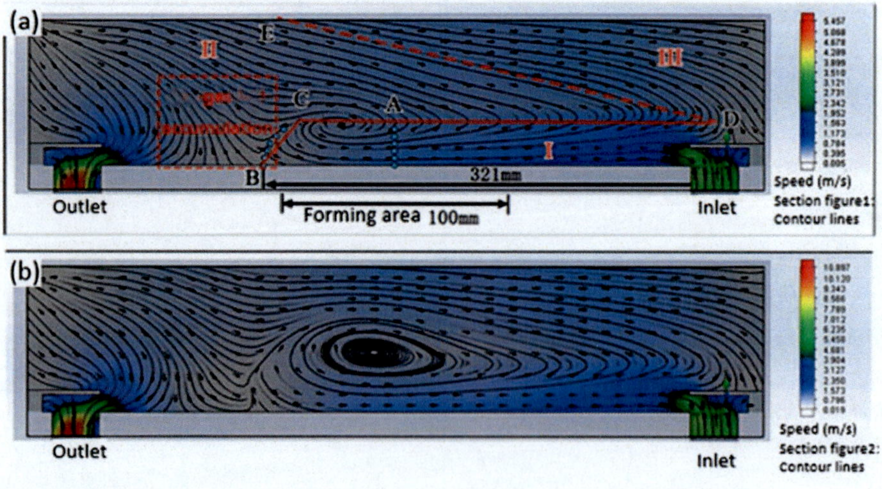

Fig. 5.36 Flow field distribution of chamber formation: **a** chamber flow field distribution at a volume flow rate of 10 m^3/h; **b** chamber flow field distribution at a volume flow rate of 20 m^3/h

of the horizontal laminar flow area, due to the gradual flow characteristics of the horizontal laminar flow, the decline of the flow performance at the head of area I and the confrontation of the airflow accumulation process in area II become important factors that hinder the development of the chamber flow performance. Under the simulation condition of trying to improve the volume flow, the flow field distribution shown in Fig. 5.36b is obtained, it is found that it is not feasible to increase the average flow rate of the protective gas in the horizontal laminar flow area by simply increasing the fan frequency to combat the gas accumulation effect in Area II, the enhanced gas flow performance improves the gas flow in the head of area I, but also enhances the degree of gas accumulation, finally, it led to more severe convection and vortex effect at the head of region I, which on the contrary caused more serious containment of the fluidity in the horizontal laminar flow region.

5.5.2.2 Influence of Building Chamber Height

According to the chamber model shown in Fig. 5.32, the height of the building chamber is doubled to 200 mm, and the other model parameters and boundary conditions remain unchanged to obtain the flow field distribution diagram of the building chamber shown in Fig. 5.37. As shown in Fig. 5.37a, when the protective gas is pumped into the building chamber, the horizontal laminar flow area I will still be formed above the forming area. There is also a large amount of gas that can't directly enter the air outlet due to the diffusion angle and rises along the wall of the building chamber at the air outlet side. However, due to the increase of the height of the building chamber, the upward flowing shielding gas finally fully develops and forms the large-scale vortex (region II) shown in Fig. 5.37a, b at the top of the building chamber. The new air flow distribution shows that the original area III disappears due to large-scale vortex flow encroachment in area II, and the horizontal vortex in Fig. 5.33d also migrates due to the replacement of area II and loses the ability to restrict the vertical height of area I.

This section briefly analyzes the distribution characteristics of each flow field inside the building chamber and the flow performance of the shielding gas. However, due to the complexity of the spatter process, the diversity of the forming dimensions and the gas circulation system of the equipments, the researches and optimization technology of the surrounding fluidity of the protective atmosphere of the building chamber for the purpose of filtering out impurities such as spatter and smoke is still in its infancy at present. Some researchers, aiming at the deposition of by-products in the LPBF process, carried out optimization research from the flow rate of shielding gas, inlet/outlet design and other aspects to enhance the average flow rate and fluidity in the horizontal laminar flow area to achieve a longer distance transport and removal process of by-products in laser processing. But in fact, due to the randomness of the movement of the spatter particles, simply enhancing the average velocity in the horizontal laminar flow area can not completely remove the particles, but tends to make their distribution more uniform. Therefore, a more comprehensive gas purification strategy may be required in the optimization of gas

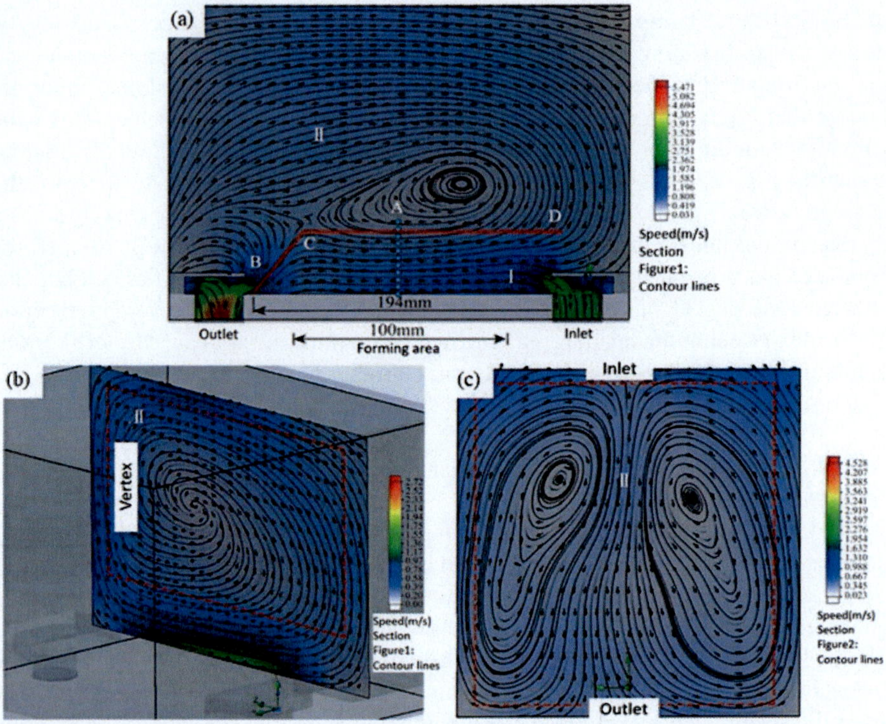

Fig. 5.37 Flow field distribution of building chamber: **a** side view of building chamber flow field; **b** vortex movement of air flows; **c** horizontal flow field distribution at 0.1 mm height

purification efficiency in the future. In the study of the influence of the height of the building chamber on the flow field distribution and fluidity, it is found that the top vortex structure makes the upper part of the horizontal laminar flow obtain a flow trend away from the powder bed. If the upward continuous entrainment can lead to the upward traction of the spatters and extend their dead time, the purification efficiency can be improved, which provides an optimization idea for the future researches on flow field optimization and by-product purification of the building chamber.

References

1. Wang D, Wu S, Fu F et al (2017) Mechanisms and characteristics of spatter generation in SLM processing and its effect on the properties. Mater Des 117:121–130
2. Vladimir S, Akira M (1997) The role of recoil pressure in energy balance during laser materials processing. J Phys D Appl Phys 30(18):2541–2552
3. Zhang MJ, Chen GY, Zhou Y et al (2013) Observation of spatter formation mechanisms in high-power fiber laser welding of thick plate. Appl Surf Sci 280:868–875

4. Li S, Hui D, Yan Z et al (2016) Material gasification and molten pool behavior induced by high power laser irradiation. Electr Weld 46(10):7–13
5. Cheng Y, Jin X, Li S et al (2012) Fresnel absorption and inverse bremsstrahlung absorption in an actual 3D keyhole during deep penetration CO_2 laser welding of aluminum 6016. Opt Laser Technol 44(5):1426–1436
6. Andani MT, Dehghani R, Karamooz-Ravari MR et al (2017) Spatter formation in selective laser melting process using multi-laser technology. Mater Des 131:460–469
7. Gunenthiram V, Peyre P, Schneider M et al (2018) Experimental analysis of spatter generation and melt-pool behavior during the powder bed laser beam melting process. J Mater Process Technol 251:376–386
8. Bidare P, Bitharas I, Ward RM et al (2018) Fluid and particle dynamics in laser powder bed fusion. Acta Mater 142:107–120
9. Ly S, Rubenchik AM, Khairallah SA et al (2017) Metal vapor micro-jet controls material redistribution in laser powder bed fusion additive manufacturing. Sci Rep 7(1):1–12
10. Zhao C, Fezzaa K, Cunningham RW et al (2017) Real-time monitoring of laser powder bed fusion process using high-speed X-ray imaging and diffraction. Sci Rep 7(1):1–11
11. Nakamura H, Kawahito Y, Nishimoto K et al (2015) Elucidation of melt flows and spatter formation mechanisms during high power laser welding of pure titanium. J Laser Appl 27(3):032012
12. Ordas N, Ardila L-C, Iturriza I et al (2015) Fabrication of TBMs cooling structures demonstrators using additive manufacturing (AM) technology and HIP. Fusion Eng Des 96–97:142–148

Chapter 6
Surface Characteristics and Roughness of Laser Powder Bed Fusion Processed Parts

6.1 Theoretical Calculation of Surface Roughness

Although the laser powder bed fusion (LPBF) technology has incomparable advantages over traditional manufacturing technology in manufacturing complex parts, the LPBF formed parts have a poor surface roughness. The poor surface roughness of LPBFed parts can be due to the surface defects such as balling, powder adhesion, connection of coarse particles and discontinuity of melting tracks. In view of the fact that each surface of the part is formed by overlapping continuous melting tracks, the shape of melting track is an important factor affecting the surface roughness. Therefore, to deeply understand the surface roughness of LPBFed parts, in this section the surface (divided into upper surface and side surface) roughness of LPBFed parts was theoretically studied from the microscopic forming angle of laser melting powder to form melting track and body, that is from one dimension to two dimensions and three dimensions. The evaluation parameters of surface roughness include the arithmetic mean deviation *Ra* of contour and the evaluation width *Rsm* of contour element. Since the surface is formed through the overlap of the melting track during the LPBF forming process, the surface will form an undulating surface shape. The surface quality of the LPBFed parts can be comprehensively reflected by using the arithmetic mean deviation *Ra* and the maximum contour height *Rz*. Thus *Ra* and *Rz* are selected as the evaluation parameters of the surface roughness in this study [1].

© National Defense Industry Press 2024

D. Wang et al., *Laser Powder Bed Fusion of Additive Manufacturing Technology*, Additive Manufacturing Technology, https://doi.org/10.1007/978-981-99-5513-8_6

6.1.1 Analysis of Single-Track

In order to study the theoretical value of surface roughness of LPBFed parts, first, it is necessary to know the shape of a single-track. A series of experiments are needed to study the shape parameters of single-track, including the track width and the upper surface shape of the track. In the process of LPBF manufacturing single-tracked molten pool, the molten pool is divided into upper and lower parts. Since the surface roughness of LPBFed parts is mainly affected by the shape of the melting track above the matrix, and the track below the matrix only plays the role of bonding different layers, the main object of study is the shape of the melting track above the matrix.

In Chap. 3, the single-track forming is studied deeply. In single-track manufacturing, high laser energy input can obtain smooth and continuous single-track, but the material will vaporize and blow away the powder around the molten pool, which is not conducive to the formation of subsequent tracks. Therefore, the ideal single-track shape is not only smooth and continuous, but also needs to ensure that the powder around the track is in the original position as far as possible. By observing the cross-sectional shape of an ideal single-track, it can be found that the shape of the track above the matrix is approximately a circular curve. The reason is that the liquid molecules in the liquid melting track attract each other, and the molecules on the liquid surface are attracted by the internal molecules more than the external gas molecules, which makes the molecules on the liquid surface shrink inward. Macroscopically, the molten liquid track shrinks into a circular curve shape under the effect of surface tension.

The width of single-track is an important index reference value in the actual forming process, which has a great influence on the surface roughness of parts, and is also the basis for the theoretical research on the surface roughness of LPBFed parts. The width of a single-track is closely related to laser power and scanning speed. Therefore, this chapter studies the relationship between laser power, scanning speed and track width through experiments, and the results are shown in Fig. 6.1. It can be seen that the width of a single-track can be roughly determined after fixing the laser power and scanning speed.

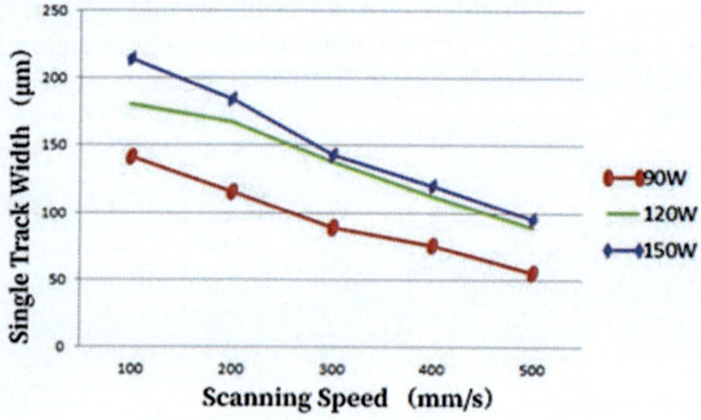

Fig. 6.1 Relation curve among scanning speed, laser power and melting track width (layer thickness: 35 μm)

6.1.2 *Upper Surface Roughness of the Part*

In order to calculate the theoretical surface roughness value of the upper surface, the following assumptions should be made:

(1) The cross-sectional shape of the melting track above the matrix is a circular curve.
(2) Each melting track has the same shape.
(3) The thermal expansion of the re-melting zone is ignored during the track lap.

After making the above assumptions, the function model of the melting track curve can be established.

Figure 6.2 shows the cross-sectional shape of a single-track above the matrix, where the shape of the track is a circular curve with radius r, the width of the track is a, and the height of the track above the matrix is h. It is deduced that the height of the track above the matrix is equal to the thickness of the powder layer (slice layer thickness). As such, we can figure out that $r = \frac{a^2}{8h} + \frac{h}{2}$.

The shape curve of the single-track was put into the plane rectangular coordinate system for modeling, and the shape equation of the melting track was established as follows:

$$y = f(x) = \sqrt{r^2 - (x - a/2)^2} + h - r(0 \leq x \leq a) \tag{6.1}$$

where y is height, x is width, r is track radius, a is track width, h is track height.

Figure 6.3 shows the lap diagram between different tracks, s is the hatch space, and the lap ratio is:

$$\delta = \frac{(a - s)}{a} \tag{6.2}$$

where δ is lap ratio, a is track width, s is hatch space.

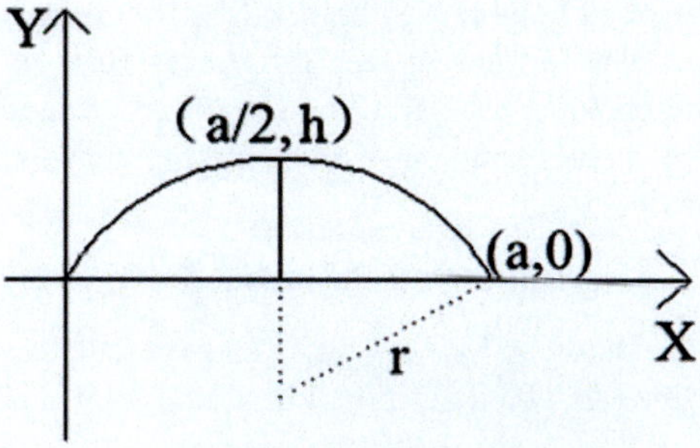

Fig. 6.2 Single-track patterns

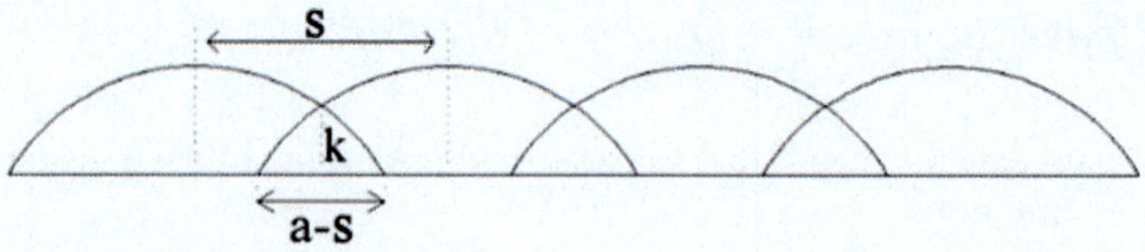

Fig. 6.3 The lapping diagram of melting tracks

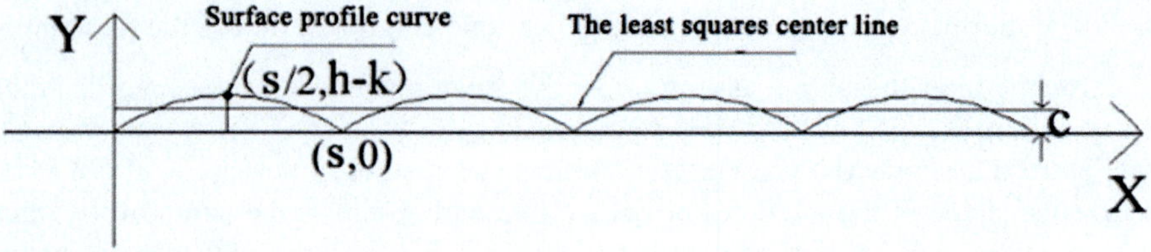

Fig. 6.4 The upper surface profiles

The depth of the re-melting zone is:

$$k = f\left(\frac{a+s}{2}\right) \tag{6.3}$$

where k is the depth of the re-melting zone, a is track width, s is hatch space.

Figure 6.4 shows the roughness curve of the upper surface of the formed part. The current roughness curve can be obtained by translating the previous two figures downward by k units and left by $(a - s)/2$ units in the plane coordinate system.

The surface roughness curve is a periodic function of period s, that is, $f'(x + s) = f'(x)$.

The function equation in a single period is:

$$y' = f\left(x + \frac{a-s}{2}\right) - k = \sqrt{r^2 - \left(x - \frac{a}{2}\right)^2} + h - k - r \quad (0 \le x \le b) \tag{6.4}$$

where y' is height, x is width, r is track radius, s is hatch space, a is track width, h is track height, k is the depth of the re-melting zone.

The formed surface roughness curve is a periodic curve, thus it is unnecessary to consider the sampling length and evaluation length in the theoretical calculation of surface roughness, but only the curve within one period needs to be selected for calculation.

1. The arithmetic mean deviation *Ra* of the contour.

As shown in Fig. 6.4, in order to calculate *Ra*, the position of the least squares center line must be determined first, and let $\int_0^s (y' - c)^2 dx$ take the minimum value, then

$$c = \frac{\pi r^2 \sin^{-1}\frac{s}{2r}}{180s} - r + k - \frac{h}{2} \tag{6.5}$$

where c is the height of the least squares center line, π is pi, r is track radius, s is hatch space, h is track height, k is the depth of the re-melting zone.

Finally, the expression of Ra is derived as:

$$\begin{aligned} Ra &= \frac{\int_0^s |y' - c| dx}{s} \\ &= \frac{\left(\frac{a^2}{8h} + \frac{h}{2}\right)^2}{180s}\left[90\cos^{-1}\left(\frac{4sh}{a^2 + 4h^2}\right) - \pi \sin^{-1}\left(\frac{4sh}{a^2 + 4h^2}\right)\right] \\ &\quad - \frac{\sqrt{\left(\frac{a^2}{8h} + \frac{h}{2}\right)^2 - \frac{s^2}{4}}}{4} \end{aligned} \tag{6.6}$$

where Ra is the arithmetic mean deviation of the contour, y' is height, x is width, c is the height of the least squares center line, s is hatch space, π is pi, a is track width, h is track height.

2. The maximum contour height Rz.

From the definition of Rz, Rz can be expressed as:

$$Rz = h - k = \frac{h}{2} - \sqrt{\left(\frac{a^2}{8h} + \frac{h}{2}\right)^2 - \frac{s^2}{4}} + \frac{a^2}{8h} \tag{6.7}$$

where Rz is the maximum contour height, h is track height, k is the depth of the re-melting zone, a is track width, s is hatch space.

From the above expression of surface roughness parameters, it can be seen that the surface roughness of the formed parts is theoretically affected by three factors: melting track width, hatch space and powder layer thickness. The track width is mainly controlled by laser power and scanning speed. Therefore, in order to improve the surface roughness of the forming parts, four factors should be considered: laser power, scanning speed, hatch space and powder layer thickness.

In order to calculate the surface roughness of the upper surface of LPBFed parts, it was assumed that the laser power used in manufacturing is 150 W and the scanning speed is 400 mm/s. By comparing the relationship between track width, laser power and scanning speed, it was found that the track width $a = 120$ μm. Assuming that hatch space $s = 80$ μm, powder layer thickness $h = 35$ μm. After inputting the calculation formulas and parameter values into Matlab software for calculation, the final theoretical calculation value of surface roughness is $Ra = 3.21$ μm, $Rz = 12.79$ μm.

6.1.3 *Side Surface Roughness of Part*

In the following, the side surface of the LPBFed part is formed, and the surface roughness of the side surface is theoretically analyzed.

As shown in Fig. 6.5, the dotted line is the theoretical contour of the inclined surface, the slice layer thickness is h, and the inclination angle is set as α. The theoretical contour curve of the inclined surface is put into the plane coordinate system for modeling.

It can be seen from Fig. 6.6 that the roughness curve equation is a periodic function, and the equation in a single period is as follows:

$$y = \begin{cases} x \cdot \cot\alpha \quad (0 \le x \le h \cdot \sin\alpha) \\ -x \cdot \tan\alpha + \frac{h}{\cos\alpha} \quad \left(h \cdot \sin\alpha \le x \le \frac{h}{\sin\alpha}\right) \end{cases} \tag{6.8}$$

where y is height, x is width, α is inclination angle, h is slice layer thickness.

The formed surface roughness curve is a periodic curve, thus it is unnecessary to consider the sampling length and evaluation length in the theoretical calculation of surface roughness, but only the curve within one period needs to be selected for calculation.

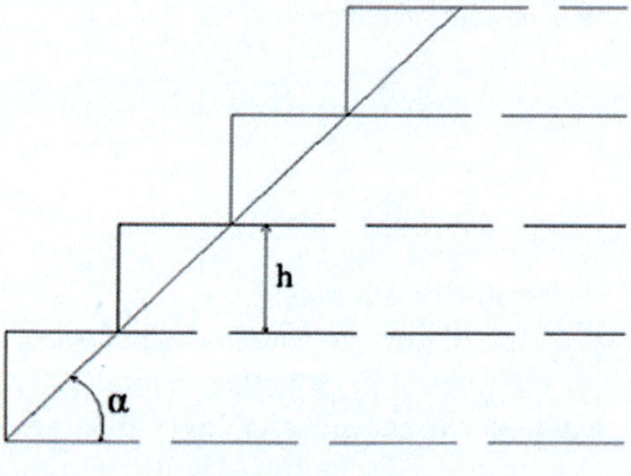

Fig. 6.5 The theoretical profile of inclined surface

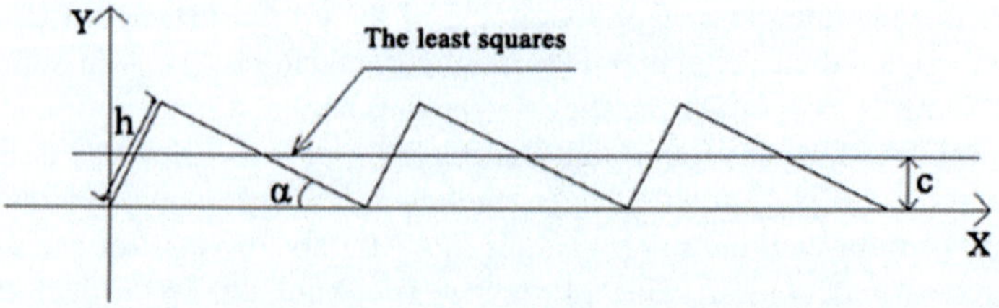

Fig. 6.6 The theoretical profile of side surface

1. The arithmetic mean deviation *Ra* of the contour.

As shown in Fig. 6.6, in order to calculate *Ra*, the position of the least squares center line must be determined first, and let $\int_0^{\frac{h}{\sin\alpha}}(y-c)^2dx$ take the minimum value, then $c=\frac{h\cdot\cos\alpha}{2}$.

Finally, the expression of *Ra* is derived as:

$$Ra=\frac{\int_0^{\frac{h}{\sin\alpha}}|y-c|dx}{\frac{h}{\sin\alpha}}=\frac{h\cdot\cos\alpha}{4} \tag{6.9}$$

where *Ra* is the arithmetic mean deviation of the contour, *y* is height, *x* is width, α is inclination angle, *h* is slice layer thickness, *c* is the height of the least squares center line.

2. The maximum contour height *Rz*.

$$Rz=h\cdot\cos\alpha \tag{6.10}$$

where *Rz* is the maximum contour height, α is inclination angle, *h* is slice layer thickness.

It can be seen from Eqs. (6.9) and (6.10) that the surface roughness of inclined plane is theoretically determined by two factors: inclination angle and slice layer thickness. From the theoretical analysis of surface roughness, it can be concluded that the larger the inclination angle, the smaller the slice layer thickness and surface roughness of the inclined plane. Therefore, in order to obtain better surface quality, smaller slice layer thickness is needed to reduce the "step effect". At the same time, it is necessary to adjust the inclination angle of the manufacturing machine in the processing to make the inclination angle as large as possible to reduce the impact of manufacturing defects on the surface quality.

In order to calculate the theoretical value of surface roughness of inclined side, it was assumed that powder layer thickness $h = 35\ \mu\text{m}$, and side inclination angle $\alpha = 45°$. After inputting the calculation formulas and parameter values into Matlab software for calculation, and the final theoretical calculation value of surface roughness is $Ra = 6.19\ \mu\text{m}$, $Rz = 24.75\ \mu\text{m}$.

6.1.4 Comparison of Theoretical and Measured Surface Roughness of Part

When the corresponding process parameters were selected for the theoretical analysis of upper surface roughness as follows: laser power of 150 W, scanning speed of 400 mm/s, powder layer thickness of 35 μm, hatch space of 80 μm, the LPBF manufacturing machine was used to process a small cube with side length of 10 mm. When the corresponding process parameters were selected for the theoretical analysis

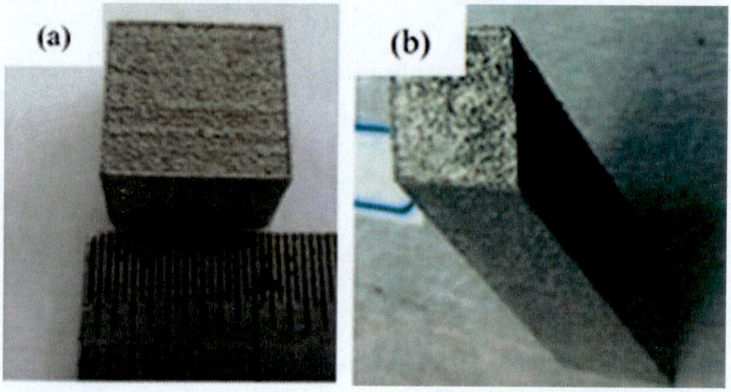

Fig. 6.7 The macroscopic surface morphology of the formed parts: **a** small block; **b** inclined long block

of side surface roughness as: powder layer thickness of 35 μm, laser power of 150 W, scanning speed of 400 mm/s, hatch space of 80 μm, the LPBF manufacturing machine was used to process an inclined part with inclination angle of 45°. The LPBF processed small cube and inclined part are shown in Fig. 6.7.

Then, the surface roughness of the upper surface of the cube and the side of the inclined part are measured respectively. The comparison results between measured values and theoretical calculated values are shown in Table 6.1.

By comparing the statistical results of surface roughness in Table 6.1, it can be found that the measured value is larger than the theoretical value, and the measured value is about twice the theoretical value.

The reasons for the difference between the theoretical value and the measured value of the surface roughness of the parts are analyzed as follows:

(1) LPBF is a complex and changeable process. It is impossible to completely simulate the single-track shape and track lap during the manufacturing process when conducting the roughness theoretical analysis, but only an idealized assumption close to the actual process can be made. In the theoretical calculation, the shape of the melting track is assumed to be a regular circular curve, but in the actual process, the molten pool is unstable, and there will be small balls on both sides of the track. The thermal influence of the re-melting zone is ignored in the theoretical calculation, while the thermal expansion of the re-melting zone

Table 6.1 The surface roughness comparison between theoretical and measured values

Type of surface roughness		*Ra* (μm)	*Rz* (μm)
Upper surface roughness	Theoretical value	3.21	12.79
	Measured value	7.36	40.01
Side surface roughness	Theoretical value	6.19	24.75
	Measured value	14.25	38.34

Fig. 6.8 Zoom-in image of the part's surface

will exist in the actual processing process, which will affect the accuracy of the theoretical calculation.

(2) As can be seen from the enlarged phase diagram of the part's surface in Fig. 6.8, the surface quality of the lap area of the melting track is poor, and there are balls and bulges on both sides of the track, which are powder particles that adhere to the surface of the track and fail to melt completely. These defects affect the quality of the latter track, resulting in discontinuity and balling of the subsequent track and undulation of the track surface. For the inclined side surface, since the outer side of the side surface is powder area, the melting track on the side surface is easy to absorb powder during the cooling process, resulting in more unmelted powders adhering to the side surface, which further deteriorates the roughness of the side surface.

(3) In the actual process of LPBF, there are some defects such as balling and warping. These defects will accumulate with the increase of the number of layers, and the surface roughness of the parts will deteriorate with the increase of the number of layers.

Based on the above reasons, the measured value of surface roughness of parts is larger than that of theoretical calculation.

6.2 Surface Characteristics and Influencing Factors of Surface Roughness

Through the background investigation of metal parts manufactured by LPBF technology and the research status of surface roughness of additive manufacturing (AM) parts at home and abroad, it can be found that there are dozens of factors affecting the

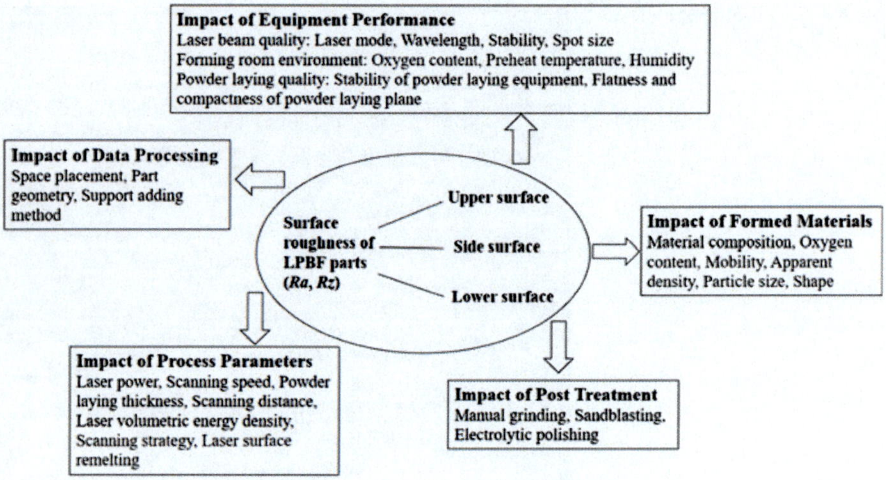

Fig. 6.9 Influencing factors of surface roughness of the LPBFed parts

surface roughness of LPBFed parts. These influencing factors can be summarized into five parts: equipment performance, feedstock material, data processing, process parameters and post-processing, as shown in Fig. 6.9.

Among the above influencing factors, the influencing factors of process parameters have huge room for improvement with practical significance. Therefore, combined with the theoretical analysis of the previous section, this section focuses on the influence of process parameters on the surface roughness of LPBFed parts. However, during a series of process experiments, it is necessary to control the other four influencing factors except the influence of process parameters, so as to reduce their influence on the surface roughness of LPBFed parts and improve the credibility of the experimental results. To this end, the following measures have been taken:

(1) As for the influencing factors of equipment performance, the experiment used the LPBF equipment DiMetal-100 independently developed by the laboratory. The laser beam was of good quality and stability by using continuous fiber laser. In the manufacturing cavity environment, the high purity inert gas protection was used in the manufacturing process to ensure that the oxygen content in the manufacturing cavity meets the manufacturing requirements. The dust purifier can timely remove the dust floating in the manufacturing cavity and ensure the clean gas in the manufacturing cavity.

(2) As for the influencing factors of manufacturing materials, researchers had done a lot of research in order to find out more suitable metal powder materials for AM. It was found that water atomized powder has high oxygen content and poor sphericity, which was not suitable for metal AM. The gas atomized powder had low oxygen content and good sphericity, which was very suitable for metal AM. Therefore, the mature 316L stainless steel gas atomized powder was selected as the manufacturing material.

(3) As for the influencing factors of data processing, the square model with regular shape and simple shape was selected for forming. Thus it is unnecessary to consider the influencing factors of data processing, such as spatial placement and support addition.
(4) As for the influencing factors of post-treatment, it has no influence on the process experiment since it is carried out after the forming of the parts. Moreover, the post-treatment is to further improve the surface roughness of the forming parts, thus the influencing factors of post-treatment will not affect the experimental results.

On the premise of taking the above measures, the process optimization experiment was carried out. The effects of laser power, scanning speed, laser energy density and scanning strategy on the surface roughness of LPBFed parts were studied. According to the theoretical research results of surface roughness of LPBFed parts, the effects of laser power, scanning speed, hatch space and laser energy on the upper surface roughness of forming parts were mainly studied. For the side surface of the parts, the influence of the thickness and inclination angle of the powder on the roughness was studied. In order to further improve the overall surface quality of the parts, the influence of different scanning strategies on the surface roughness was also studied.

6.2.1 *Analysis of Upper Surface Characteristics and Roughness*

In order to study the effect of process parameters on the surface roughness of the parts, two plates of 10 mm × 10 mm × 5 mm small blocks were processed. The process parameters are shown in Fig. 6.10, in which the scanning strategy adopts S-shaped orthogonal interlayer interleaved scanning.

Figure 6.11 shows the surface morphology of the small blocks after forming.

As meeting the requirements of relative density is the premise of LPBF processing, relative density should also be considered when optimizing the surface roughness. Therefore, the relative density and surface roughness of the square were measured simultaneously in this study for comprehensive consideration. The measurement results are listed in Tables 6.2 and 6.3.

1. Effect of scanning speed on upper surface roughness

According to the experimental results and measured data, the influence of scanning speed on the roughness of the upper surface of the forming block was drawn, as shown in Fig. 6.12.

As can be seen from the Fig. 6.12, the technological parameters of power of 150 W, hatch space of 80 μm, powder layer thickness of 30 μm, scanning speed of 400 mm/s correspond to the minimum surface roughness. When the scanning speed was less than 400 mm/s, the roughness of the upper surface decreased with the increase of the

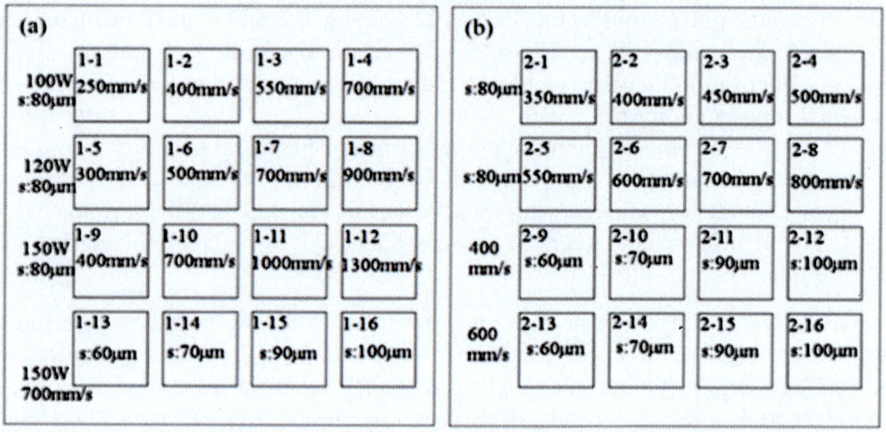

Fig. 6.10 The design of upper surface process experiment: **a** the process parameters of Experiment 1 (powder layer thickness of 30 μm); **b** the process parameters of Experiment 2 (laser power of 150 W, powder layer thickness of 30 μm)

Fig. 6.11 The macroscopic appearance of the LPBF-processed samples: **a** macroscopic appearance of the experimental sample in Experiment 1; **b** macroscopic appearance of the experimental sample in Experiment 2

scanning speed. When the scanning speed was greater than 400 mm/s, the roughness of the upper surface increased with the increase of the scanning speed.

The morphology of the upper surface of the formed small block was photographed with an optical camera, as shown in Fig. 6.13.

It can be seen from Fig. 6.13 that the upper surfaces of the three blocks with scanning speed of 400, 450 and 500 mm/s are relatively flat. In particular, the upper surfaces of the blocks with scanning speed of 400 mm/s are the smoothest, without

Table 6.2 The measurement results of the sample in Experiment 1

Number	The process parameters (*P*, *V*, *S*)	Volume energy density (J/mm^3)	Relative density (%)	*Ra* (μm)
1-1	100 W, 250 mm/s, 80 μm	166.67	89.31	21.2435
1-2	100 W, 400 mm/s, 80 μm	104.17	91.24	14.4619
1-3	100 W, 550 mm/s, 80 μm	75.76	85.93	18.0475
1-4	100 W, 700 mm/s, 80 μm	59.52	81.93	13.9109
1-5	120 W, 300 mm/s, 80 μm	166.67	91.35	24.6544
1-6	120 W, 500 mm/s, 80 μm	100.00	91.72	8.8209
1-7	120 W, 700 mm/s, 80 μm	71.43	83.74	12.0262
1-8	120 W, 900 mm/s, 80 μm	55.56	81.27	13.7635
1-9	150 W, 400 mm/s, 80 μm	156.25	91.93	4.7635
1-10	150 W, 700 mm/s, 80 μm	89.29	92.23	6.2869
1-11	150 W, 1000 mm/s, 80 μm	62.50	81.78	13.8497
1-12	150 W, 1300 mm/s, 80 μm	48.08	79.53	13.8864
1-13	150 W, 700 mm/s, 60 μm	119.05	91.84	5.8069
1-14	150 W, 700 mm/s, 70 μm	102.04	93.68	6.6108
1-15	150 W, 700 mm/s, 90 μm	79.37	89.53	11.2448
1-16	150 W, 700 mm/s, 100 μm	71.43	89.50	10.1152

any defects, and have metallic luster. The upper surface of the block at other speeds is relatively rough. The surface of the block at 350 mm/s speed shows uneven morphology. There are some small convex spheres on the surface. And there is a slight overmelting phenomenon. The upper surfaces of the blocks with the scanning speed greater than 500 mm/s present the appearance of scattered sand and no metallic gloss. Especially for the blocks with the speed of 800 mm/s, the scattered sand particles on the upper surface are larger and exhibit obvious incomplete melting and sintering state.

Table 6.3 The measurement results of the sample in Experiment 2

Number	The process parameters (*P*, *V*, *S*)	Volume energy density (J/mm^3)	Relative density (%)	*Ra* (μm)
2-1	150 W, 350 mm/s, 80 μm	178.57	94.12	10.558
2-2	150 W, 400 mm/s, 80 μm	156.25	99.35	6.0259
2-3	150 W, 450 mm/s, 80 μm	138.89	98.23	6.3237
2-4	150 W, 500 mm/s, 80 μm	125.00	97.87	7.6237
2-5	150 W, 550 mm/s, 80 μm	113.64	96.61	8.434
2-6	150 W, 600 mm/s, 80 μm	104.17	96.13	8.2149
2-7	150 W, 700 mm/s, 80 μm	89.29	94.21	10.2749
2-8	150 W, 800 mm/s, 80 μm	78.13	93.13	12.4864
2-9	150 W, 400 mm/s, 60 μm	208.33	93.16	24.1587
2-10	150 W, 400 mm/s, 70 μm	178.57	94.45	9.285
2-11	150 W, 400 mm/s, 90 μm	138.89	98.76	9.7863
2-12	150 W, 400 mm/s, 100 μm	125.00	97.87	8.7042
2-13	150 W, 600 mm/s, 60 μm	138.89	98.77	9.3675
2-14	150 W, 600 mm/s, 70 μm	119.05	96.64	6.5157
2-15	150 W, 600 mm/s, 90 μm	92.59	95.56	11.7433
2-16	150 W, 600 mm/s, 100 μm	83.33	93.26	13.2258

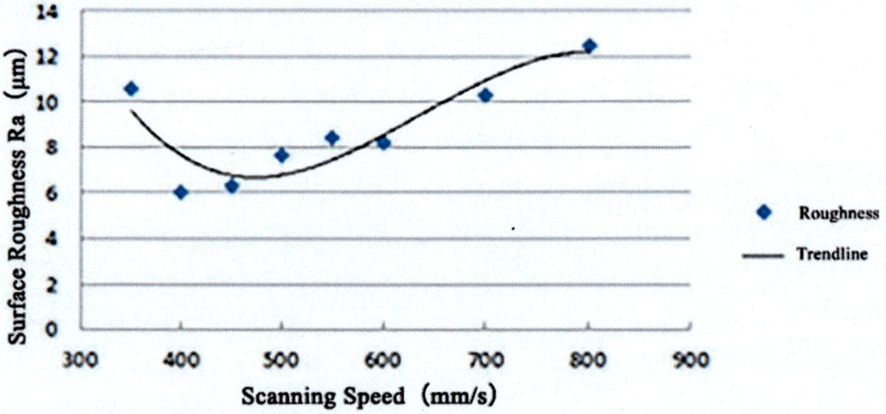

Fig. 6.12 The upper surface roughness versus laser scanning speed (P: 150 W, S: 80 μm, h: 30 μm)

In order to further study the effect of scanning speed on the upper surface roughness of the part, stereomicroscope images were taken of the upper surface of the part. Representative pictures are selected for display and illustration, as shown in Fig. 6.14.

Fig. 6.13 The upper surface morphology of samples processed by different scanning speeds

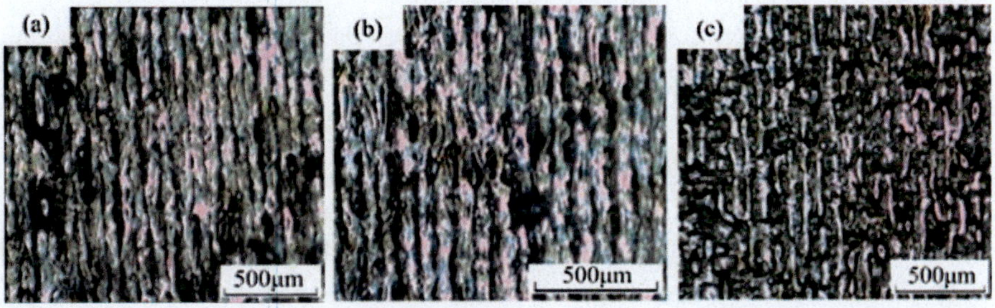

Fig. 6.14 The upper surface micromorphology of the part: **a** scanning speed of 350 mm/s; **b** scanning speed of 400 mm/s; **c** scanning speed of 800 mm/s

It can be found from Fig. 6.14a that the melting track on the upper surface of the part is continuous and overlaps tightly, but the track is not smooth enough, and there are obvious raised balls on both sides of the track. The reason for producing bulged globules is analyzed: under the process parameters of 150 W and 350 mm/s, the energy density is high, the melting condition is good, and thus the continuous melting track can be formed. However, the low speed of 350 mm/s leads to a significant increase in the temperature of the machining center and a sharp increase in the melting amount of powder, which makes the track wider and the cooling and solidification time of the track longer. In this process, the melting track is easy to absorb the powders on both sides. Although the adhesive powder on both sides of the track does not produce obvious defects during the processing of the initial layer, the adhesive powder will affect the powder laying quality of the next layer. The processing quality deteriorates as the number of layers increase until the surface accumulates into a raised ball. It can be seen from Fig. 6.14b that when the scanning speed is

400 mm/s, the melting track of the formed block is continuous and smooth. The lap joint is tight, and there are fewer raised balls on both sides of the track. It can be seen from Fig. 6.14c that when the scanning speed is 800 mm/s, the upper surface of the formed block cannot form a continuous track. The surface presents the shape of a worm, which is rough with a serious balling phenomenon.

Based on the above research, it can be concluded that the lower scanning speed and higher scanning speed will deteriorate the roughness of the upper surface. If the scanning speed is too low, the laser energy density will be too high, which is easy to produce overmelting phenomenon. Moreover, if the scanning speed is too low, the melting track will be widened and the cooling and solidification time will be longer. This results in serious powder adhesion. When the scanning speed is too high, the melting state is not good, and the continuous track cannot be formed. The track presents a striped worm shape, and the melting depth and width are small, which will also increase the tendency of balling. Therefore, when selecting the scanning speed, a moderate scanning speed should be selected while considering the other process parameters, so as to improve the roughness of the upper surface.

2. Effect of laser power on upper surface roughness

According to the experimental results and measured data, the relationship diagram of the influence of laser power on the roughness and relative density of the upper surface of the shaped block was drawn, as shown in Fig. 6.15.

It can be seen from Fig. 6.15 that under the quantitative process parameters of scanning speed of 700 mm/s, hatch space of 80 μm, and powder layer thickness of 30 μm, the upper surface roughness of the block first decreases and then increases, and the relative density of the block first increases and then decreases with the increase of laser power. In order to further study the cause of this effect, stereographic micrographs were taken on the upper surface of the formed block. Select representative pictures for demonstration and illustration, as shown in Fig. 6.16.

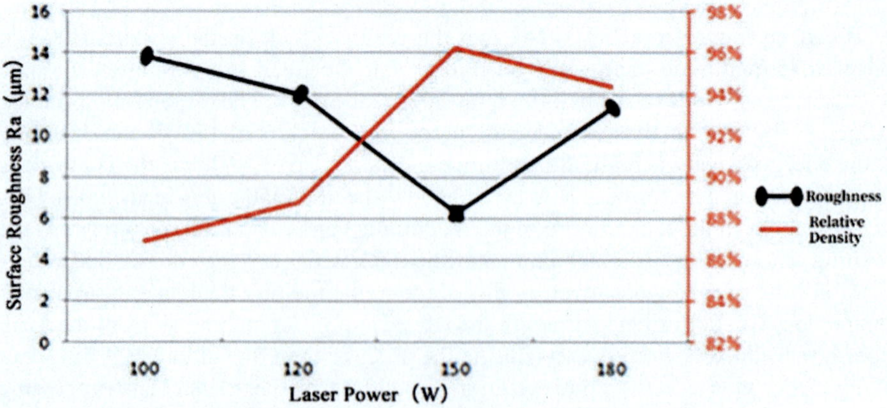

Fig. 6.15 Effect of laser power on the relative density and surface roughness (V: 700 mm/s, S: 80 μm, h: 30 μm)

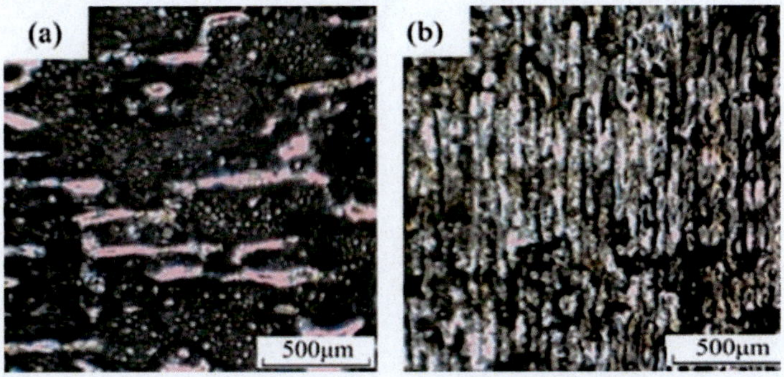

Fig. 6.16 The upper surface micro topography of the block: **a** laser power of 100 W; **b** laser power of 180 W

As can be seen from Fig. 6.16a, when the laser power is 100 W, the laser energy is low. More powders on the upper surface of the formed block are not completely melted, and uneven sintering state appears on the surface without continuous melting track. Due to the poor wettability of melted powder on solid matrix, balling is more serious. In this case, the relative density and surface quality of the parts are poor. As can be seen from Fig. 6.16b, when the laser power is 180 W, the laser energy is large. Although there are continuous melting tracks on the upper side, the melting tracks are not smooth and the lap is too tight without slight overmelting phenomenons. When the laser power is too high, on the one hand, the surface area of the molten pool will be larger, leading to longer cooling and solidification time and more serious powder adsorption phenomenon. On the other hand, the liquid molten pool may spatter and flow around before cooling, and solidify to form a rough and irregular surface.

Therefore, both too low and too high laser power is not conducive to the improvement of surface roughness of LPBFed parts. In the process adjustment, laser power should cooperate with scanning speed and hatch space to make the volume energy density in the smooth manufacturing area, so as to achieve the effect of improving surface roughness.

3. Effect of hatch space on upper surface roughness

According to the experimental results and measurement data, the influence diagram of hatch space on the roughness of the upper surface of the formed block was drawn, as shown in Fig. 6.17.

As can be seen from the Fig. 6.17, under the technological parameters of power of 150 W, scanning speed of 700 mm/s, powder layer thickness of 30 μm, when hatch space is 80 μm, the upper surface roughness is minimum. When the hatch space in the range of less than 80 μm increases, the upper surface roughness increases as the hatch space significantly decreases. When hatch space is greater than 80 μm, the surface roughness increases as the hatch space and slowly increases. Figures 6.18 and 6.19 show the macro topography and stereo micrograph taken on the upper surface of the formed block.

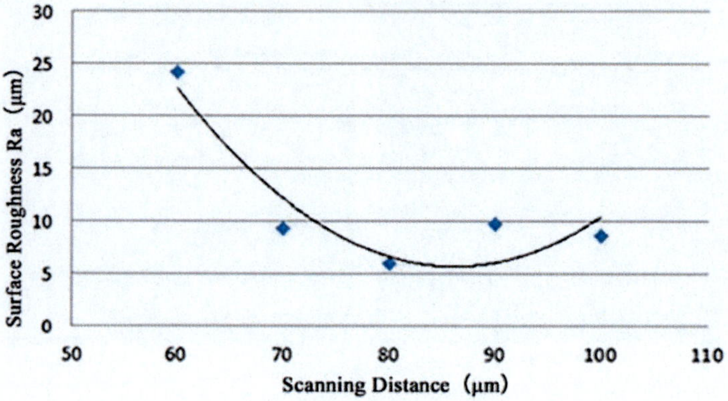

Fig. 6.17 The upper surface roughness trend affected by hatch space (P: 150 W, V: 700 mm/s, h: 30 μm)

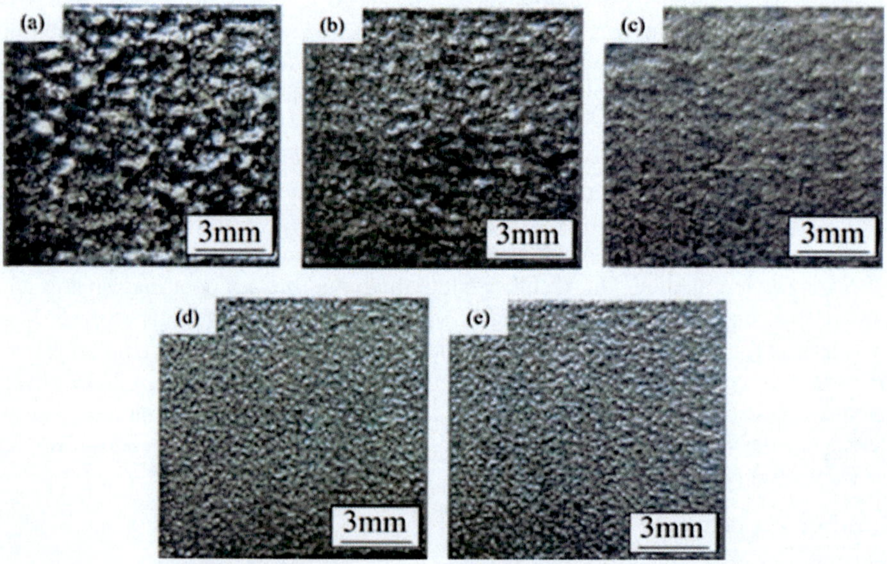

Fig. 6.18 The macro topography of upper surface at different hatch spaces: **a** 60 μm; **b** 70 μm; **c** 80 μm; **d** 90 μm; **e** 100 μm

As can be seen from the Fig. 6.18, the upper surface of the square corresponding to the hatch space of 80 μm is the smoothest, without any defects, and has metallic luster. The upper surface of the square corresponding to the hatch space of 60 μm presents a serious overmelting pattern, and the overmelting phenomenon makes the upper surface exhibit a rough and uneven pattern. The upper surface of the block corresponding to 70 μm hatch space is relatively flat, but there are many small bumps. The upper surfaces of the squares corresponding to the hatch space of 100 μm are all

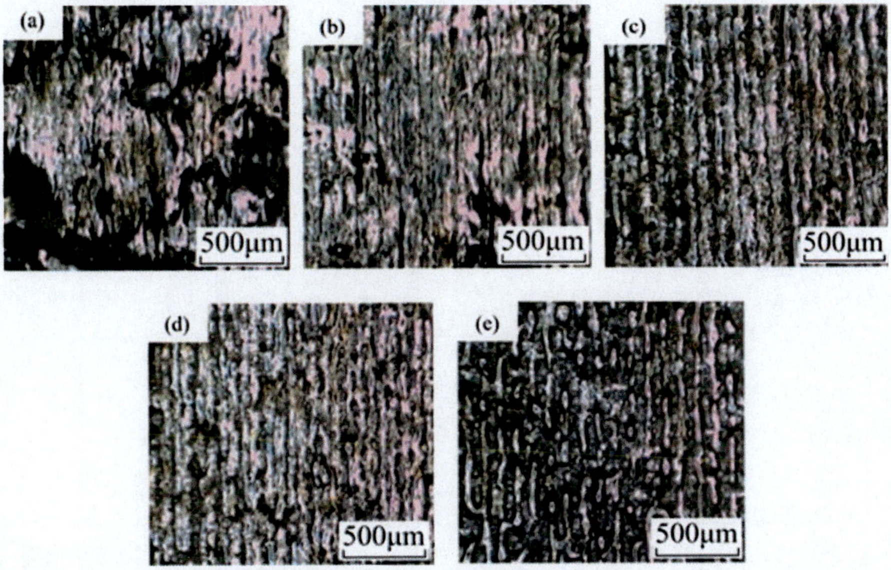

Fig. 6.19 The stereoscopic micrograph of upper surface at different hatch spaces: **a** 60 μm; **b** 70 μm; **c** 80 μm; **d** 90 μm; **e** 100 μm

coarse and granular. As can be seen from Fig. 6.19, when the hatch space is 60 and 70 μm, the melting track lap is relatively close, and the re-melting area is wide with large heat accumulation, resulting in overmelting, uneven surface, and serious powder adhesion. In particular, the upper surface of the block with a hatch space of 60 μm has some defects, such as distortion and depression, due to the serious overmelting. The block tracks at 80 and 90 μm hatch space are smooth and continuous, with good lap connection and no defects. However, when the hatch space increases to 100 μm, the melting track lap is not close enough, the gap between the melting track is wide with insufficient heat accumulation, and the melting track appears intermittent. Moreover, due to the large distance between the wave crest and the wave trough, the surface ripple effect is obvious, resulting in a rough surface.

From the above analysis, it can be seen that the influence of hatch space on the roughness of the upper surface is mainly influenced by the lap ratio between tracks. Therefore, it is necessary to further study the effect of melting track lap ratio on the roughness of the upper surface. The overlap rate of the melting track refers to the ratio of the lap width between the tracks to the width of a single-track. The calculation method is as follows:

$$\delta = \frac{(a - s)}{a} \tag{6.11}$$

where δ is lap ratio, a is track width, s is hatch space.

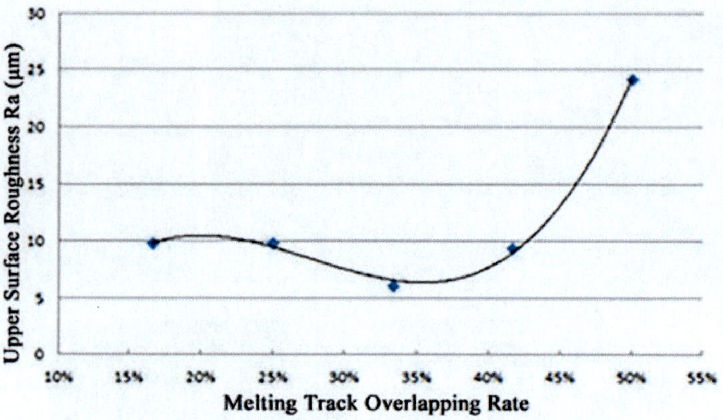

Fig. 6.20 The upper surface roughness affected by overlapping rate

According to the previous research results on the width of single-track, when the laser power is 150 W and the scanning speed is 400 mm/s, the width of single-track is about 120 μm. Therefore, the effect of the overlap rate on the roughness of the upper surface can be studied by calculating the overlap rate between tracks, as shown in Fig. 6.20.

As can be seen from Fig. 6.20, too low or too high overlap rate will have adverse effects on the surface roughness of the forming parts. When the overlap rate is kept at 30–40%, the roughness of the upper surface is minimum. When the overlap rate is 33.3%, the roughness *Ra* of the upper surface is 6.2 μm.

When the hatch space is larger than the width of a single-track, the overlap rate is zero. The two molten pools cannot be connected or the connection is not firm after solidification, thus generating the phenomenon of "disassembly". In this case, the parts have low relative density and are difficult to form metallurgically combined metal parts. Therefore, in order to improve the surface roughness of LPBFed parts, the laser power and scanning speed should be considered comprehensively when selecting the hatch space value. Because there are different track widths under different laser energy densities, different hatch space should be selected under different energy densities to keep the overlap rate between melting tracks at the ideal value of 30–40%.

4. Effect of volume energy density on upper surface roughness

According to previous studies, it is not difficult to find that the laser power, scanning speed, hatch space and other process parameters in the manufacturing process of LPBF are interrelated and affect each other. Therefore, when conducting process optimization experiments to study a single process parameter, the influence of other process parameters should not be ignored, and these mutually influencing process parameters should be considered comprehensively. The four process parameters, e.g. laser power, scanning speed, hatch space and powder layer thickness, are integrated into one parameter, "volume energy density". Its physical meaning is the energy input per unit volume in unit time. Through the optimization, the range of volume energy density of metal parts can be found, so that the whole control of laser melting process and forming quality is possible. The calculation formula of volume energy density is as follows:

$$\omega = \frac{P}{vsh} \tag{6.12}$$

where P is laser power, v is scanning speed, s is hatch space, h is powder layer thickness.

According to the previous experimental results and measured data, the corresponding volume energy density of each forming block was calculated, and the influence of volume energy density on the surface roughness and relative density of the formed block was drawn, as shown in Fig. 6.21.

It can be seen from Fig. 6.21 that there is a close relationship between volume energy density and the quality of LPBFed parts. The forming characteristics of LPBFed parts are different under different volume energy density. According to different volume energy density corresponding to different forming characteristics, the LPBF in the figure is divided into 5 types: A is the incomplete melting zone; B is the low energy density balling zone; C is the successfully manufacturing zone; D is the high energy density balling zone; E is the overmelting zone.

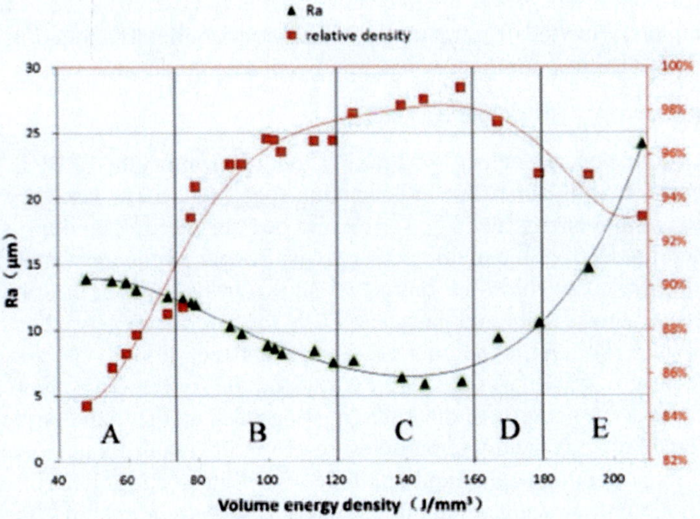

Fig. 6.21 Effect of volume energy density on relative density and surface roughness of LPBFed parts

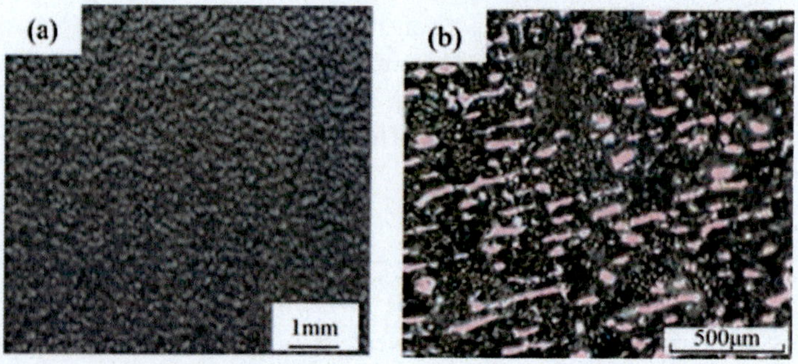

Fig. 6.22 The typical sample of the incompletely melting zone: **a** low magnification micrograph; **b** high magnification micrograph (*P*: 100 W, *V*: 700 mm/s, *s*: 80 μm, *h*: 30 μm)

(1) The incomplete melting zone.

When the volume energy density $\omega < 75$ J/mm^3, the LPBF is in the incomplete melting zone. In this forming area, the volume energy density is too small to completely melt the metal powder in the scanning area. The melting track shape cannot be formed, and some powders are in the sintering state. In the incomplete melting area, the relative density of the LPBF sample is low, and the surface is sandy, rough and without metallic luster. Figure 6.22 shows a typical sample in the incomplete melting zone, with volume energy density of 59.52 J/mm^3, relative density of 86.93%, and surface roughness *Ra* of 13.91 μm. As can be seen from the figure, the melting track of the sample is discontinuous. There are many black pellets on the surface, and the powder is not completely melted. The metal powder is not completely melted in this case, the LPBF sample has poor quality, low relative density and rough surface.

(2) The low energy density balling zone.

When the volume energy density 75 J/mm^3 $< \omega <$ 120 J/mm^3, the LPBF is in the low energy density balling zone. Too fast scanning speed or too low laser power results in too low volume energy density, so that the powder melting amount is reduced, and the melting depth of the melting track is insufficient with poor wettability of the liquid melting track to the solid matrix. As shown in Fig. 6.23, F1 is the interface energy between solid state and gas state, F2 is the interface energy between solid state and liquid state, F3 is the interface energy between liquid state and gas state. Due to the poor wettability of the melting track on the solid matrix, the force of the solid particles at the position of the liquid melting track on the liquid particles is less than the force between the liquid particles, resulting in greater F2 than F1. The total resultant force direction of the liquid particle at the intersection point of gas, liquid and solid three phases points to the liquid interior, so that the melting track wetting angle θ is greater than 90°. Under the action of interfacial tension, the liquid surface shrinks into the molten pool and forms a sphere, and the metal sphere appears on the surface after manufacturing. This is the reason of balling phenomenon when the laser volume energy density is low in the process of LPBF. Figure 6.23 shows a typical

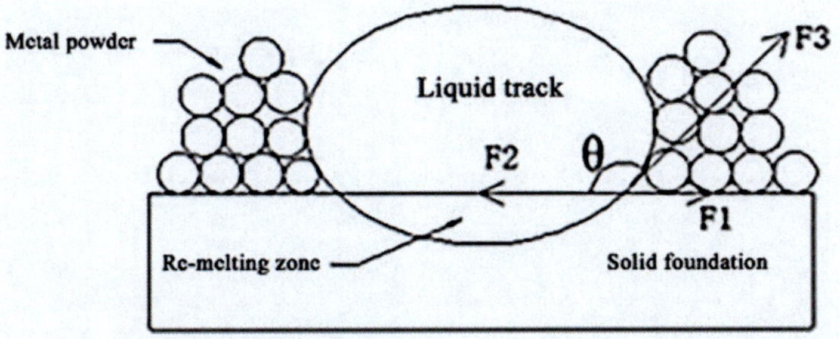

Fig. 6.23 Schematic diagram of balling

sample in the low energy density balling zone, with volume energy density of 79.37 J/mm^3, relative density of 95.53%, and surface roughness *Ra* of 11.24 μm. It can be seen from the figure that the sample has serious balling, discontinuous melting track, poor melting condition, and rough surface without metallic luster. Balling results in the existence of pores between adjacent tracks, which affects the relative density of the LPBFed parts. Balling also affects the smoothness of the next layer, deteriorates the subsequent forming, and greatly increases the surface roughness of the LPBFed parts [2] (Fig. 6.24).

(3) The successfully manufacturing zone.

When the volume energy density 120 J/mm^3 < ω < 160 J/mm^3, the LPBF is in the successfully manufacturing zone. In this area, the laser can melt the metal powders to form continuous smooth tracks, and the depth of the track is suitable, which has better wettability on the solid matrix. The wetting angle is small, which greatly alleviates the trend of balling. The track width is reasonable and the track lap is ideal, thus the relative density is high, which can reach more than 97%, and the surface is smooth. Figure 6.25 shows a typical sample in the successfully manufacturing zone, with volume energy density of 156.25 J/mm^3, relative density of 98.93%, and surface roughness *Ra* of 6.03 μm. It can be seen from the figure that the track of the sample

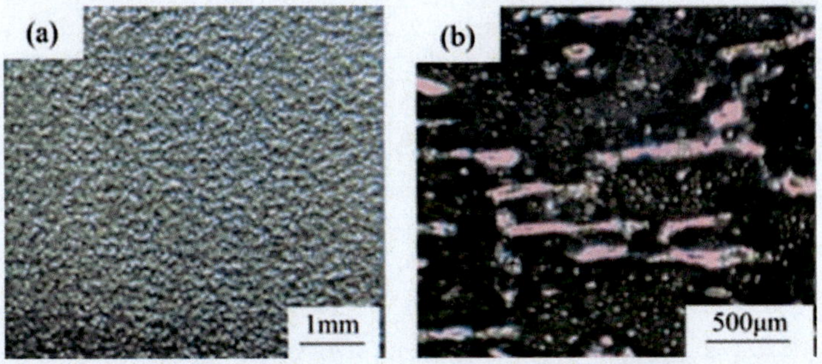

Fig. 6.24 The typical sample of the low energy density balling zone: **a** low magnification micrograph; **b** high magnification micrograph (*P*: 150 W, *V*: 700 mm/s, *s*: 90 μm, *h*: 30 μm)

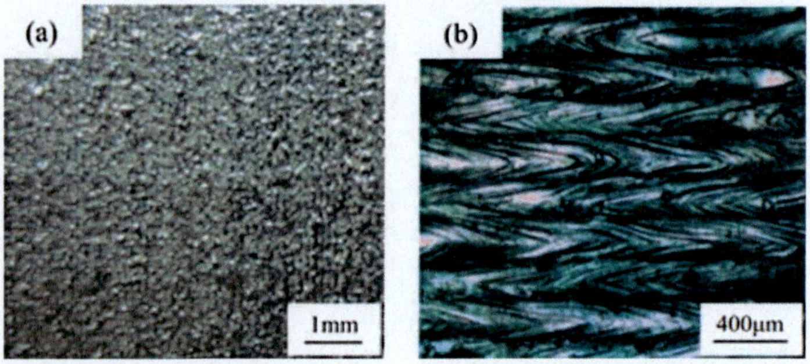

Fig. 6.25 The typical sample of the successfully forming zone: **a** low magnification micrograph; **b** high magnification micrograph (*P*: 150 W, *V*: 400 mm/s, *s*: 80 μm, *h*: 30 μm)

is continuous and smooth, with ideal lap joints, few black bumps, and basically no balling.

(4) The high energy density induced balling zone.

When the volume energy density 160 J/mm^3 < ω < 180 J/mm^3, the LPBF is in the high energy density balling zone. Too low scanning speed, too high laser power or too small hatch space (high overlap rate) will result in too high volume energy density, and balling phenomenon will also occur in this case. The high volume energy density can increase the sintering temperature significantly, and lead to the sharp increase of powder melting amount and the increase of molten pool surface area. On the one hand, the excess liquid phase significantly reduces the melt viscosity. The interfacial tension of the solid–liquid interface is relatively large in the larger molten pool, and the solution is prone to balling. On the other hand, the lower scanning speed makes the liquid phase exist longer, and makes the cooling and solidification time longer. The melt overheat tendency is obvious, and the Marangoni effect is enhanced. Thus, there is more time for balling to occur, which increases the possibility of oxidation. Under these conditions, with the decrease of liquid surface energy, the melt will split into a large number of tiny spheres when the liquid column breaks. Figure 6.26 shows a typical sample in the high energy density balling zone, with volume energy density of 178.57 J/mm^3, relative density of 95.12%, and surface roughness *Ra* of 10.56 μm. It can be seen from the figure that although the molten passage of the sample is continuous and overlapped closely, it is not smooth enough. Moreover, there are raised balls on both sides of the track and balling in some areas.

(5) The overmelting zone.

When the volume energy density ω > 180 J/mm^3, the LPBF is in the overmelting zone. In this manufacturing area, the processing temperature is too high and the amount of powder melting is too much. The liquid melting track flows violently, making the liquid melt spatter seriously. There are serious overmelting phenomena and defects in this area, such as distortion, cracks and depressions. Figure 6.27 shows a typical sample in the overmelting zone, with volume energy density of 208.33 J/mm^3, relative

Fig. 6.26 The typical sample of the high energy density balling zone: **a** low magnification micrograph; **b** high magnification micrograph (*P*: 150 W, *V*: 350 mm/s, *s*: 80 μm, *h*: 30 μm)

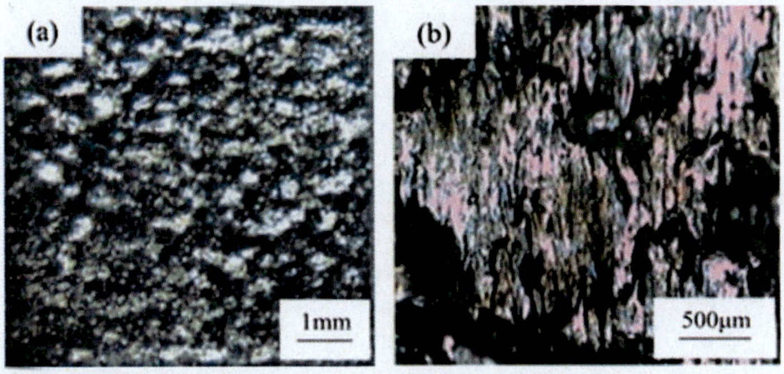

Fig. 6.27 The typical sample of the overmelting zone: **a** low magnification micrograph; **b** high magnification micrograph (*P*: 150 W, *V*: 400 mm/s, *s*: 60 μm, *h*: 30 μm)

density of 93.16%, and surface roughness *Ra* of 24.16 μm. As can be seen from the figure, the surface of the sample is depressed and has large black objects.

In a word, too high or too low volume energy density will cause balling effect, which has adverse effects on the relative density and surface roughness of LPBFed parts. Therefore, in order to eliminate the balling as much as possible and ensure the quality of LPBFed parts, the volume energy density should be controlled in the successfully manufacturing zone. At the same time, the process parameters should also be in a reasonable range.

6.2.2 *Analysis of Characteristics and Roughness of Side Surface*

It can be seen from the theoretical research results of the side surface roughness of LPBF formed parts that the side surface roughness is theoretically affected by the powder layer thickness and the inclination angle. However, from the experimental study, it is found that the scanning speed and laser volume energy density are also important factors affecting the side surface roughness. Therefore, in order to reduce the interference factors, by controlling the laser power and scanning speed and other process parameters, the laser volume energy density is controlled within the range of successfully manufacturing zone. The influence of the powder layer thickness and inclination angle on the side surface roughness is mainly studied.

In the research of side surface roughness, the forming overhang mechanism is needed. However, there is a ultimate forming angle when forming the overhang structure. When the inclination angle is less than the ultimate forming angle, it cannot be smoothly formed, and the overhanging surface will produce serious slag scraping or warping defects, or even collapse, which cannot continue processing. In this case, support must be added to the overhanging surface to prevent forming defects. When the inclination angle is greater than the ultimate forming angle, the overhanging surface can be formed smoothly without adding support. Under different process parameters, there are different ultimate forming angles in forming drape structures. By increasing the scanning speed to reduce the laser energy density, the ultimate forming angle of the forming overhang mechanism will also decrease. Therefore, the laser power was set as 150 W, the scanning speed was 600 mm/s, and the hatch space was 80 μm. The experiments on the influence of the powder layer thickness and the inclination angle on the side surface roughness were carried out. Under these parameters, the ultimate forming angle is 27°.

Because this chapter studies the influencing factors of the roughness of the side sloping surface without adding support, it aims at the influence of the inclination angle on the roughness of the side sloping surface. Under the conditions of inclination angles of 30°, 35°, 40°, 45°, 50°, 60°, 70°, 80° and 90°, the inclined block was manufactured. The side surface roughness was measured, and the influence of inclination angle on the side surface roughness was analyzed. In order to study the influence of powder layer thickness on the surface roughness of the side, the corresponding inclined blocks were fabricated under the thickness of 25 μm, 30 μm and 35 μm respectively. The side surface roughness was measured and the influence of powder layer thickness on the surface roughness was analyzed. The processing results of partial inclined blocks are shown in Fig. 6.28.

The formed regular inclined block was cut from the substrate, and the surface macro topography and microscopic magnification images of the side surface were taken. The surface roughness of the upper side and the lower side was measured respectively. The influence of inclination angle and powder layer thickness on the surface roughness of the side was shown in Figs. 6.29 and 6.30.

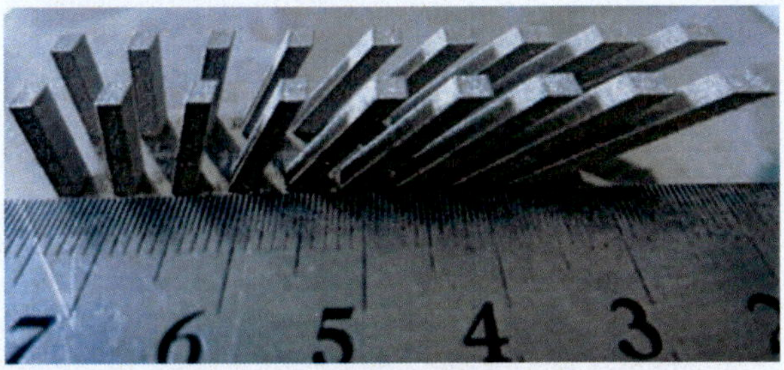

Fig. 6.28 The processed inclined blocks with different angles (powder layer thickness of 25 μm)

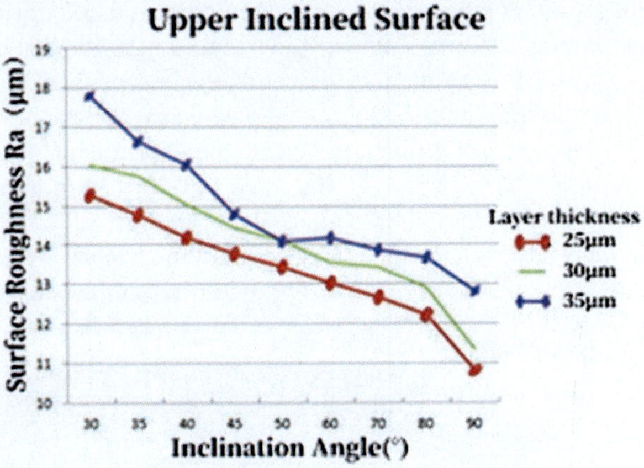

Fig. 6.29 Effect of inclination angle and powder layer thickness on the roughness of upper inclined surface

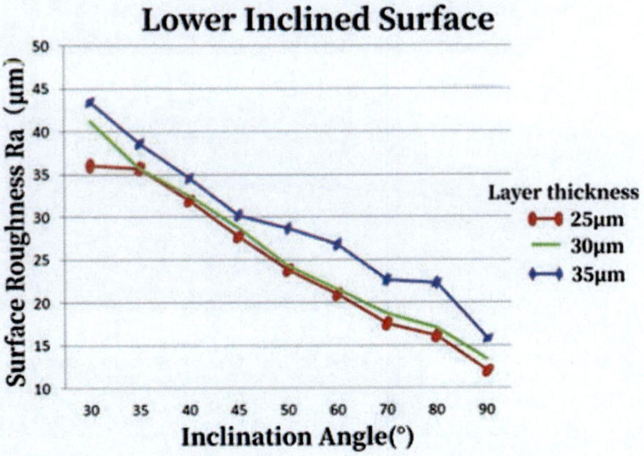

Fig. 6.30 Effect of inclination angle and powder layer thickness on the roughness of lower inclined surface

It can be seen from Figs. 6.29 and 6.30 that the surface roughness of both the upper and lower inclined sides tend to gradually decrease with the increase of the inclination angle in the range higher than the ultimate forming angle. This is also consistent with the results of the previous theoretical analysis of side surface roughness, which proves that the theoretical analysis process is correct. Moreover, when the powder layer thickness is 25 μm, three inclined blocks with inclination angles of 35°, 45° and 70° are selected. The surface morphology of the upper inclined surface is photographed, as shown in Fig. 6.31. When the inclination angle is 35°, the side surface of the inclined part is very rough, showing small intervals of concave and convex ups and downs, with obvious ladder shape. When the inclination angle is 45°, the side surface of the inclined part is still rough, showing a concave and convex shape, but the stepped surface shape is significantly improved compared with that of the 35° inclined block. When the inclination angle increased to 70°, the lateral surface is smooth. Although there are also bumps and downs, the surface don't have the ladder shape. It can also be concluded that in order to improve the roughness of the side surface of the LPBFed parts, the position of the parts should be adjusted during the manufacturing process, and the inclination angle of the forming parts should be improved as far as possible.

According to the theoretical analysis on side surface roughness in the previous part, the side surface roughness increases when the powder layer thickness increases. This conclusion is also proved by the experiment. Here, the surface morphology of the upper side of the 30° inclined part was photographed under the powder layer thickness of 25 and 35 μm, as shown in Fig. 6.32. The surfaces of the two inclined blocks have obvious ladder shape, but the lateral surface of the inclined block with a layer thickness of 35 μm is significantly rougher than that of the inclined block with

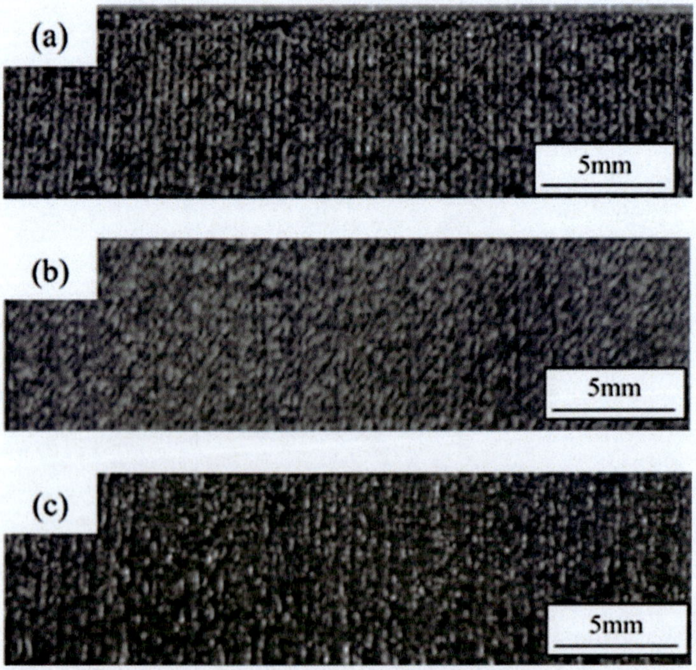

Fig. 6.31 The macroscopic upper side surface morphology in different inclination angles: **a** inclination angles of 35°; **b** inclination angles of 45°; **c** inclination angles of 70° (powder layer thickness of 25 μm)

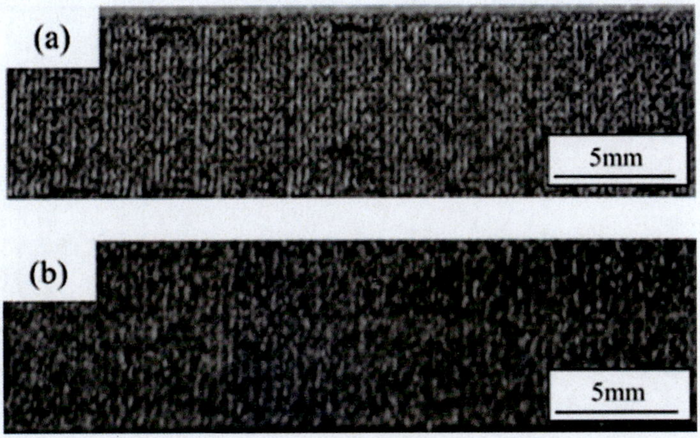

Fig. 6.32 The macroscopic upper side surface morphology in different thickness of powder layer: **a** powder layer thickness of 25 μm; **b** powder layer thickness of 35 μm (inclination angle of 30°)

a layer thickness of 25 μm. The edge of the melting track is exposed to the edge of the side surface, on which a series of distinct raised lines are formed. The convex lines on the lateral surface of the inclined block with a thickness of 25 μm are thin and continuous, and the convex lines are closely connected. The convex lines on the lateral surface of the inclined block with a thickness of 35 μm are discontinuous, and the distance between the convex lines is wide. Therefore, in order to improve the roughness of the side surface of the LPBFed parts, the thickness of the powder layer should be kept as small as possible while ensuring the forming efficiency.

In addition, it can be found from Figs. 6.29 and 6.30 that for the same inclined block, the surface roughness value of the lower side is much larger than that of the upper side. Figure 6.33 shows the macroscopic surface topography of the upper and lower surfaces of a 45° inclined block with powder layer thickness of 25 μm. It can be seen from Fig. 6.33 that the upper surface is smooth, with continuous convex lines and tight lap joints, and there is no hanging slag phenomenon on the surface. But the surface of the lower side is rough and granular, there is obvious powder adhesion phenomenon without continuous convex line.

The reason why the lower side of inclined block is rougher than the upper side is analyzed. As shown in Fig. 6.34, when the laser beam scans the solid support area (point a), the heat conductivity is high. When the laser beam is scanned to the edge of the lower side of the inclined block, the laser beam will incident on the powder support area (point b), at which time the heat conductivity is only 1/100 of the corresponding solid material. Therefore, under the same conditions of laser processing parameters, the energy input in the powder support area is much larger than that in the solid support area, resulting in a large molten pool in the powder support area. The molten pool sinks into the powder due to the action of gravity and surface stress. In this case, the laser deep penetration effect will occur, and more powders will adhere to the molten pool. These reasons lead to the appearance of

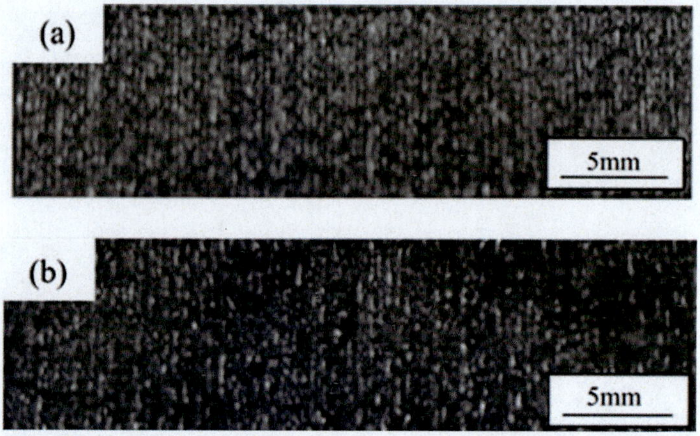

Fig. 6.33 The macroscopic surface morphology on upper side surface and lower side surface: **a** upper side; **b** lower side (powder layer thickness of 25 μm; inclination angle of 45°)

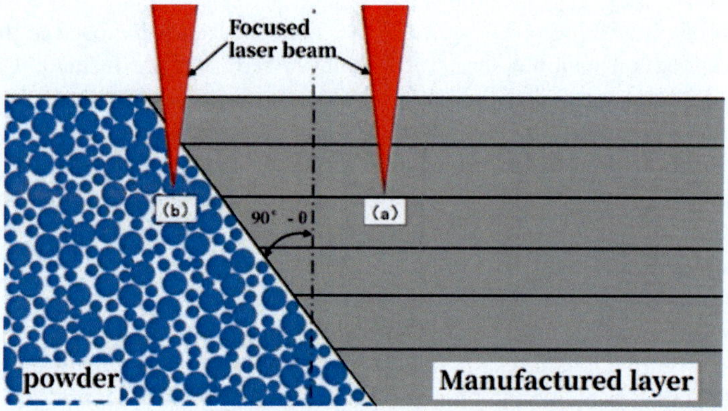

Fig. 6.34 Schematic diagram of overhanging surface scanning by laser beam

overhang when forming the lower side of the inclined block, which deteriorates the surface quality and dimensional accuracy of the lower side. And the smaller the inclination angle, the more obvious the laser deep penetration effect. The upper side of the inclined block is not affected by laser deep penetration. Therefore, under the same process conditions, the lower side of the inclined block is usually rougher than the upper side. In order to improve the surface roughness of the lower side of the inclined block, the influence of laser deep penetration on the lower side should be reduced as much as possible. The effect of laser deep penetration can be alleviated by increasing the laser scanning speed.

Table 6.4 The number and scanning strategy of each block

Block 1 Z-shaped X direction scan	Block 2 S-shaped X direction scan	Block 3 Z-shaped orthogonal scan	Block 4 S-shaped orthogonal scan
Block 5 Z-shaped X direction interlayer interleaved scan	Block 6 S-shaped X direction interlayer interleaved scan	Block 7 Z-shaped orthogonal interlayer interleaved scan	Block 8 S-shaped orthogonal interlayer interleaved scan
Block 9 S-shaped orthogonal interlayer interleaved scan with before raster	Block 10 S-shaped orthogonal interlayer interleaved scan with after raster	Block 11 S-shaped orthogonal interlayer interleaved scan with before raster in higher scanning speed	Block 12 S-shaped orthogonal interlayer interleaved scan with after raster in higher scanning speed
Block 13 Contour offset scan	Block 14 Interlayer interleaved contour offset scan	Block 15 Partition scan	Block 16 Partition scan with after raster

6.2.3 *Effect of Scanning Strategy on Surface Characteristics and Roughness of Forming Parts*

In the LPBF process, the laser beam scans the manufacturing area of each layer section, and fills the powders in the forming area to form a part with a thick powder layer. Therefore, the scanning strategy of laser beam has a great impact on the forming quality. Different scanning strategies have a direct impact on the dimensional accuracy and surface roughness of LPBFed parts. When forming the upper surface, different scanning strategies affect the relative density and surface roughness of scanning layer. When forming side surface, the scanning strategy is closely related to the surface roughness and dimensional accuracy of parts. Therefore, the scanning strategy has an effect on the roughness of both the upper surface and the side surface.

In order to study the influence of different scanning strategies on the surface roughness of the forming parts, 16 small blocks of 10 mm × 10 mm × 5 mm were processed with different scanning strategies. The influence of scanning strategies was analyzed by the surface quality of the forming blocks. The number of blocks and their scanning strategies are listed in Table 6.4.

A schematic of the various scanning strategies are shown in Fig. 6.35.

The 16 blocks in Table 6.4 were formed using laser power of 150 W, powder layer thickness of 30 μm, hatch space of 80 μm and scanning speed of 400 mm/s. Among them, the raster scanning speed of Block 11 and Block 12 is 600 mm/s, and that of Block 13 and Block 14 is 450 mm/s. The forming results of the blocks are shown in Fig. 6.36.

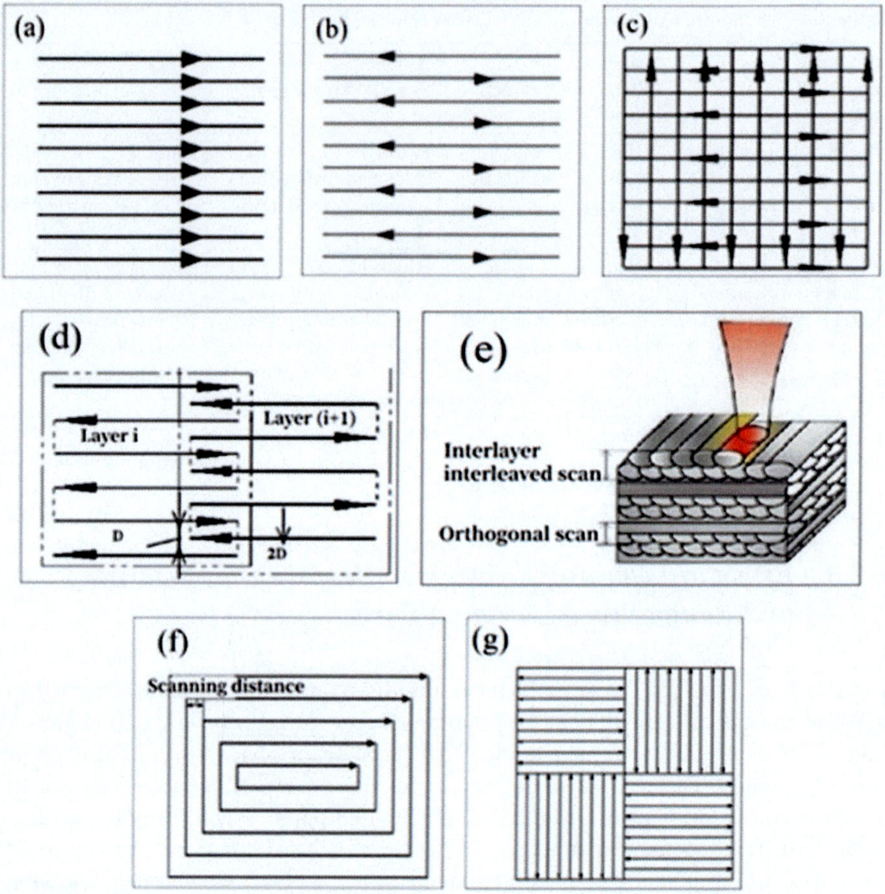

Fig. 6.35 Schematic diagram of different scanning strategies: **a** Z-shaped single direction scan; **b** S-shaped round-trip scan; **c** orthogonal scan; **d** interlayer interleaved scan; **e** orthogonal interlayer interleaved scan; **f** contour offset scan; **g** partition scan

1. Z-shaped single direction scan and S-shaped round-trip scan

As shown in Fig. 6.37, by comparing the macroscopic morphology of the surface of the two blocks, it can be found that there is an obvious raised line at the starting point of the scanning line on the left edge of the Z-shaped single direction scanning part. This is because of the high energy density at the beginning of the scanning line, which will quickly melt the powder at the beginning of the scanning line. As a result, a large temperature difference is formed between the liquid tracks in this area, and the viscosity of the liquid tracks increases. Before they have time to spread out, these melting tracks shrink into small balls under the action of surface tension. Since each scanning line starts on the left, a series of edge balls accumulate to form the raised line

Fig. 6.36 The macroscopic morphology of blocks in different scanning strategies

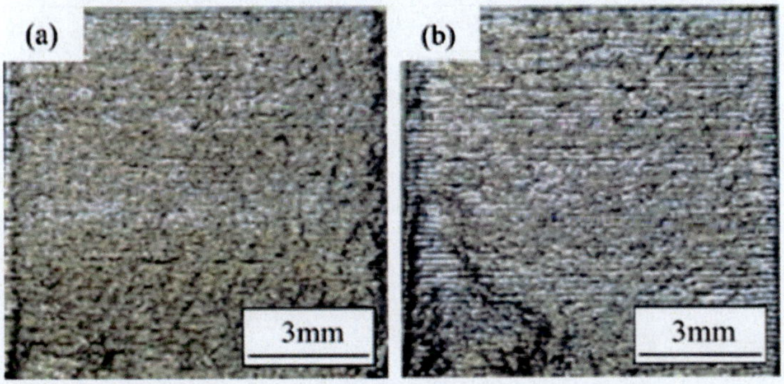

Fig. 6.37 The macro morphology comparison between Z-shaped single direction scan and S-shaped round-trip scan: **a** Block 1—Z-shaped single direction scan; **b** Block 2—S-shaped round-trip scan

shown here. However, the scan trajectory of S-shaped round-trip scan is relatively continuous. After scanning a line, a scan interval is offset and the scan is returned. In this case, the next scanning line can make use of the molten pool and heat of the previous one, thus there is no obvious edge raised line on the surface of the block. Compared with Z-shaped scan, S-shaped round-trip scan has better surface quality. S-shaped scan can make the starting point of the track more continuous, reducing the starting point bump effect and alleviating the convex edge of the forming part.

2. Orthogonal scan and interlayer interleaved scan

As shown in the schematic diagram of scanning strategy (Fig. 6.35), orthogonal scan means that when the previous layer scans along the X direction, the next layer scans along the Y direction. The scanning direction between adjacent layers is 90° orthogonally. Interlayer interleaved scan means that, some spacing is staggered to scan after scanning a layer, so that the lower scanning line falls between the two upper scanning lines. The orthogonal interlayer interleaved scan is to combine the interlayer interleaved scan and the orthogonal scan. Two layers of interleaved scanning are carried out first, and then the other two layers of interleaved scanning are carried out in the orthogonal direction, so that the cycle accumulates.

In order to study the advantages of interlayer interleaved scanning and orthogonal scanning, the surface morphologies of parts corresponding to four different scanning strategies were compared, and the comparison results are shown in Fig. 6.38.

The surface roughness of the upper surface of the four parts in Fig. 6.38 was measured, and the column shape comparison diagram was drawn, as shown in Fig. 6.39.

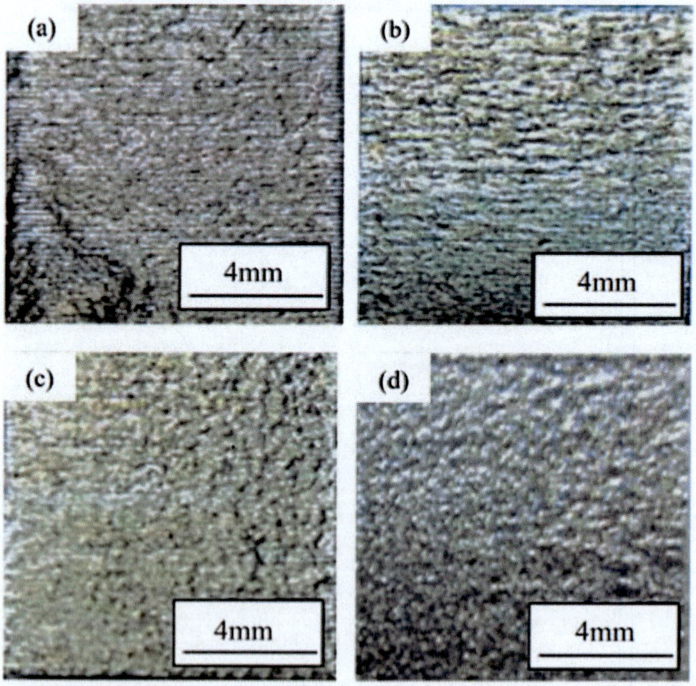

Fig. 6.38 Surface morphology of parts in four different scanning strategies: **a** S-shaped X direction scanning without interlayer interleaved; **b** S-shaped orthogonal scanning without interlayer interleaved; **c** S-shaped X direction scanning with interlayer interleaved; **d** S-shaped orthogonal scanning with interlayer interleaved

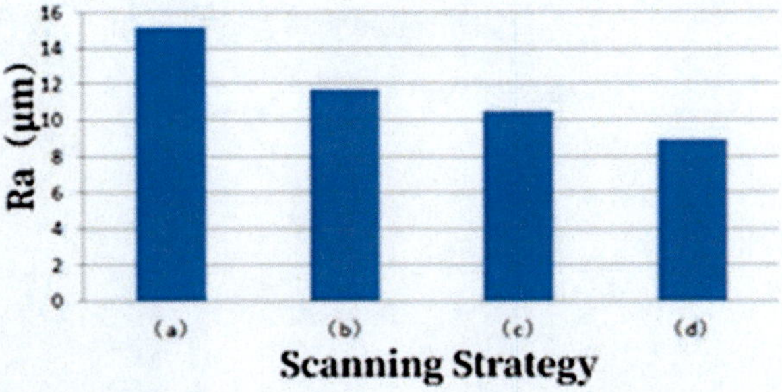

Fig. 6.39 The *Ra* variation in different scanning strategies

By comparing Fig. 6.38a and b, it can be found that the surface of parts scanned in orthogonal direction is relatively flat compared with the surface topography of parts scanned in single direction, and there is basically no bump along the edges. This is because the S-shaped orthogonal scan can not only eliminate the bump effect along the edge, but also reduce the ripple effect of the surface melting track during the orthogonal scanning and repair the surface defects of the previous layer. By comparing Fig. 6.38c and d, it can be found that the surface quality is significantly improved. The surface roughness is greatly reduced, and the surface is more compact and brighter after using layer interleaved scanning. Because the staggered scanning strategy between layers can repair the forming defects of the previous layer. The liquid metal is easy to wet the matrix at the gullies of the previous layer, so as to make the lap between layers closer and alleviate the ripple topography. Therefore, interlayer interleaved scanning between layers can make the parts more compact and improve the surface quality. The part shown in Fig. 6.38d adopts the S-shaped orthogonal interlayer interleaved scanning strategy, and its surface is the flattest and smoothest among the four parts. Therefore, the combination of orthogonal scan and interlayer interleaved scan can make use of the advantages of the two scanning strategies to further improve the surface roughness.

3. Raster operation

The surface morphology of the formed parts under different raster operations was compared (Fig. 6.40) to study the effect of raster operations on the surface roughness of the parts. It can be seen from Fig. 6.40 that the raised defects along the edge of Block 9 with before raster operation along the edge are very serious. And the area near the edge is depressed. This is because the before raster operation will make the edge track absorb a lot of powder after laying a layer of powder, and suck away the powder around the edge. With the accumulation of layers, the bulge along the edge is serious, and the area near the edge is depressed. The surface quality of Block 10 with after raster operation in the same scanning speed is better than that of Block 9. There is a slight bulge on the edge, but it is not serious. The surface quality of Block 11 with before raster in higher scanning speed is better than that of Block 9, but there is still a bulge on the edge. The surface quality of Block 12 with after raster operation in higher scanning speed is very good, and there is basically no edge bulge. It can be seen that the unreasonable raster operation cannot improve the surface quality of the

Fig. 6.40 Surface morphology of parts formed by different raster: **a** Block 9—S-shaped orthogonal interlayer interleaved scan with before raster operation; **b** Block 10—S-shaped orthogonal interlayer interleaved scan with after raster operation; **c** Block 11—S-shaped orthogonal interlayer interleaved scan with before raster in higher scanning speed; **d** Block 12—S-shaped orthogonal interlayer interleaved scan with after raster in higher scanning speed

edge of the parts, and even make the surface quality of the parts worse. The before raster operation will aggravate the bulge of the edge of the part, and the after raster operation is equivalent to the re-melting operation of the contour of the part, which can improve the quality of the edge of the part. Therefore, after raster operation in higher scanning speed is the reasonable operation of raster.

4. Contour offset scan

Figure 6.41 shows the part surface macroscopic topography obtained by contour offset scanning. It can be seen from the figure that the surface of the parts formed by contour offset scanning has overmelting phenomenon, and the bulge along the edge is more serious. The reason for the poor morphology is that the scanning line of contour offset scanning is too long, and the heat accumulates too much, which is easy to cause the overmelting phenomenon. Using contour offset scanning, we can get a nearly ideal part profile and a relatively flat surface. However, the length of the scanning line and the amount of heat accumulation are easy to cause edge warping and overmelting. Thus, the appropriate energy density and hatch space should be

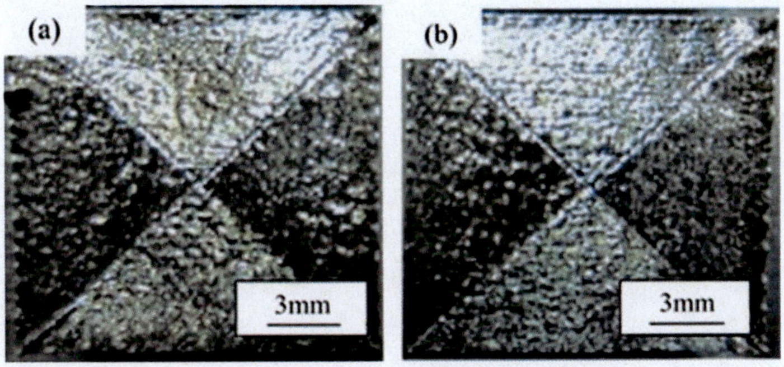

Fig. 6.41 The macro surface morphology of contour offset scanning part: **a** Block 13—contour offset scan; **b** Block 14—interlayer interleaved contour offset scan

selected. Moreover, the scanning speed and the hatch space should be increased when the contour offset scanning is carried out.

5. Partition scan

Partition scan is the most effective method to reduce the thermal deformation. The larger the length of the scanning line, the greater the thermal stress, which is easier to cause warping. The length of scanning line can be reduced by using partition scanning, and the warping of LPBFed parts can be effectively reduced. However, the overlap between different areas is a difficult problem in the partition scan, which is easy to produce the phenomenon of weak connection between different areas. Figure 6.42 shows the macro topography of the part surface obtained by the partition scanning. As can be seen from the figure, the surface quality in the small area of the part is very good. However, the connection part between Block 15 area is rough, with an obvious indented boundary. After raster operation is adopted in Block 16, and the roughness of the area connection is improved. Therefore, when using partition scan, the amount of spot compensation should be reasonably set, and the contour lines of each area should adopt after raster operation.

6.3 Measures to Improve the Surface Roughness of Parts

Many applications have strict requirements on the surface roughness of parts, because the surface roughness of each individual part has a non-negligible impact on the whole assembly. The surface roughness will affect the matching properties between parts, thus affecting the sealing and wear resistance of the assembly, as well as the performance and service life of the assembly. Therefore, this section puts forward measures to improve the surface roughness of parts from the perspective of process, data processing, post-processing and process parameter optimization.

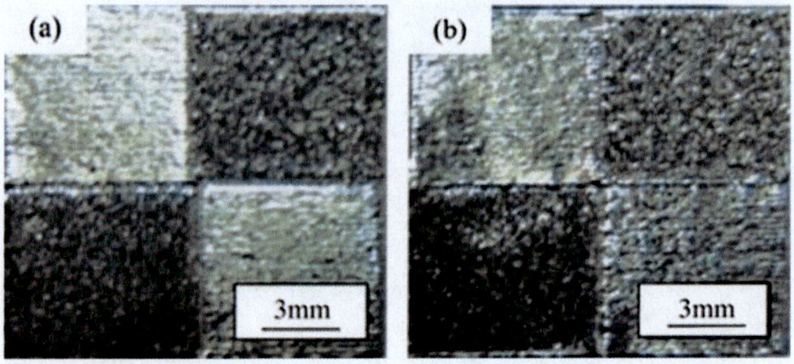

Fig. 6.42 The macro surface morphology of the partition scanning part: **a** Block 15—Partition scan; **b** Block 16—Partition scan with after raster

6.3.1 Laser Surface Re-melting

Laser surface re-melting is an important method to improve surface roughness in LPBF process [3–5], and its principle is shown in Fig. 6.43. After scanning a layer by using laser, before placing a new powder layer, the same surface would be scanned again by laser offset at a certain distance to make it remelted and smooth, meanwhile removing some defects. However, if each layer is remelted in the process of LPBF, the forming time will inevitably increase. Furthermore, this method can be used when the relative density of 98–99% still cannot meet the requirements. The process of re-melting each layer can greatly reduce the internal micropores and improve the mechanical properties of the forming parts [6]. Laser surface re-melting can also be applied only to the last layer or to the outer surface of the part if only with the purpose of improving the surface quality of the part.

In order to study the improvement effect of laser surface re-melting on the surface quality of LPBFed parts, three small blocks of 10 mm × 10 mm × 5 mm were prepared. The laser power was 150 W, the layer thickness was 80 μm, the scanning speed was 600 mm/s, and the scanning strategy was ordinary raster scanning. The re-melting method and the prepared sample are shown in Fig. 6.44. The samples shown in Fig. 6.44a were not remelted, while the samples shown in Fig. 6.44b, c adopted different re-melting strategies.

The surface roughness and surface profile were measured with a roughness measuring instrument, and the comparison map shown in Fig. 6.45 was obtained. As

Fig. 6.43 The schematic of laser surface re-melting

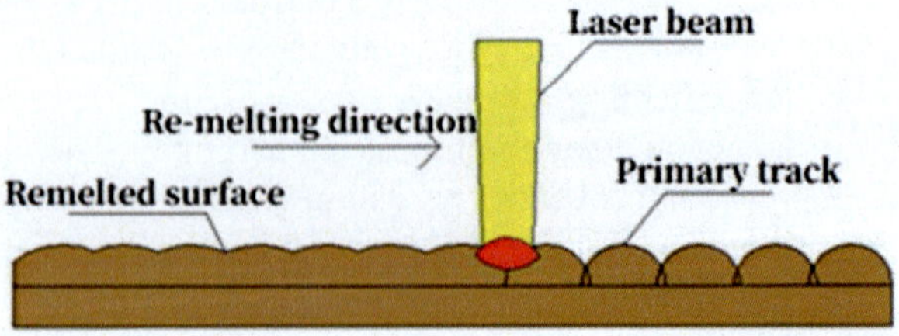

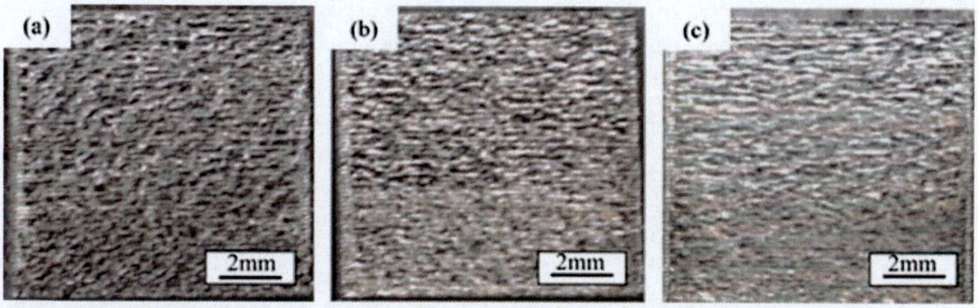

Fig. 6.44 Surface topography of 3 blocks: **a** non re-melting; **b** surface orthogonal re-melting 2 times; **c** surface orthogonal re-melting 5 times

can be seen from Fig. 6.45, the surface of the block without re-melting is uneven, with severe surface undulations. The distance between the peak and trough is large, and the surface microscopic contour is extremely irregular. The surface contour of the block remelted 2 times by the laser becomes smooth, the peak height and peak valley distance in the contour element are reduced, and the contour is more regular. The surface profile of the block remelted 5 times by laser will become smoother, without steep peaks and troughs, and the contour is very regular. Among them, the surface roughness (*Ra*) of the block without re-melting is 14.33 μm, and the surface roughness (*Ra*) of the block with laser re-melting for 5 times is 3.34 μm. The surface roughness decreases greatly.

By comparing the surface of the above three blocks, it is found that laser surface re-melting can obviously improve the surface roughness of LPBFed parts. And the surface roughness decreases with the increase of re-melting times. It should be noted that the increase of re-melting time also means the increase of manufacturing time, thus appropriate re-melting strategies should be adopted according to the actual needs to improve the surface roughness of parts.

6.3.2 Data Processing

After the parts are designed in the 3D software, they are saved as STL files, which need to be processed before processing. Data processing is very important for the smooth machining of parts, especially for the parts with complex structure. Good data processing is beneficial to the successfully manufacturing of parts. It can not only improve the manufacturing quality of parts, but also improve the surface roughness of parts. The process of data processing includes determining the placement of parts, adding supports for overhanging surfaces, etc.

The position of the parts has a significant effect on the surface roughness of the parts with complex shapes. Taking the transverse cylinder manufacturing experiment shown in Fig. 6.46a as an example, two groups of transverse cylinders were formed (group A and group B). The placement angles of the two groups were different. The

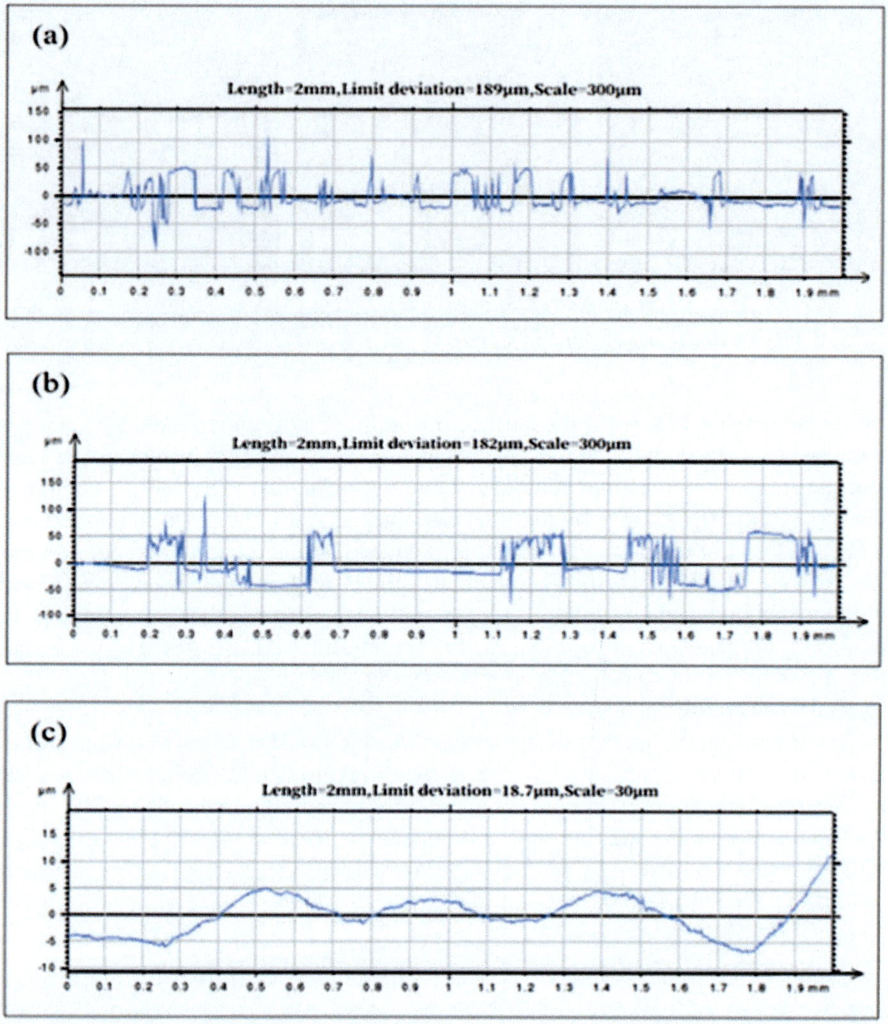

Fig. 6.45 Surface morphology of 3 blocks: **a** non re-melting; **b** surface orthogonal re-melting 2 times; **c** surface orthogonal re-melting 5 times

included angle between the curved generatrix of group A transverse cylinder and X axis was 90°, and that between the curved generatrix of group B transverse cylinder and X axis was 45°. The surface roughness of the inner and outer curved surfaces of the two groups of transverse cylinders was measured at different test angles. The schematic diagram of the test angle is shown in Fig. 6.46b, and the measurement result is shown in Fig. 6.47.

It can be seen from Fig. 6.47 that the surface roughness curves of group A and group B are obviously different regardless of the outer curved surface or the inner

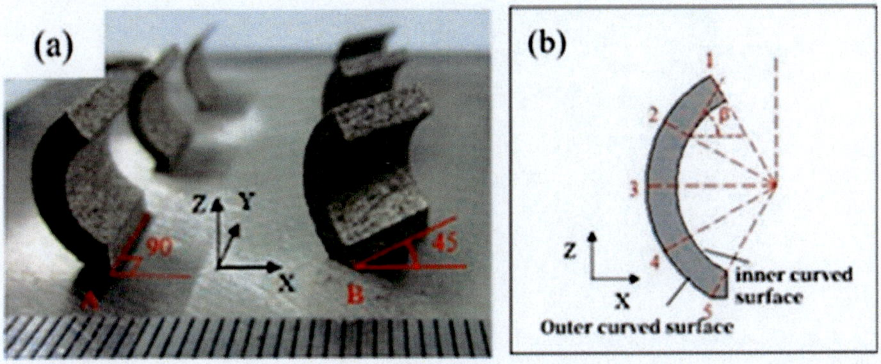

Fig. 6.46 Transverse cylinder: **a** the forming results; **b** schematic diagram of test angle

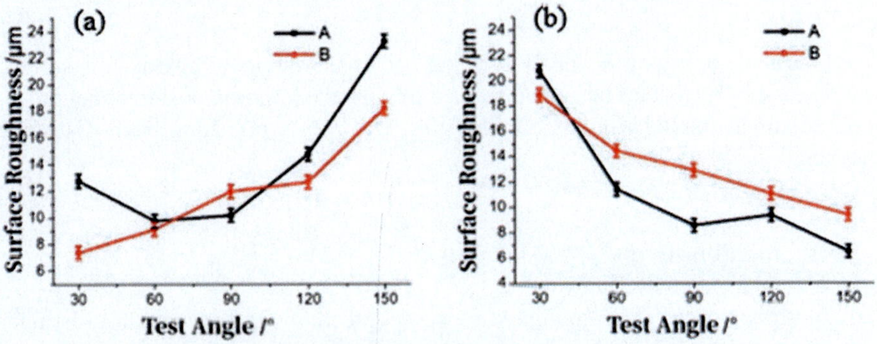

Fig. 6.47 Effect of placement angle on surface roughness of transverse cylinder: **a** outer curved surface; **b** inner curved surface

curved surface. This indicates that changing the placement angle of the parts will have a significant impact on the surface roughness of the parts. Moreover, the surface roughness of different test angles of the same curve is also significantly different, indicating that the inclination angle of the part surface has an important influence on the surface roughness, which is verified by the previous theoretical analysis.

Therefore, the surface roughness of the parts can be improved by adjusting the space position of the parts reasonably. In addition, setting appropriate support for the hanging surface of the parts can prevent warping, balling, hanging slag and other problems in the process of processing, so as to prevent the surface of the parts from showing poor surface quality due to defects, and obviously improve the surface roughness of the parts [7].

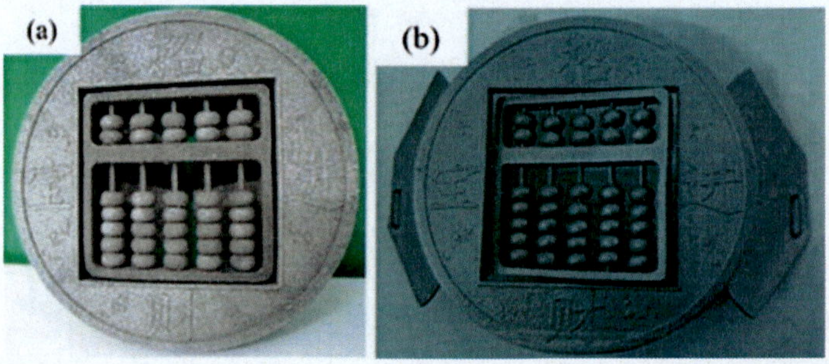

Fig. 6.48 Comparison of LPBFed parts before and after sandblasting: **a** before sandblasting; **b** after sandblasting

6.3.3 *Post Treatment*

If the surface quality of the LPBF forming parts has high requirements, the surface roughness can be further improved by post treatment. Common post treatment techniques include manual grinding, sandblasting, electrolytic polishing and hot isostatic pressing.

6.3.3.1 Sandblasting

Sandblasting is a very common post processing technology for parts. It uses high-pressure air to form high-speed jet beam to spray the material onto the surface of the treated part, and improves the cleanliness and roughness of the surface of parts through the impact and cutting effect of the abrasive on the surface. Sandblasting is a general, rapid and efficient cleaning method, and can be arbitrarily selected in terms of different roughness.

Figure 6.48 shows the comparison of the LPBFed parts before and after sandblasting. It can be seen that the surface quality of copper money abacus has been significantly improved after sandblasting treatment. After sandblasting, the surface of the parts is brighter, exhibiting metallic luster. And the surface is smoother, without the form of scattered sand. It can be seen that sandblasting has a good improvement effect on the surface quality of LPBFed parts, which can remove the oxide layer on the surface of parts and the adhesive powder.

6.3.3.2 Electrolytic Polishing

In the process of electrolytic polishing, the part is usually located at the anode, and the cathode is usually a lead plate. After electrification, the electrolytic solution will dissolve the bulge in the anode part. A slime layer will appear on the surface of the part to fill the concave part on the surface of the part, so that the part becomes smooth

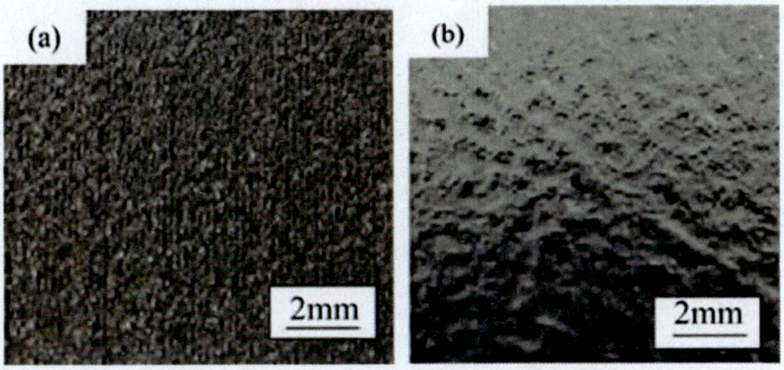

Fig. 6.49 Comparison of LPBFed parts before and after electrolytic polishing: **a** before electrolytic polishing; **b** after electrolytic polishing

and bright. Electrolytic polishing has the advantages of high production efficiency, low equipment investment, continuous use of electrolyte, and lower processing cost than mechanical polishing.

Figure 6.49 shows the surface morphology before and after electrolytic polishing. In the process of electrolytic polishing experiment, YQ series high-frequency switch electrolytic polishing equipment is used. The electrolytic polishing liquid is 316L stainless steel special electrolyte. In the process of electrolytic polishing, the polishing mode of constant voltage and variable current is adopted, in which the voltage is set as 8 V, the polishing time is 7 min, and the distance between cathode and anode is 50 mm.

It can be seen from Fig. 6.49 that the surface of the parts before polishing is rough and presents a thin linear shape. After the electrolytic polishing, the surface quality of the parts is obviously improved. After treatment, the surface of the formed part becomes very smooth, exhibiting a bright metallic luster, and the original thin linear shape disappears.

The surface roughness measurement results before and after electrolytic polishing of LPBFed parts are listed in Table 6.5. After the electropolishing treatment, the roughness of the part is greatly reduced. *Ra* and *Rz* are respectively reduced by 311.97 and 233.60%. The *Ra* of the part after electrolytic polishing is 2.34 μm. The surface quality can reach the level of ordinary machining.

Table 6.5 Results of surface roughness measurement before and after electrolytic polishing

Processing of forming parts	*Ra* (μm)	*Rz* (μm)
Before electrolytic polishing	9.64	24.72
After electrolytic polishing	2.34	7.41
Roughness reduction effect (%)	311.97	233.60

References

1. Liu R (2014) The study on surface roughness of metal parts fabricated by selective laser melting and the application on non-assembly mechanisms. South China University of Technology, Guangzhou
2. Wu W, Yang Y, Wang D (2010) Balling phenomenon in selective laser melting process. J South China Univ Technol (Nat Sci Ed) 05:110–115
3. Yasa E, Kruth JP (2011) Application of laser re-melting on SLM parts. Adv Prod Eng Manag 6(4):259–270
4. Yasa E, Kruth JP, Deckers J (2011) Manufacturing by combining selective laser melting and selective laser erosion/laser re-melting. Manuf Technol 60:263–266
5. Yasa E, Kruth JP (2011) Microstructural investigation of selective laser melting 316L stainless steel parts exposed to laser re-melting. Procedia Eng 19:389–395
6. Lu J (2011) Design optimization and process study on directly manufacturing of customized precise metal parts by selective laser melting. South China University of Technology, Guangzhou
7. Mai S (2016) Study on the forming processes and properties of customized CoCr alloy crowns and fixed bridges manufactured by selective laser melting. South China University of Technology, Guangzhou

Chapter 7
Technology of Quality Detection and Feedback in Laser Powder Bed Fusion Process

As the laser powder bed fusion (LPBF) processing process involves more than 130 processing parameters, the processing process and manufacturing quality have strong instability and relatively poor repeatability. The research on the quality detection and feedback technology for the LPBF process is the key content to explore the manufacturing and defect mechanism, as well as the inevitable stage to realize the quality control and backtracking of the LPBF process and improve the LPBFed part quality and level.

At present, the mainstream LPBF equipments still adopts the open-loop control mode, and they are still unable to achieve high-level quality detection and backtracking in the manufacturing process. The research and application of LPBF processing detection on improving the part precision and quality are still at the initial stage. Furthermore, the general technique is still offline detection on process parameter testing and optimization.

The detection technologies for the LPBF process can be divided into online detection and offline detection technology. Online detection has high real-time performance, which can feed back information to the control system in time. The light, heat, sound, electricity and vibration signals accompanying the LPBF processing are mainly detected. Offline detection usually has high accuracy, which is convenient for comprehensive quality detection. It is an irreplaceable supplement for online detection. The main detection object is the surface and internal defect of the part, such as surface roughness, porosity, tensile property and other performance.

© National Defense Industry Press 2024

D. Wang et al., *Laser Powder Bed Fusion of Additive Manufacturing Technology*, Additive Manufacturing Technology, https://doi.org/10.1007/978-981-99-5513-8_7

7.1 Classification of Quality Detection and Feedback Technology

7.1.1 Online Detection

According to the characteristics of LPBF technology, the mainstream online detection technologies can be classified into a light signal, temperature signal, acoustic signal and ultrasonic signal.

7.1.1.1 Optical Signal

1. Molten pool detection

With the intense heat transfer process in the interaction between laser beam and powders, photothermal radiation signal has been widely used in online detection of laser processing. In the LPBF process, the metal powder undergoes temperature rise, melting and vaporization under the action of the laser beam to form a micro-molten pool structure. The molten liquid and metal vapor of the high-temperature molten pool radiate high-brightness light signals. The researchers studied the thermal behavior of the molten pool and explored the processing technology by detecting the local light intensity of the molten pool.

The light intensity detection system based on photodiodes can convert the collected light intensity into current signals, establish the mapping mechanism between radiated light signals and processing technology by an optical system and signal processing system, and obtain the influence mechanism of different process parameters on the thermal radiation behavior of the molten pool to optimize the LPBF process. Figure 7.1 shows a solution for detecting the light intensity of the molten pool based on photodiodes [1].

In addition to the detection methods for the light intensity signal of the molten pool, high-speed imaging technology can capture information about the molten pool, spatter particles and plasma plume by its sensitivity to near-infrared and visible light bands. Through the designed optical system and digital image processing scheme, processing information such as the shape and size of the molten pool, spatter particle size, and spatter behavior can be obtained. Compared with the detection of light intensity information of molten pool, the coaxial or non-coaxial detection system can not only obtain more complete processing information but also have a strong anti-interference ability. Figure 7.2 shows the structure of the coaxial LPBF process detection system based on infrared and visible light [2].

2. Appearance detection of the formed layer

As the LPBF belongs to the laminated manufacturing technology, powder spreading quality directly affects the final manufacturing quality of the parts. Inappropriate processing parameters will result in the defects of the track and the overlap defects

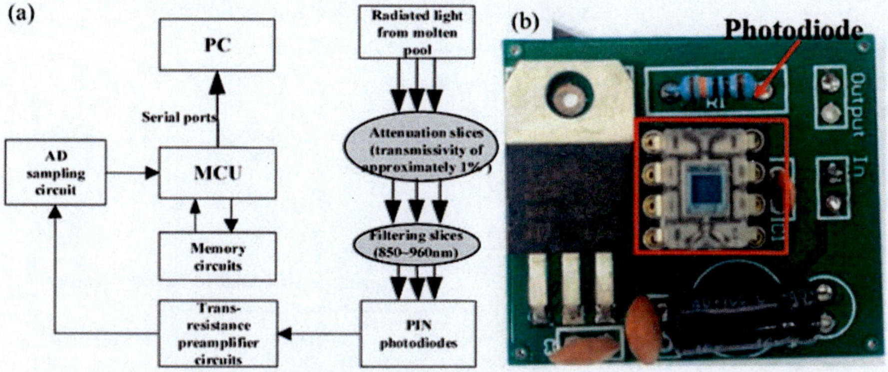

Fig. 7.1 **a** The principle of the molten pool data acquisition system with photodiodes; **b** the circuit board of the photodiode monitoring system. Reprinted with permission from a work by Zhang et al. [1]

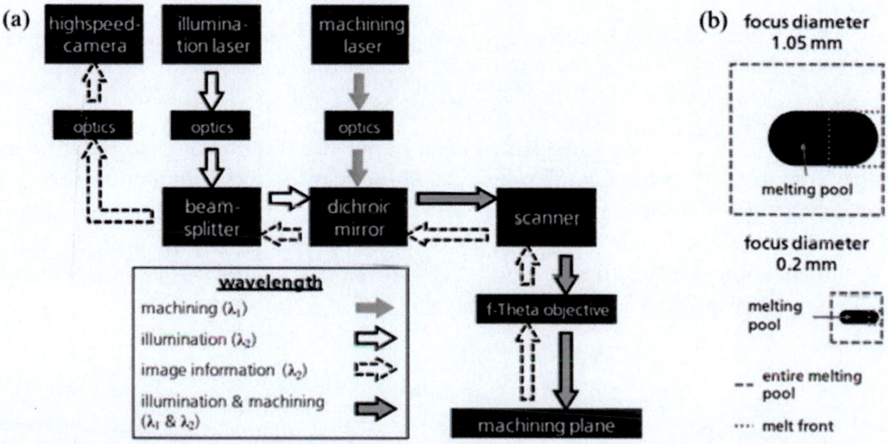

Fig. 7.2 Coaxial process detection systems in LPBF: **a** basic structure of the process detection system, **b** illustration of the different imaging sizes. Reprinted with permission from a work by Lott et al. [2]

of the tracks, including spheroidization, bulge, porosity, etc., which will eventually appear in the formed layer. In addition, defects such as bulges on the formed layer will damage the powder-spreading device and reduce the powder-spreading quality, resulting in powder-spreading defects such as insufficient powder supply and streaky powder-spreading morphology.

Based on light signals, machine vision can collect, identify and locate powder-spreading defects and formed layer surface defects, which can accurately and quickly detect the morphology and powder-spreading quality of the formed layer. Compared with the photoelectric signals acquisition, it has better robustness. In addition, the

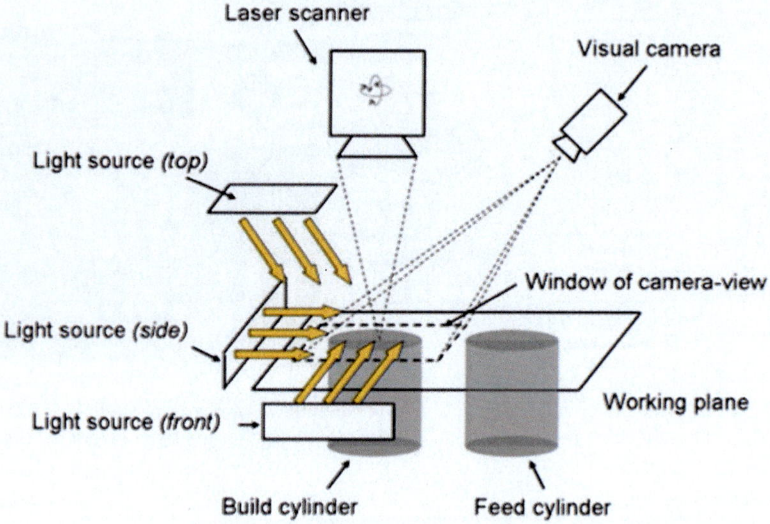

Fig. 7.3 Principle of the visual detection system for monitoring the powder layer surface. Reprinted with permission from a work by Craeghs et al. [3]

optical detection process of the formed layer or powder layer belongs to intermittent processing quality detection. Through the subsequent defect elimination process, the defects of the powder layer or the formed layer are repaired. Although there is a certain lag in the timeliness compared with the molten pool detection, the technical realization is less difficult. Figure 7.3 shows the structure of the optical visual detection of the powder layer (formed layer) [3].

7.1.1.2 Temperature Signal

In the L-PBF process, the complex thermodynamic process causes serious heat accumulation and residual stress accumulation in the formed parts, which directly affect the microstructure and mechanical properties of the parts. The measurement and evaluation of the temperature of the molten pool and the formed layer using infrared thermography equipment is helpful to study the solidification process of the molten pool and the residual stress distribution of the workpiece. The process and scanning strategy can be optimized according to the temperature distribution characteristics measured by the infrared thermography equipment. Figure 7.4 shows the principle of thermal imaging detection of the powder bed in LPBF [4], and Fig. 7.5 shows the non-coaxial measuring equipment of infrared thermography and the temperature of the heat-affected zone [5].

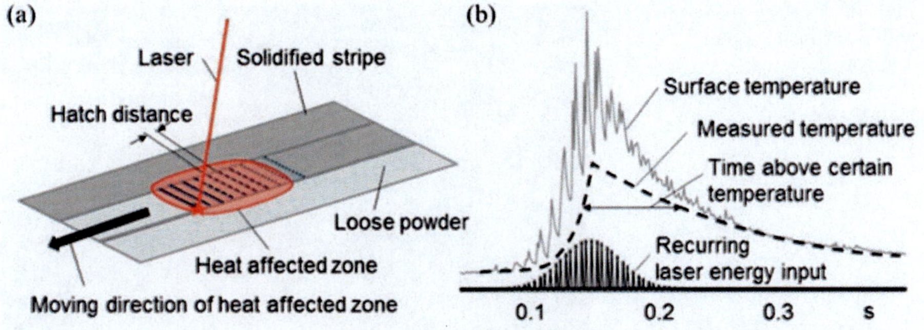

Fig. 7.4 Principle of thermal imaging detection of the powder bed in L-PBF: **a** schematic diagram, **b** temperature profile. Reprinted with permission from a work by Krauss et al. [4]

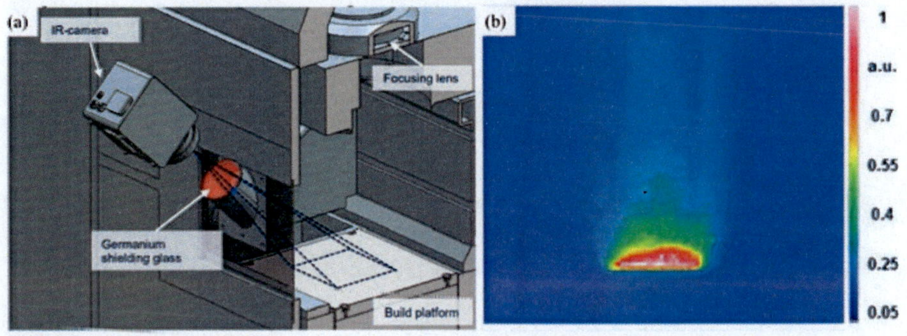

Fig. 7.5 **a** The non-coaxial measuring equipment of infrared thermography; **b** a sample thermogram of the heat-affected zone. Reprinted with permission from a work by Krauss et al. [5]

7.1.1.3 Acoustic Signal

1. Passive acoustic signal detection

The acoustic signal sensor has a simple structure, low cost and small amount of data acquisition, which is suitable for real-time acquisition and processing of signals. Acoustic signal detection is an effective method to realize online detection of weld quality in the welding process. The acoustic signal is generated during the interaction of laser and powder. However, the sensitivity of acoustic signals to light, heat, plasma and other signals is not enough to realize the state detection in the LPBF process. Researchers try to establish the potential relationship between sound information and laser power, scanning speed, surface quality and other parameters by monitoring the acoustic signal [6, 7]. Figure 7.6 shows an online acoustic signal detection device.

2. Active acoustic signal detection

In the LPBF process, active ultrasonic signal detection is carried out for the formed parts through ultrasonic emission and echo reception, and the amplitude-frequency

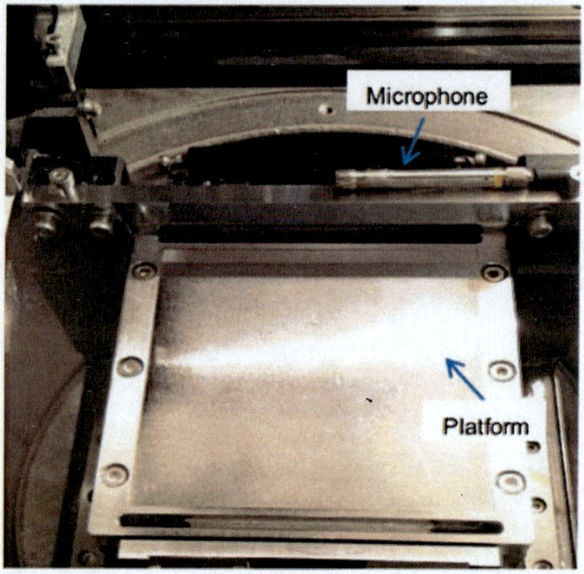

Fig. 7.6 Experiment setups with an acoustic signal acquisition system. Reprinted with permission from a work by Ye et al. [7]

characteristics of echo are used to detect and identify the surface quality, internal defects and residual stress of formed parts [8, 9]. Active ultrasonic signal detection is a non-contact and nondestructive technology that can be applied to the LPBF process.

7.1.1.4 Electronic Signal

At present, it is possible to detect the process of metal powder melting and vaporizing to produce plasma under the continuous radiation of a high-energy laser beam. There are a large number of free electrons and positive ions moving upward in an electrically neutral plasma. Because free electrons are lighter and have a faster moving speed, the plasma structure will have an electric potential difference in the vertical direction. It has been found that the change of free electrons in the traditional welding process can judge the change in melting depth [10], which is helpful to identify and locate holes or bulges. Figure 7.7 shows the schematic of the plasma charge signal detection system in the traditional welding process [10].

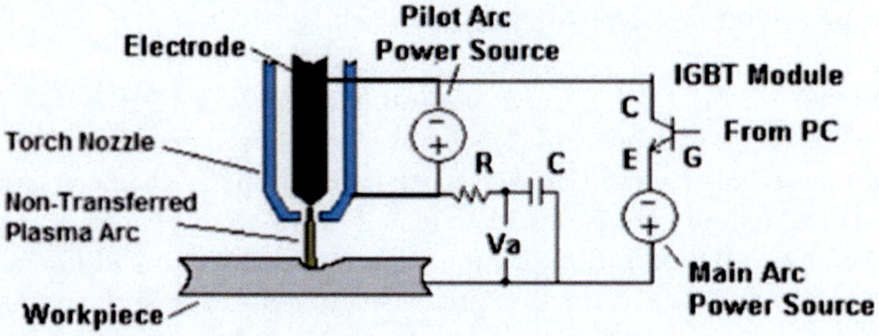

Fig. 7.7 The schematic of the non-transferred plasma charge sensor system. Reprinted with permission from a work by Lu et al. [10]

7.1.2 *Offline Detection*

7.1.2.1 Micro-CT Offline Detection Technology

Micro-CT technology has a resolution of micrometer level and can nondestructively detect the structures with a size of 1 μm inside the part. At present, advanced micro-CT technology has been used to analyze the internal defects of LPBFed parts, and the characterization of the three-dimensional shape of internal cracks, pores and other defect structures is more refined [11]. Compared with the two-dimensional analysis methods (optical microscope, scanning electron microscope, X-ray flaw detection, ultrasonic flaw detection), the defected information is more sufficient, which is conducive to accurate analysis of the LPBF process and defect formation mechanism [12].

7.1.2.2 Laser-Induced Breakdown Spectroscopy

Laser-induced breakdown spectroscopy (LIBS) is a kind of atomic emission spectroscopy used for qualitative and quantitative analysis of chemical multi-element. LIBS ablates and atomizes the sample material by focused laser to obtain the plasma. High-energy plasma melts the nanoparticles of an atomized sample. The atoms are excited to emit light, which can be captured by the detector and recorded as a spectrum. Through the analysis of the spectrum, information on the element types in the sample can be obtained, and further qualitative. Moreover, quantitative analysis of the spectrum can be carried out through software algorithms. In principle, LIBS technology can detect all elements, and the detection range is not limited by the physical state of the sample, but only limited by the laser power, sensitivity and wavelength range of the spectrometer and detector. In essence, LIBS is a method for material analysis. For single metals, ceramic materials, multiple and gradient metals, and ceramic and other composite materials formed by LPBF, LIBS technology can characterize the properties of these materials to evaluate processes [13].

7.1.2.3 Other Offline Detection Methods

The traditional material testing and analysis methods for LPBFed parts include scanning electron microscope (SEM), energy dispersive spectrometer (EDS), electron backscattered diffraction (EBSD), mechanical property characterization, electrochemical testing, and laser ultrasonic testing technology. These detection technologies belong to the offline detection of the microstructure and internal defects of the LPBFed parts, which can comprehensively detect the surface morphology, microstructure, crystal structure, macro and micro mechanical properties, electrochemical properties, and internal surface defects of the LPBFed parts. This kind of detection method is quite popular, but it is difficult to operate. Therefore, special personnel is required to operate relevant instruments.

7.2 Quality Feedback Technology

7.2.1 Online Detection

In 2017, to optimize the process of LPBFed complex ceramic $A1_2O_3$ parts, Zhang et al. [1] used array photodiodes to improve the accuracy of molten pool data collected by photodiodes during LPBF. Based on this system, they studied the influence of laser power on the behavior of molten pools in single-track and analyzed the corresponding relationship between molten pool information and several defects. In the experiment of Al_2O_3 single-track, they verified the corresponding relationship between the fluctuation of the light intensity signal and the stability of the molten pool under the laser power, as shown in Fig. 7.8. The scanning delay, edge effect and unstable temperature field are discussed based on the comparison of the light intensity data of the molten pool.

In 2010, based on the coaxial detection architecture, Chivel et al. [14] used a pyrometer and a photodiode to collect optical information to achieve online monitoring of the temperature distribution of the molten pool in the L-PBF and Selective Laser Sintering (SLS) processes. They studied the Rayleigh Taylor instability of the liquid interface in the LPBF. In 2011, Lott et al. [2] introduced the design of an optical system for monitoring the flow dynamics of the molten pool at a high scanning speed. They used a dichroic mirror and spectroscope to collect the visible and infrared light information in the molten pool area of the LPBF process in situ by building a coaxial optical path.

Gökhan Demir et al. [15]. conducted online monitoring on the LPBF process of 18Ni300 maraging steel by building a multi-band optical coaxial detection system, as shown in Fig. 7.9. The characterization ability of the light intensity signal radiated from the molten pool and the near-infrared wave band light signal to the manufacturing process is discussed. The test results and statistical data prove that the accuracy of the light signal of the molten pool reflecting the actual manufacturing

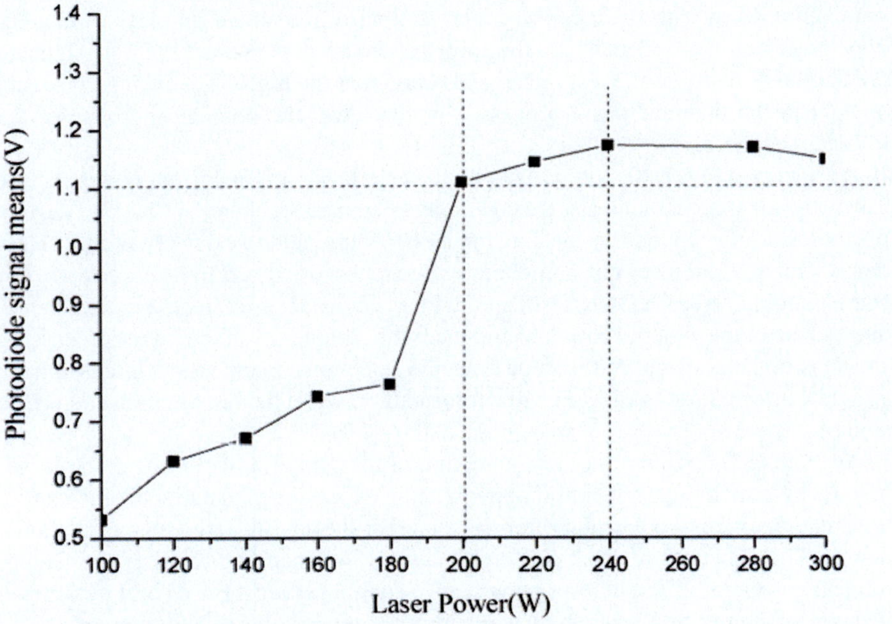

Fig. 7.8 The average values of photodiode signals in a single track with different laser powers. Reprinted with permission from a work by Zhang et al. [1]

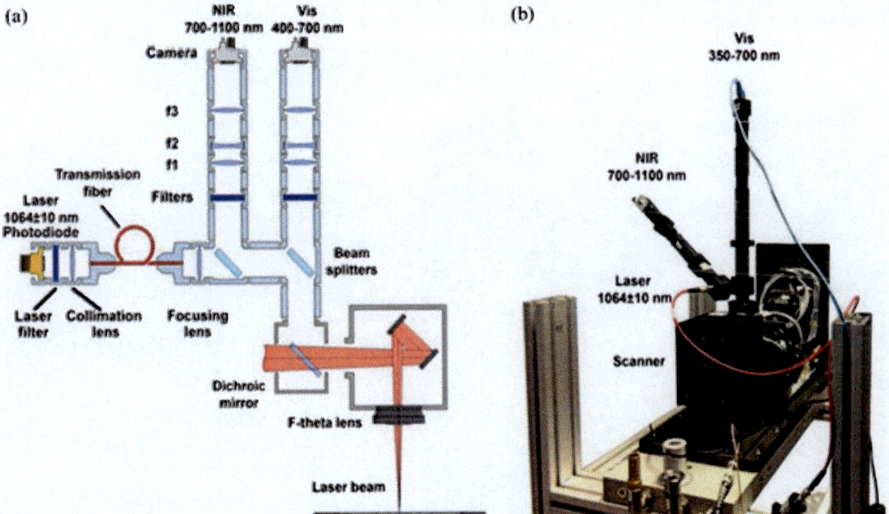

Fig. 7.9 **a** Schematic representation of the monitoring module; **b** the implemented system. Reprinted with permission from a work by Gökhan Demir et al. [15]

state is improved with the increase of the number of manufacturing layers. In addition, they discussed and analyzed the ability of the average intensity of light signal to predict the porosity of the test piece, and found that the higher the average intensity of light in the manufacturing process was, the lower the obtained porosity of the formed piece was.

As shown in Fig. 7.10, compared with the coaxial structure, the paraxial structure has many advantages, such as a simple optical path and mechanical structure design. It is not necessary to modify the existing LPBF equipment to carry out online detection research. Therefore, at present, many researchers use this framework to carry out corresponding research [16]. Compared with the coaxial structure, the optical detection system of the paraxial structure enhances the complexity of the image processing and detection algorithm. At the same time, the detection content is converted from the top view information to the side view information, which is flexible in the detection angle.

Ye et al. [17] carried out online detection of the spattering behavior in the LPBF process by establishing a paraxial optical detection system, and extracted the feature information of the plume and spattering image for the near-infrared image, as shown in Fig. 7.11. Then, they discussed and analyzed the changing trend of the feature information under different laser powers and scanning speeds. Based on 17 extracted features of plume and spatter, they combined neural network and support vector machine technology to realize the classification detection of five melting states in the LPBF, with an accuracy rate of 71.02% to establish the mapping relationship between plume, spatter behavior and melting manufacturing quality. Their research results verified the feasibility of the online detection mechanism of the LPBF processing state on the information of the paraxial molten pool.

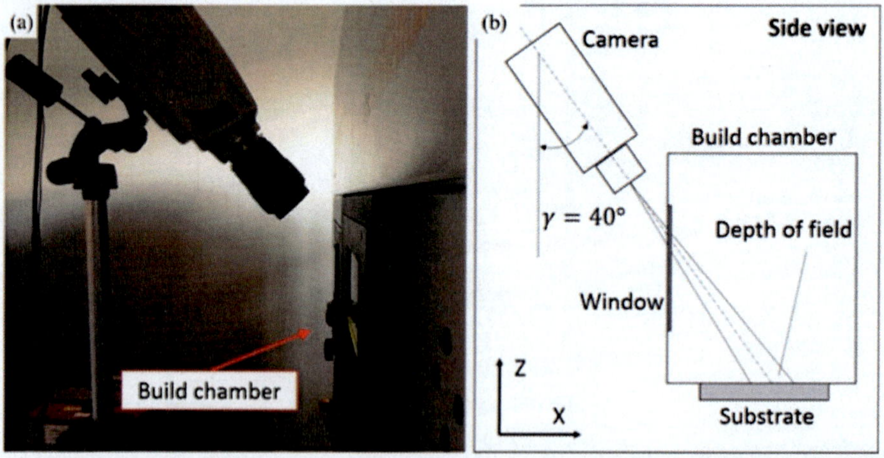

Fig. 7.10 Experimental setup with the high-speed camera outside the build chamber: **a** side view picture and **b** schematic representation of the same view. Reprinted with permission from a work by Grasso et al. [16]

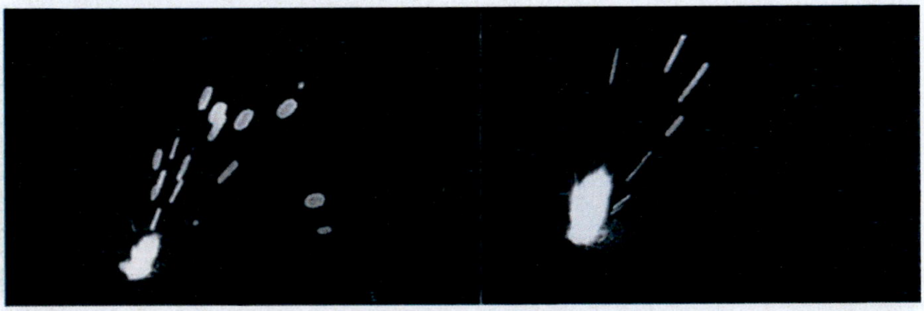

Fig. 7.11 Near-infrared image of plume and spatter. Reprinted with permission from a work by Grasso et al. [17]

In 2014, Karuss et al. [4]. evaluated the stability of the LPBF process and the quality of parts by collecting the temperature distribution data of layers during the LPBF process. They show that the heat distribution varies with the length of the scanning vector, laser power, layer thickness, hatch space and other parameters. The solidification and melting processes were monitored and evaluated by the integration of a paraxial uncooled temperature detector, which helps to identify hot spots early in the solidification process and contributes to the continuity of the manufacturing process. The underlying quality indicators are derived from the spatial analytic measurement data and are related to the generated part characteristics. By measuring the response of the material under different heat inputs, they proposed a heat dissipation model. The results show that it is feasible to use thermal imaging technology to monitor the finite section of the manufacturing platform.

In 2018, Ye et al. [17] verified the feasibility of the manufacturing process detection system based on sound signals according to the generation principle and dynamic features of sound signals in the LPBF process and analyzed the change rule and the cause of power spectral density of sound signals by changing laser frequency, laser power and scanning speed. Through packet analysis and feature statistical analysis, they decomposed the frequency band and extracted the feature parameters of the acoustic signals, and classified the five melting states with Fisher linear classification dimension reduction and linear support vector machine classification model, with an accuracy of 73.02%.

7.2.2 Offline Detection

In 2017, Grasso et al. [16] aimed at the phenomenon that the accumulation of defects in the upper layer would lead to the manufacturing failure of the other layer in the LPBF (Fig. 7.12). To overcome the problem that traditional micro-CT and ultrasonic nondestructive testing can only detect the defects offline and cannot eliminate the defects in the processing process in real-time, a lateral structure was used to study the

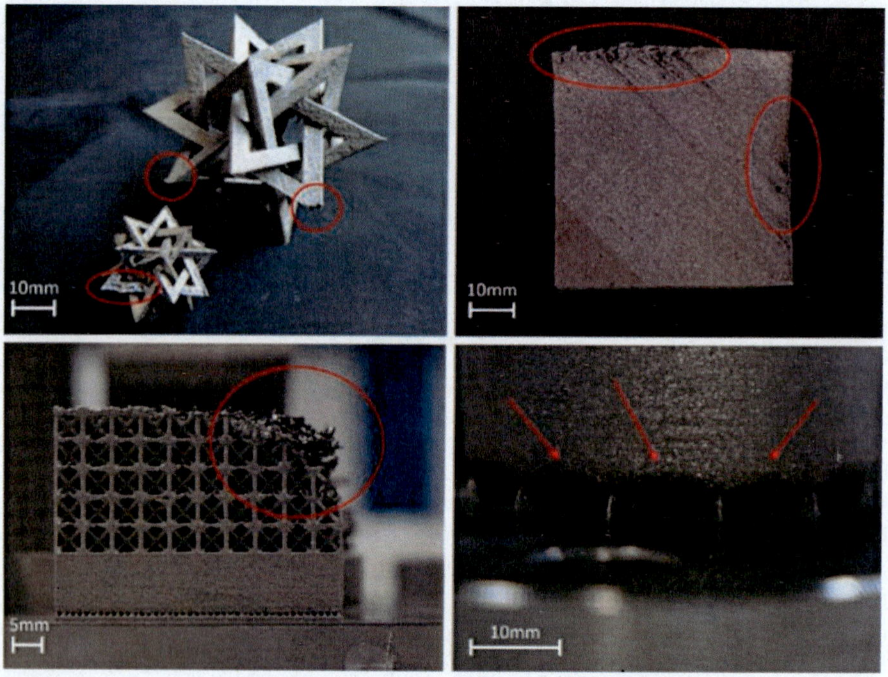

Fig. 7.12 Processing failure of the rear layer caused by defects in the front layer during L-PBFF. Reprinted with permission from a work by Grasso et al. [16]

in-situ monitoring method for rapid defect detection and location in the LPBF. This book conducts a relatively in-depth study of specific statistical detection technologies.

In 2011, Van Bael et al. [11] manufactured complex porous parts with controlled structures by LPBF technology, but there may be differences between the design and production in the morphological characteristics. Therefore, the production robustness and controllability of porous Ti6Al4V structure were improved by optimizing the process parameters of the LPBF, and the mismatching of morphology and mechanical properties between design and production was reduced by detection and compensation. In the first round, they designed LPBFed porous Ti6Al4V structures with different pore sizes. Morphological parameters of the formed parts were analyzed by micro-CT offline detection techniques, such as pore size, pillar thickness, porosity, surface area and volume. Compared with the original design and based on the difference between the design and measured properties, the compensation is integrated into the second LPBF process, which significantly improves the quality of the LPBFed porous structure. Zhou et al. [12] also carried out basic research on microscopic CT in LPBF detection and characterization. In 2019, Bobel et al. [18] also studied an in-situ synchrotron radiation X-ray imaging technology applied to LPBF, which could detect the manufacturing state of the single-track and was verified by SEM

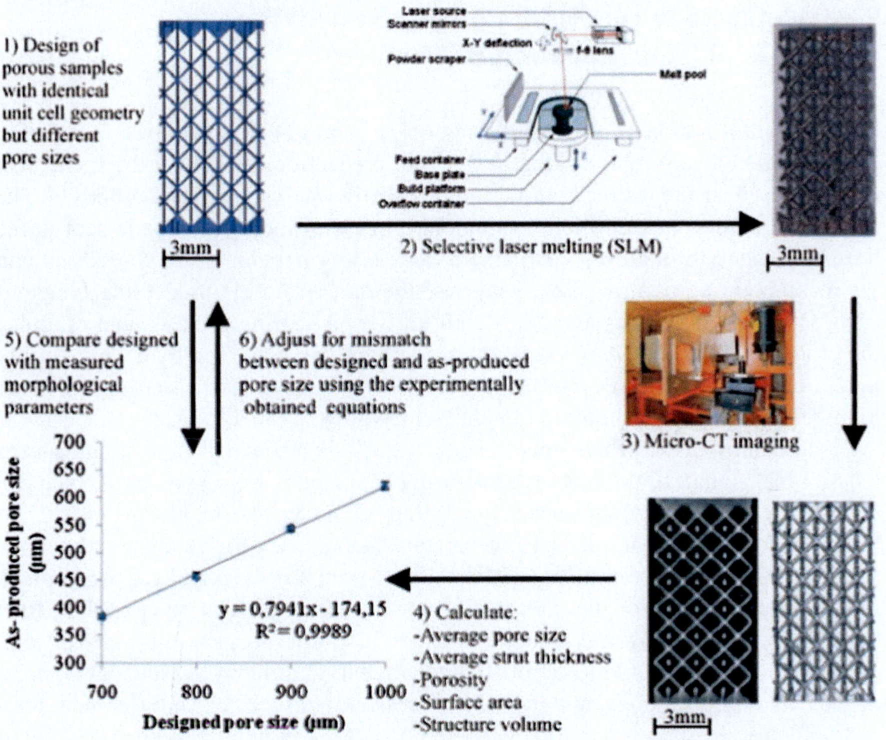

Fig. 7.13 The offline detection of micro-CT to characterize the internal void structure of the LPBFed parts. Reprinted with permission from a work by Van Bael et al. [11]

and EBSD. Figure 7.13 shows the offline detection of micro-CT to characterize the internal void structure of the LPBFed parts.

In 2018, Vrábel et al. [13] first tried to use LIBS technology in the field of additive manufacturing (AM) to introduce the possibility of using LIBS technology to improve the LPBF process and measure the substrate, powder and as-built samples using a desktop-level LIBS system. They also simulated the interference and influence of plasma generated by LIBS on the LPBF process under low irradiance. The obtained data are analyzed by multivariate data, which provides a basis for material analysis under LPBF technology.

7.3 Metal AM Process Monitoring System Based on Real-Time Shooting HD Camera

To obtain information on the melting region morphology of each layer and quality of the formed layer, a monitoring device and realization method of the metal AM process based on a real-time high-definition camera was designed and proposed. The scene information of the whole machining process of the workpiece is sent to the terminal device through the HD camera. Processing personnel can directly obtain the manufacturing quality of each layer of the parts from the processing image of each part, timely analysis and adjustment of the processing process, and optimize the process parameters. At present, the scheme has been granted by a China utility model patent, a metal 3D printing process monitoring device based on real-time high-definition camera shooting (patent No. CN201821180360.4).

In the control of the LPBF process, some teams carried out real-time monitoring of the physical signals of the molten pool during the manufacturing process, which can be fed back to the control system and quickly adjust the processing parameters such as laser power and scanning speed. However, this method has a high technical threshold and puts forward high requirements for the real-time processing effect of the metal AM system. The size of the molten pool under the metal AM, such as the LPBF technology, is generally only about 100 μm, which improves the difficulty of real-time monitoring of the molten pool. To overcome the shortcomings and deficiencies of the existing technology, a metal AM process monitoring device based on a real-time shooting HD camera is provided. The operator can find the defective parts of the formed parts during the processing, improve the processing technology, and thus improve the performance of the formed parts.

7.3.1 Hardware Scheme of a Real-Time Shooting Monitoring System

7.3.1.1 Basic Composition

As shown in Fig. 7.14, it is a metal AM process monitoring device based on the real-time shooting of a high-definition camera, including a sealed manufacturing chamber, in which a scanning galvanometer is used for LPBF the metal powder layer laid on the manufacturing plane. A high-definition camera is arranged on the top wall on the right side of the manufacturing chamber, and the front end of the high-definition camera is provided with an adjustable focusing head, which extends into the manufacturing chamber through an image acquisition cylinder arranged on the inner side of the top wall of the manufacturing chamber, and is used for real-time photographing the manufacturing appearance of the melting layer during the metal AM powder melting process. The electrical signal of a high-definition camera is connected to an external terminal device, and the formed shape data captured

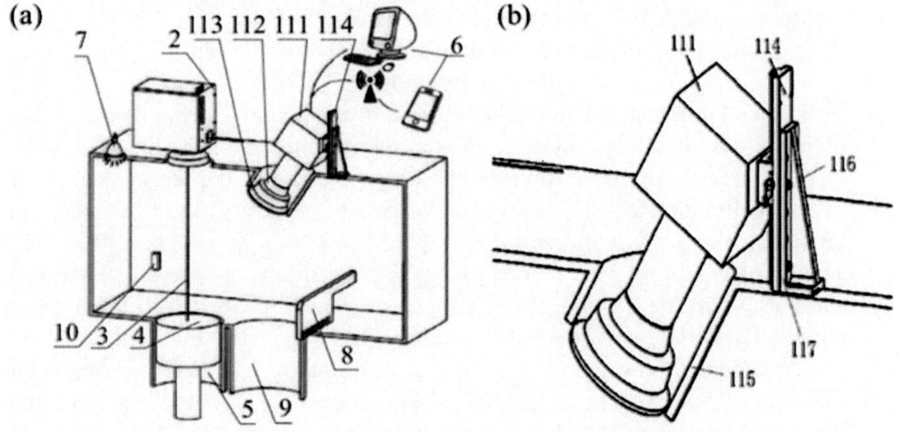

Fig. 7.14 Schematic diagram of real-time shooting monitoring system: **a** process monitoring device; **b** HD camera assembly. 1-HD camera assembly; 2-scanning galvanometer; 3-laser; 4-manufacturing plane; 5-manufacturing cylinder; 6-terminal equipment; 7-halogen lamp; 8-scraper; 9-powder cylinder; 10-left limit switch; 111-HD camera; 112-adjustable focusing head; 113-filter; 114-sliding strip; 115-image collecting cylinder; 116-reinforcing rod; 117-L-shaped support back plate

is transmitted to the terminal device. The high-definition camera is installed on the height fine-tuning assembly, which is fixed on the right top wall of the manufacturing chamber.

7.3.1.2 Implementation Mode

(1) There is a certain angle between the high-definition camera and the processing area of the workpiece, which can be adjusted according to the actual processing situation of the workpiece. Adjust the height of the high-definition camera by using the height fine-tuning component to ensure that the focus position is accurately focused on the manufacturing surface to ensure that the photos taken are clear and complete.

(2) The tilt angle of the image-collecting cylinder is about 40°–45°. If the entire image-collecting cylinder is made of silica gel, the adjustment angle is larger and more flexible, and the sealing performance may be better.

(3) The high-definition camera can be a CCD high-speed camera, and the recorded image area covers the entire manufacturing surface. The CCD high-speed camera shall not be less than 1024 × 1024 pixel resolution, and the number of frames can reach 300 frames/s. The minimum exposure time of the overall shutter is 1 μs. The dynamic range is 120 dB. The spectral range is 400–950 nm, with 8-bit sampling resolution.

(4) A filter is installed in front of the adjustable focusing head, which only transmits light of 440–950 nm wavelength to avoid damage to the sensitive parts inside the camera caused by a high-intensity light.
(5) A halogen lamp is installed on the left side of the manufacturing chamber. The halogen lamp is always on during the exposure process of the high-definition camera or is always on during the whole processing of the workpiece. It can be manually adjusted, or the trigger switch can be turned on or off when the manufacturing surface drops.
(6) The focus position of the halogen lamp is concentrated on the manufacturing surface. When the manufacturing surface descends, the switch of the halogen lamp is triggered synchronously to make it light up. After the high-definition camera completes the exposure, the switch of the halogen lamp is triggered again to make it go out, that is, the luminous time of the halogen lamp is greater than the exposure time of the high-definition camera to ensure that there is enough light when the high-definition camera takes photos.
(7) During the manufacturing process, the laser is focused on the manufacturing surface through the scanning galvanometer to melt the paved metal powders. After the melted powder on the manufacturing surface condenses and the manufacturing base platform drops by a layer thickness, the high-definition camera takes a picture of the manufacturing area. At the same time when the camera is exposed, the halogen light source pulse lights up (fill light). After the exposure, the halogen light goes out, and the manufacturing condition of this layer is recorded. The high-definition camera transmits the image information to the computer in real-time and stores it for the processing personnel to view the manufacturing situation layer by layer. Then, the metal AM system processes the next layer of data, processes layer by layer and captures the image of the formed layer through the high-definition camera.

7.3.2 Realization of a Real-Time Shooting Monitoring System

Based on the above scheme, the author team has adapted the equipment based on the self-developed Di-Metal 100F AM equipment and realized the metal AM process monitoring system based on the real-time shooting of high-speed and high-definition cameras, as shown in Fig. 7.15. Based on this system, AM technical engineers can obtain a better viewing angle and more timely observation means, find the defective parts of the formed parts during the processing, remelt the formed layer according to the type and location of the defects, and improve the processing technology to avoid manufacturing failure and improve the performance of the formed parts.

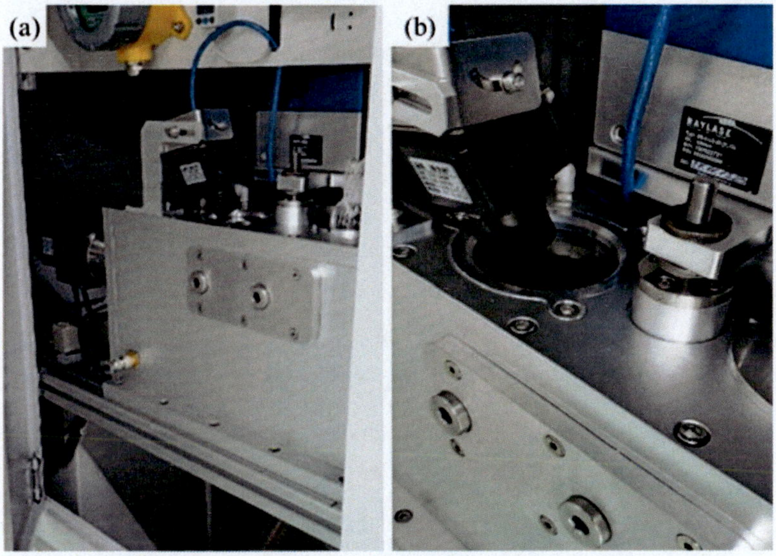

Fig. 7.15 Real-time shooting monitoring system

7.4 Coaxial Monitoring and Quality Feedback Technology

To realize the tracking and monitoring of the molten pool in the LPBF process, this book designs and proposes a coaxial monitoring and quality feedback solution and device for the molten pool information, and upgrades the independently developed Di-Metal AM equipment to achieve the whole-process, real-time monitoring and closed-loop control of the LPBF process. At present, the solution has been granted a Chinese invention patent, a coaxial monitoring method and a device for a selective laser melting process (patent No. CN201710244822.8).

A coaxial monitoring method for the LPBF process is proposed, that is, in the LPBF process, high-speed camera imaging technology and photodiode detection technology are used for coaxial monitoring of the molten pool shape and light intensity signal, and finally a point-by-point and layer-by-layer online monitoring technology is realized and used for the quality control of the processing process. This technology has a higher local resolution of molten pool and faster acquisition rate of molten pool information due to the sharing of optical detection optical path and laser processing optical path.

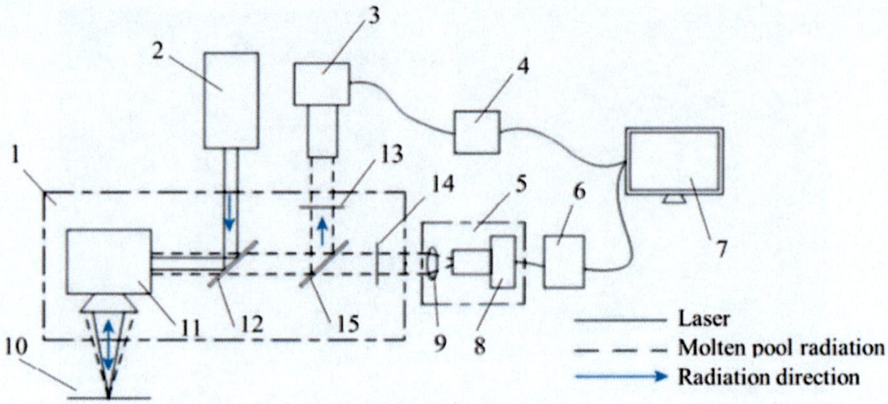

Fig. 7.16 Schematic diagram of coaxial monitoring system hardware scheme. 1-Optical path module; 2-laser head; 3-COMS high-speed camera; 4-camera controller; 5-photodiode module; 6-diode controller; 7-computer; 8-photodiode; 9-focusing lens; 10-manufacturing plane; 11-scanning galvanometer; 12-semi-transparent and semi-reflective mirror; 13-first filter; 14-second filter; 15-beam splitter

7.4.1 *Hardware Composition of the Coaxial Monitoring System*

7.4.1.1 Basic Composition

As shown in Fig. 7.16, this scheme uses two detectors, a high-speed camera and a photodiode, which are distributed on the same plane. They share the same optical system with the laser. The coaxial monitoring system is based on modular design and construction. The implementation process is shown in Fig. 7.16, including the optical path module, laser head, COMS high-speed camera, camera controller, photodiode module, diode controller, and computer. The optical path module comprises a scanning galvanometer, a semi-transparent and semi-reflective mirror, a first filter, a second filter and a beam splitter. The photodiode module includes a photodiode and a focusing lens.

7.4.1.2 Module Composition

1. The camera controller module is composed of an image acquisition module, an image conversion module, an image filtering module, an threshold segmentation module, an data transmission module, etc.

   (1) Image acquisition module: It is used to control COMS high-speed camera to acquire real-time image data of the molten pool and save it in memory.

(2) Image conversion module: The color image fed back to the COMS high-speed camera is displayed as a gray image, and the perspective coordinate system and the processing plane coordinate system are converted.
(3) Image filtering module: It uses a median filter to filter the gray image to smooth the image and remove noise.
(4) Threshold segmentation module: Using the gray histogram, select the threshold of the histogram as the minimum value, according to the threshold, the image is binarized, and divided into the molten pool area and non-molten pool pixels.
(5) Data transmission module: Output the processed image to the computer and save it.

2. The diode controller module includes an optical signal acquisition circuit, a programmable amplifier, a low-pass filter, an AD acquisition card, a data transmission module, etc.

(1) Optical signal acquisition circuit: It is used to control the photodiode to collect the visible light signal of the molten pool.
(2) Programmable amplifier: It automatically changes its gain according to the size of the input signal, so that its output voltage is always within the range of the full range value.
(3) Low pass filter: Since the output signal contains high-frequency noise, use a low-pass filter to suppress high-frequency noise.
(4) AD acquisition card: It is used to collect the analog signals output by the sensor and convert them into digital signals that can be recognized by the computer, and then send them to the computer for corresponding calculation and processing according to different needs to obtain the required data.
(5) Data transmission module: Output the processed data to the computer and save it.

7.4.1.3 Electrical Characteristic Index

(1) The pixel resolution of the COMS high-speed camera shall not be lower than 1024 × 1024 pixels, which can reach 75 frames/s at full resolution, and the minimum exposure time of the overall shutter is 1 μs. The wide dynamic range is 120 dB, the spectral range is 400–950 nm, and the sampling resolution is 8-bit.
(2) The photodiode is a silicon photodiode with a diameter of 9 mm × 9 mm effective area. The sensitive area of a single Si photodiode is 3 × 3 mm, with a spectral range of 190–1100 nm. The focusing lens focuses the radiation emitted by the entire molten pool on the silicon photodiode plane.
(3) The first filter and the second filter are used to obtain the molten pool information of the required wave band. Among them, the first filter is located between the COMS high-speed camera and the beam splitter. A narrowband filter with

a central wavelength of 600–650 nm is used to ensure the high spectral sensitivity of the COMS high-speed camera. The second filter is located between the photodiode and the beam splitter. The combination of a low-pass filter with a cutoff wavelength of 950 nm and a high-pass filter with a cutoff wavelength of 780 nm is used to avoid the sensor being exposed to possible reflected laser radiation and to exclude the influence of surrounding light.

7.4.1.4 Signal Connection

(1) The photodiode is connected to the computer through the diode controller.
(2) COMS high-speed camera is connected to the computer through a camera controller.
(3) The photodiode, the focusing lens, the scanning galvanometer, the semi-transparent and semi-reflective mirror, the second filter and the beam splitter are connected in turn by an optical path.
(4) The COMS high-speed camera is connected to the beam splitter optical path through the first filter.
(5) The laser head is connected with the optical path of the semi-transparent and semi-reflective mirror.

7.4.2 Implementation Mode

Based on the coaxial detection device, the detection is performed according to the following steps.

Step 1: Start manufacturing parts. The laser beam is emitted from the laser head, reflected into the scanning galvanometer by the semi-transparent and semi-reflective mirror, and then projected onto the metal powder on the surface of the worktable substrate to manufacture LPBF-made parts.

Step 2: During the LPBF process, the optical signal radiated from the high-temperature molten pool is projected to the semi-transparent and semi-reflective mirror through the scanning galvanometer. The semi-transparent and semi-reflective mirror will reflect 100% of the 1064 nm laser wavelength, while the visible light and near-infrared light will be 100% transmitted to the beam splitter. The beam splitter deflects 30% of the emission radiation to the photodiode module and 70% to the COMS high-speed camera.

Step 3: Place the first filter on the optical transmission path between the COMS high-speed camera and the beam splitter to improve the spectral sensitivity of the COMS high-speed camera. The second filter is installed on the light transmission path between the photodiode and the beam splitter to prevent the photodiode from being exposed to possible reflected laser radiation to exclude the influence of surrounding light.

Step 4: COMS high-speed camera converts the radiation information of the molten pool into image information and transmits it to the camera controller. The photodiode module feeds back the brightness of the molten pool to the diode controller. The camera controller obtains the molten pool contour according to the image information and transmits it to the computer for storage. The diode controller transmits the digital signal of light intensity to the computer in real-time.

Step 5: The computer modifies the laser parameters, thereby improving the stability of the LPBF equipment and the quality of the workpiece, and realizing the closed-loop control of the LPBF process.

7.5 Reverse Quality Feedback Control Layer by Layer

Online coaxial monitoring technology is based on molten pool information. To achieve accurate measurement of the actual size of each melting layer and accurate acquisition of the position, three-dimensional shape and other information of internal defects during AM, this book improves and designs an AM layer-by-layer detection, part reversed model and defect positioning device and method based on coaxial monitoring. At present, the solution has been granted a Chinese invention patent (patent No. CN201710245808. X).

The part model is obtained by the follow procedure: monitor the powder melting in the LPBF process, feed it back to the computer software interface to reflect the features of the molten pool at different positions in real time, accurately measure the contour of each melting layer, and finally, obtain the model through reverse calculation. Then, the error between AMed metal part model and the original model data in size precision is gotten by comparison. At the same time, it can accurately obtain the position and three-dimensional shape of internal defects in the AM process, avoiding damage to printed parts in the later stage. The scheme of layer-by-layer reverse engineering and defect positioning device is shown below.

7.5.1 Hardware Composition

7.5.1.1 Basic Composition

Figure 7.17 shows an AM layer-by-layer detection reversed part model and positioning defect device, including a laser head, scanning galvanometer and computer, semi-transparent and semi-reflective mirror, high-speed camera, etc. The laser is reflected into the scanning galvanometer through the semi-transparent and semi-reflective mirror, and the scanning galvanometer controls the laser beam to selectively melt the metal powder paved on the working platform. At the same time, the scanning galvanometer collects the molten pool radiation and transmits it to

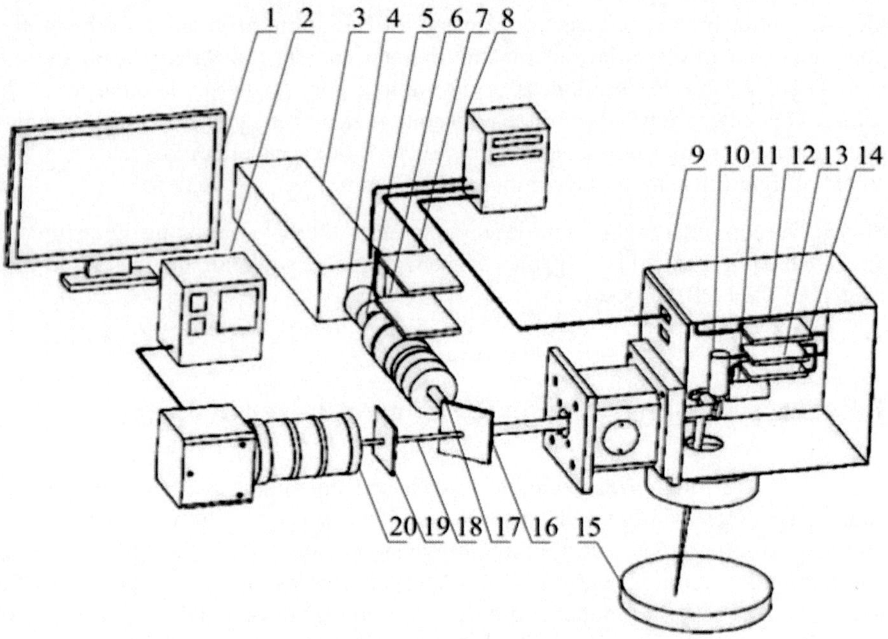

Fig. 7.17 Structure diagram of layer-by-layer reverse engineering and defect locating device. 1-Computer; 2-controller; 3-laser head; 4-beam expander of auxiliary structure; 5-three-dimensional dynamic focusing system; 6, 7-control panel; 8-galvo control card; 9-scanning galvanometer; 10-scanning motor and its lens; 11-X scanning motor and its lens; 12, 13, 14-control panel; 15-working platform; 16-semi-transparent and semi-reflective mirror; 17-laser optical path; 18-radiant optical path of the molten pool; 19-filter; 20-high-speed camera

the high-speed camera through the semi-transparent and semi-reflective mirror. The high-speed camera processes the molten pool radiation data and converts it into image information and transmits it to the controller, which is used to process the image data to determine the location of the molten pool and generate the contour of each melting layer. A filter is added on the optical path between the high-speed camera and the semi-transparent and semi-reflective mirror to filter out the acquisition band of the molten pool.

7.5.1.2 Module Composition

The controller includes an image acquisition module, an image contour extraction module and an image triangulation module. At the beginning of the working cycle, the image acquisition module collects the image information, transmits it to the image contour extraction module to extract the molten pool contour information, establishes a process file based on this information, and feeds back the processing status on the computer interface. When the processing of this layer is completed, the layer contour is extracted according to the process file. The image triangulation module obtains the complete 3D model of the workpiece according to the multi-layer contour of the workpiece and outputs the STL file.

(1) Image acquisition module: It is used to control the high-speed camera to acquire real-time image data of the molten pool in the manufacturing process of each layer of the workpiece, and save it in its memory.
(2) Image contour extraction module: the color image fed back to the high-speed camera is displayed as a gray image, and its coordinate system is established. The median filter template is used to filter the gray image to smooth the image and remove noise. The gray histogram is used to select the threshold value of the histogram as the minimum value. According to the threshold value, the image is binarized, divided into the molten pool pixels and non-molten pool pixels, and the molten pool contour is extracted.
(3) Image triangulation module: The fault contour obtained from image processing is approximated by polygons, and then connected into triangles between the vertices of adjacent fault polygons. Then the upper and lower end faces of the object are triangulated, and STL files are exported.

7.5.1.3 Electrical Characteristic Index

(1) The semi-transparent and semi-reflective mirror is used for 100% reflection of 1064 nm laser wavelength and 100% transmission of visible light and near-infrared light to the high-speed camera.
(2) Narrow band filters with a central wavelength of 600–650 nm are used to ensure the spectral sensitivity of high-speed cameras.
(3) The high-speed camera is a COMS high-speed camera with a pixel resolution of no less than 1024 × 1024 pixels, and the number of frames can reach 7000 frames/s. The minimum exposure time of the overall shutter is 1 μs. The dynamic range is 120 dB. The spectral range is 400–950 nm. The sampling resolution is 8-bit.
(4) Due to the performance constraints of the digital image processing algorithm and the systematic error of the mechanical structure, the deviation range of the final molten pool size from the standard value is about 5–15%.

7.5.1.4 Signal Connection

The high-speed camera is connected to the computer through the controller.

7.5.2 Implementation Mode

Step 1: The scanning galvanometer, semi-transparent semi-reflective mirror and filter form a coaxial optical path, through which the radiation of the molten pool is reflected and filtered to the high-speed camera.

Step 2: Establish a coordinate system with the center of the manufacturing plane of the workpiece as the origin. The high-speed camera captures the position of the molten pool on the manufacturing plane according to the plane manufacturing tracks, and records the shape of the molten pool at this position.

Step 3: The size of the molten pool is obtained through the image processing of the controller. When the size of the molten pool deviates from the deviation range of the standard value, it is recorded as an abnormal position. Otherwise, it is a normal position. The controller will feed back the position information to the real-time monitoring interface of the computer in real-time, and the corresponding position of the monitoring interface will reflect the molten pool information. If it is in a normal position, it will display green. And if it is in an abnormal position, it will display red.

Step 4: After the data processing of this layer is completed, the high-speed camera collects the manufacturing plane data of this layer. And the controller extracts the contour data of this layer of the workpiece and saves it. After the whole processing of the part is completed, the 3D model is generated according to the contour data of each layer of the workpiece, and the currently generated 3D model is embedded in the computer in advance.

By comparing and analyzing the original 3D model of the metal AM part with the original model data, the error in precision size is obtained. At the same time, the abnormal position in the model is highlighted, and the number of layers is prompted for viewing, as shown in Fig. 7.18.

7.6 Other Quality Monitoring Methods

Statistical process control (SPC) is a traditional quality control method. It applies statistical techniques to evaluate and monitor each stage of the process, establishes and maintains the process at an acceptable and stable level, and thus ensures that products and services meet the specified requirements. SPC is the basic analysis tool for online monitoring research of the LPBF process. The online monitoring algorithm of LPBF based on machine learning also includes statistical process analysis, such as dispersion graph analysis, histogram analysis, descriptive statistics analysis, correlation analysis, regression analysis, etc. In addition, the control chart of SPC is also important for the stability and quality control of the LPBF process.

In 2017, Repossini et al. [19] used high-speed image acquisition technology, coupled with image segmentation and feature extraction to estimate different statistical descriptors of spatter behavior on the laser scanning path, as shown in Fig. 7.19.

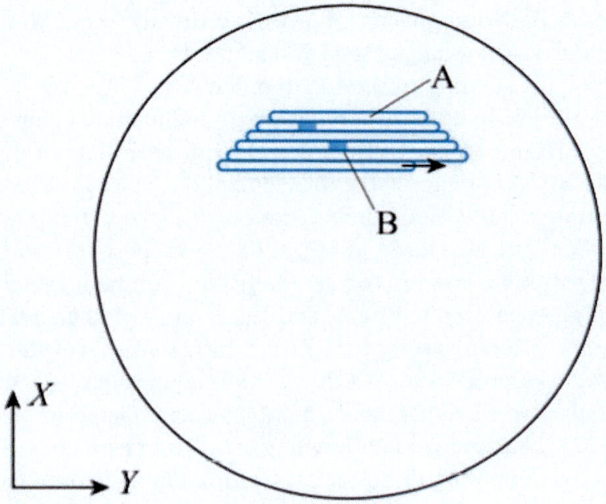

Fig. 7.18 Schematic diagram of computer interface (A represents the manufacturing layer of each layer of the part, and B represents abnormality)

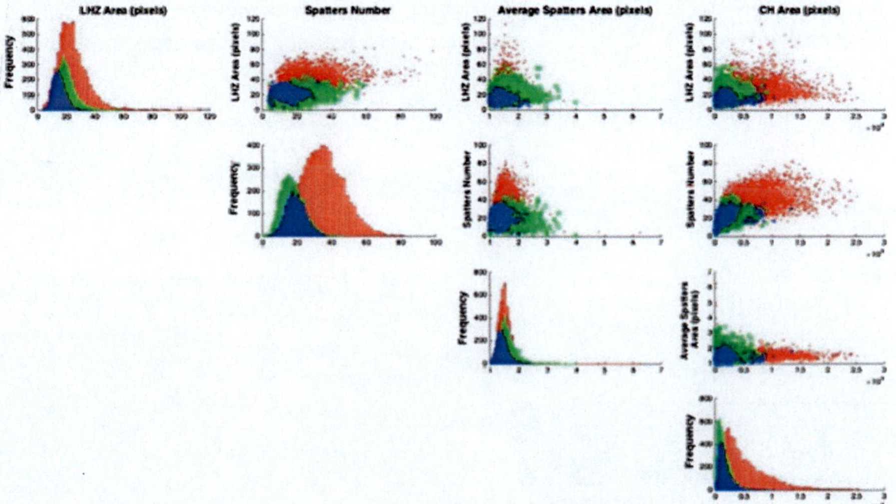

Fig. 7.19 Scatterplot of spatter and LHZ descriptors; blue crosses: under-melted, green squares: normal-melted, red circles: over-melted. Reprinted with permission from a work by Repossini et al. [19]

The logistic regression model is established, and the classification ability of the statistical descriptors of the spattering behavior to different energy densities corresponding to different mass states is determined. The results show that the detection ability under penetration and over-melting conditions can be significantly improved by using spatter as the driving factor of process characteristics. All of these indicate that spatter characteristic analysis and modeling are effective on-site monitoring means to improve the stability of the LPBF process.

Grasso et al. [20] have done a lot of research on the monitoring of molten pool, spatter and plume features of the LPBF process, proposed a monitoring and evaluation algorithm for LPBF by combining the related methods of SPC and the related model of machine learning, and verified the effectiveness and reliability of this algorithm in practical applications. As shown in Fig. 7.20, the support vector machines (SVM) classification algorithm is used to extract the features of the infrared video imaging plume of the high-speed molten pool, and the K control chart method is used to build an online monitoring system for LPBF, which forms an evaluation system for judging the quality of processing. At the same time, through the example of LPBF of zinc powder, they verified that the evaluation system is superior to other methods.

Zhang et al. [21] characterized the existing data through off-axis vision and image processing methods, monitored the features of the plume and molten pool in real-time with high-speed cameras, and made statistical analysis on the training data by using tools such as scatterplot of SPC, and further evaluated the quality of a single track in combination with the depth confidence network (DBN) model. Figure 7.21 shows the analysis results of the molten pool, spatters and spectrum of the plume.

Garmendia et al. [22] proposed a method to control the precision of formed parts online based on the data of structured light scanner for laser near net shaping

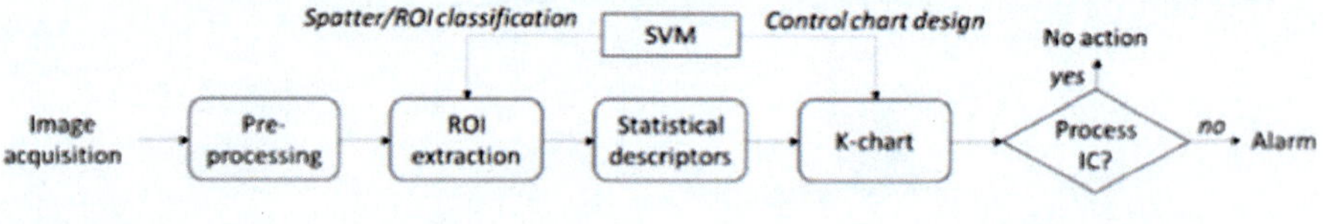

Fig. 7.20 Process control scheme based on SVM and SPC. Reprinted with permission from a work by Grasso et al. [20]

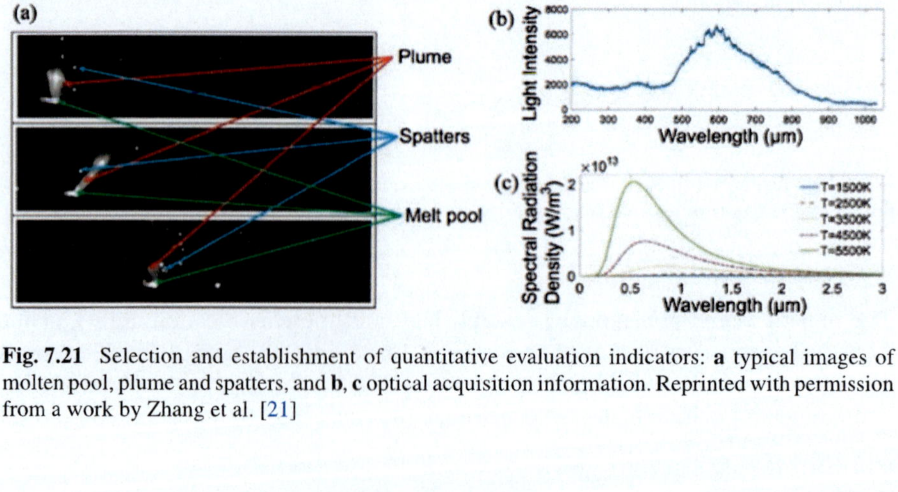

Fig. 7.21 Selection and establishment of quantitative evaluation indicators: **a** typical images of molten pool, plume and spatters, and **b**, **c** optical acquisition information. Reprinted with permission from a work by Zhang et al. [21]

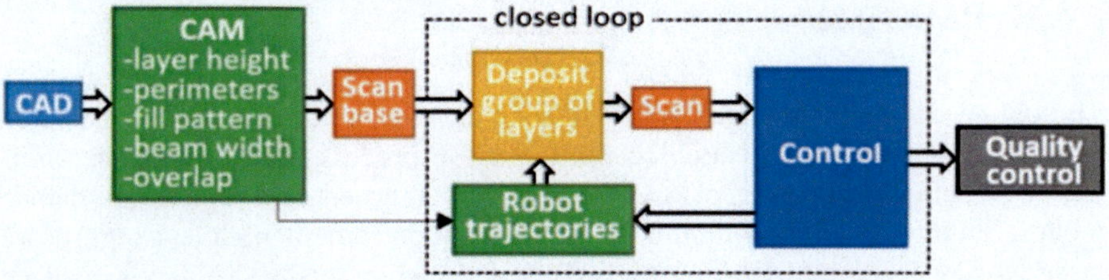

Fig. 7.22 Process scheme with the control strategy. Reprinted with permission from a work by Garmendia et al. [22]

(LNNS) technology, and took corresponding correction measures according to the measurement error, which is practical and stable in the actual LPBF process. The process control strategy is shown in Fig. 7.22, which is also a good reference for the manufacturing accuracy control of the LPBF process.

7.6.1 Combination of Online and Offline Detection Methods

Multi-sensor fusion is an important research direction of sensor science, and also an inevitable requirement for improving the detection level of LPBF. In the future, the online detection of LPBF must be based on a comprehensive offline detection method, integrating the process detection of visible light, infrared light, acoustic signal, LIBS and even spectral signal, realizing the closed-loop control and quality traceability of the LPBF technology.

7.6.2 Fusion of Simulation Analysis Methods

At present, many simulation technologies, such as the finite element method (FEM), finite volume method (FVM), and multi-physical field coupling analysis method, have a powerful role in understanding the manufacturing mechanism of LPBF that cannot be realized by online detection, which can be explained scientifically in the mathematical and physical level. In the future, the online detection of the LPBF process must be deeply combined with simulation analysis technology. At present, Wuhan University, Nanjing University of Aeronautics and Astronautics and other institutions in China have made good exploration.

7.6.3 *Real-Time*

The online detection process for LPBF may involve building an online detection simulation platform to detect defects in the LPBF process in real-time. The online defect detection simulation platform is mainly composed of a high-speed camera module, an infrared camera module, a high-resolution camera module, a high-power laser, powder feeding platform, a high-speed and high-precision XYZ axis displacement platform, a vacuum and inert gas environment, a high-speed signal processing control system, a signal data processing software and other modules. The detection process involves massive data transmission, which puts forward relatively high requirements for hardware and communication technology, and needs continuous optimization of the hardware level.

7.6.4 *Application of the Intelligent Algorithm*

Statistical quality control is the foundation and machine learning-related algorithms are effective means. The combination of the two will produce many interesting and effective things. In a word, the algorithm framework based on machine learning is quite practical in the online defect monitoring of LPBF and shows great advantages in the accuracy and reliability of discrimination. Deep learning and artificial intelligence (AI) is a global research hotspots. It has very high natural adaptability to sound, image features, pattern recognition and control decision-making, and is a development direction in the field of sensor detection. In the process of online detection for LPBF, algorithms are the soul. Therefore, the application of algorithms such as deep learning is an inevitable trend for LPBF technology. The machine learning model has a good application prospect for feature analysis of various information sources, defect identification and process stability judgment in LPBF process. At present, some teams have applied machine learning to the research of online monitoring systems for the LPBF process.

References

1. Zhang K, Liu T, Liao W et al (2018) Photodiode data collection and processing of molten pool of alumina parts produced through selective laser melting. Optik 156:487–497
2. Lott P, Schleifenbaum H, Meiners W et al (2011) Design of an optical system for the in situ process monitoring of selective laser melting (SLM). Phys Procedia 12:683–690
3. Craeghs T, Clijsters S, Yasa E et al (2011) Online quality control of selective laser melting. In: 2011 International solid freeform fabrication symposium. University of Texas at Austin
4. Krauss H, Zeugner T, Zaeh MF (2014) Layerwise monitoring of the selective laser melting process by thermography. Phys Procedia 56:64–71

5. Krauss H, Eschey C, Zaeh MF (2012) Thermography for monitoring the selective laser melting process. In: 2012 International solid freeform fabrication symposium. University of Texas at Austin
6. Ye D, Zhang Y, Zhu K et al (2017) Characterization of acoustic signals during a direct metal laser sintering process. In: Advances in energy science and equipment engineering II. CRC Press
7. Ye D, Hong GS, Zhang Y et al (2018) Defect detection in selective laser melting technology by acoustic signals with deep belief networks. Int J Adv Manuf Technol 96(5):2791–2801
8. Everton S, Dickens P, Tuck C et al (2015) Evaluation of laser ultrasonic testing for inspection of metal additive manufacturing. In: Laser 3d manufacturing II. SPIE, vol 9353, pp 145–152
9. Popovich AA, Masaylo DV, Sufiiarov VS et al (2016) A laser ultrasonic technique for studying the properties of products manufactured by additive technologies. Russ J Nondestr Test 52(6):303–309
10. Lu W, Zhang YM, Emmerson J (2004) Sensing of weld pool surface using non-transferred plasma charge sensor. Meas Sci Technol 15(5):991
11. Van Bael S, Kerckhofs G, Moesen M et al (2011) Micro-CT-based improvement of geometrical and mechanical controllability of selective laser melted Ti6Al4V porous structures. Mater Sci Eng A 528(24):7423–7431
12. Zhou X, Wang D, Liu X et al (2015) 3D-imaging of selective laser melting defects in a Co–Cr–Mo alloy by synchrotron radiation micro-CT. Acta Mater 98:1–16
13. Vrábel J, Pořízka P, Klus J et al (2019) Classification of materials for selective laser melting by laser-induced breakdown spectroscopy. Chem Pap 73(12):2897–2905
14. Chivel Y, Smurov I (2010) On-line temperature monitoring in selective laser sintering/melting. Phys Procedia 5:515–521
15. Gökhan Demir A, De Giorgi C, Previtali B (2018) Design and implementation of a multisensor coaxial monitoring system with correction strategies for selective laser melting of a maraging steel. J Manuf Sci Eng 140(4)
16. Grasso M, Laguzza V, Semeraro Q et al (2017) In-process monitoring of selective laser melting: spatial detection of defects via image data analysis. J Manuf Sci Eng 139(5)
17. Ye DS (2018) Study of in-situ monitoring methods in selective laser melting process. University of Science and Technology of China
18. Bobel A, Hector LG Jr, Chelladurai I et al (2019) In situ synchrotron X-ray imaging of 4140 steel laser powder bed fusion. Materialia 6:100306
19. Repossini G, Laguzza V, Grasso M et al (2017) On the use of spatter signature for in-situ monitoring of laser powder bed fusion. Addit Manuf 16:35–48
20. Grasso M, Colosimo BM (2019) A statistical learning method for image-based monitoring of the plume signature in laser powder bed fusion. Robot Comput-Integr Manuf 57:103–115
21. Zhang Y, Fuh JYH, Ye D et al (2019) In-situ monitoring of laser-based PBF via off-axis vision and image processing approaches. Addit Manuf 25:263–274
22. Garmendia I, Leunda J, Pujana J et al (2018) In-process height control during laser metal deposition based on structured light 3D scanning. Procedia Cirp 68:375–380

Chapter 8
Typical Geometric Shape Features Fabricated Through Laser Powder Bed Fusion

8.1 Classification of Different Geometric Features

Although there are many types and shapes of mechanical parts, the basic constituent features of mechanical parts are faces, columns, holes, angles, spheres, gaps, etc., as shown in Fig. 8.1. These features can represent the basic elements of the parts. The key factors of whether the laser powder bed fusion (L-PBF) technology can form mechanical parts can be reflected by these typical structural features. By manufacturing various typical geometric features such as thin plates, sharp corners, cylinders, round holes, square holes, spheres and gaps in Fig. 8.2, the processing capability of L-PBF was investigated in this section [1].

8.2 Defects and Formation Mechanisms in Different Feature Shapes

8.2.1 *Experimental Procedures*

The experiments are implemented on a self-developed LPBF equipment DiMetal-100, consisting of fiber laser, optical path transmission unit, sealed manufacturing chamber (including powder laying device), mechanical transmission and control system, process software, etc. The scanning speed range is 10–5000 mm/s, the processing layer thickness is 20–100 μm, the diameter of the laser focusing spot is 70 μm, and the maximum manufacturing volume is 100 mm × 100 mm × 120 mm. The optimized processing parameters were listed in Table 8.1.

The 316L stainless steel spherical powder is adopted as the raw material, and the average particle size is 17 μm, the maximum particle size is 35 μm, and the apparent density of the powder is 4.42 g/cm^2. The specific chemical composition is listed in Table 8.2.

© National Defense Industry Press 2024

D. Wang et al., *Laser Powder Bed Fusion of Additive Manufacturing Technology*, Additive Manufacturing Technology, https://doi.org/10.1007/978-981-99-5513-8_8

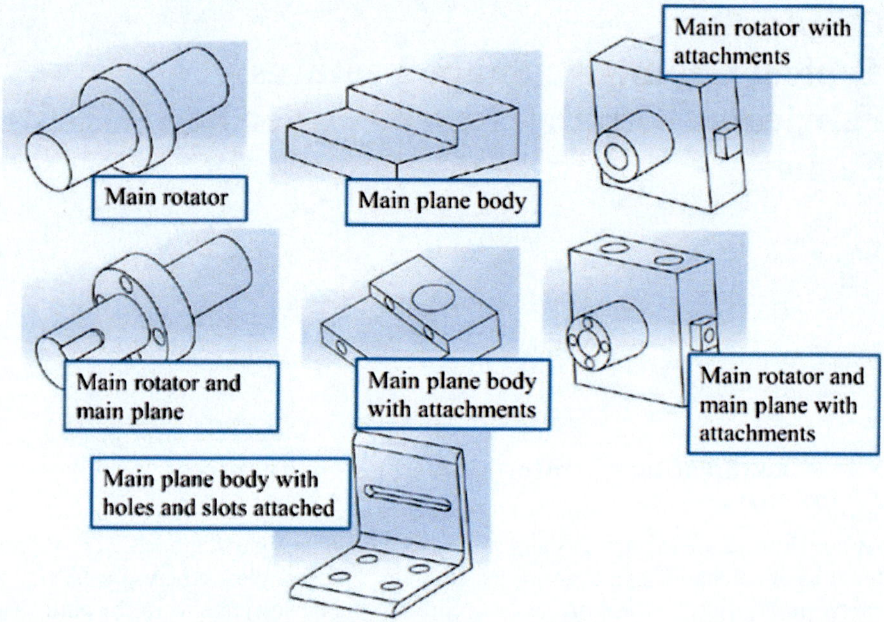

Fig. 8.1 Seven basic classifications of mechanical parts

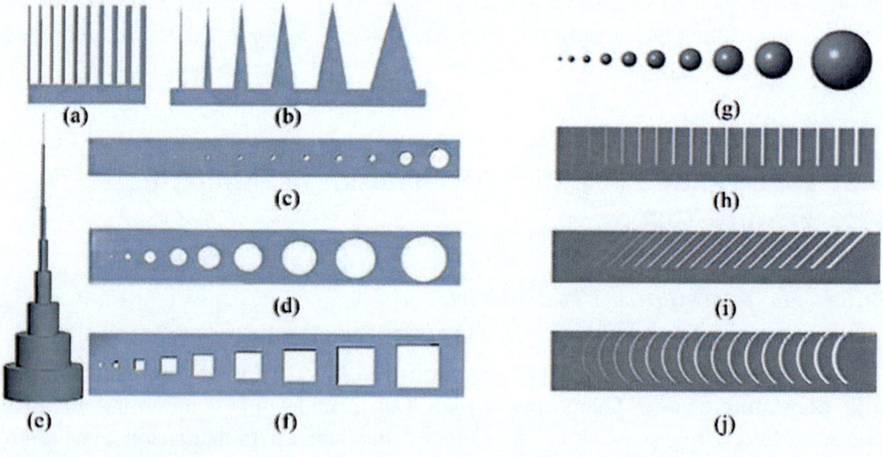

Fig. 8.2 3D model diagrams of typical geometric features: **a** thin plates; **b** sharp corners; **c** round holes parallel to Z axis; **d** round holes perpendicular to Z axis; **e** cylinders; **f** square holes perpendicular to Z axis; **g** spheres; **h** vertical gaps; **i** inclined gaps; **j** surface gaps

Table 8.1 Main processing parameters

Laser power (W)	Scanning speed (mm/s)	Layer thickness (μm)	Hatch space (mm)	Scanning strategy
170	600	30	0.08	XY orthogonal interlaminar stagger

Table 8.2 Chemical composition of 316L stainless steel powder (mass fraction, %)

C	Cr	Ni	Mo	Si	Mn	O	Fe
0.03	17.5	12.06	2.06	0.86	0.3	0.09	Bal

8.2.2 *Thin Plate*

During the experiment, 10 thin plates formed by L-PBF with a thickness of 0.05–0.5 mm (step value of 0.05 mm) were placed along the X axis, Y axis and 45° to the X axis in three directions (Fig. 8.3). As shown in Fig. 8.4, in both of the three directions, the thin plates with a thickness larger than 0.15 mm can be formed smoothly. The thin plate with a thickness of 0.1 mm can only be formed in half, while the thin plate with a thickness of 0.05 mm fail to form. As exhibited in Fig. 8.5, the absolute error of thin plates thickness increases with the increase of thin plates thickness, while the relative error decreases simultaneously, indicating that the thickness of plate is positively related to the manufacturing accuracy. When the thin plates are placed along the X axis for processing, the absolute error is 0.048–0.065 mm, the relative error is 13–48%, and the overall error is the minimum. Therefore, the thin plates processed along the X axis has the highest precision and the best quality among the three directions.

From the above analysis, the thin plates with a thickness of 0.15 mm can still be formed by LPBF. However, when the thickness of the thin plates is small to a certain extent, the constraint of laser spot on the formed part should be considered.

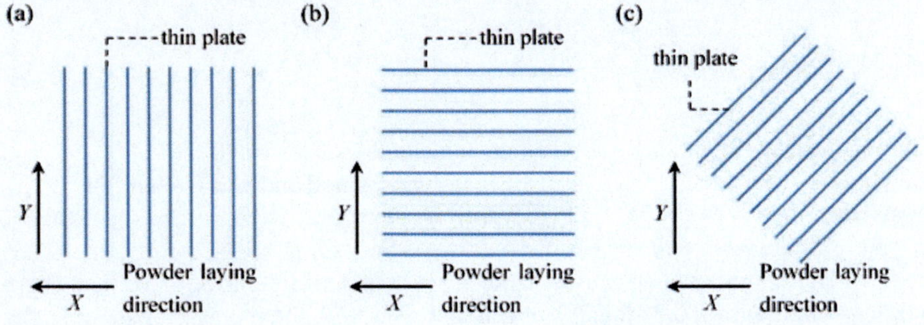

Fig. 8.3 Placement of thin plates: **a** along the X axis; **b** along the Y axis; **c** 45° to the X axis

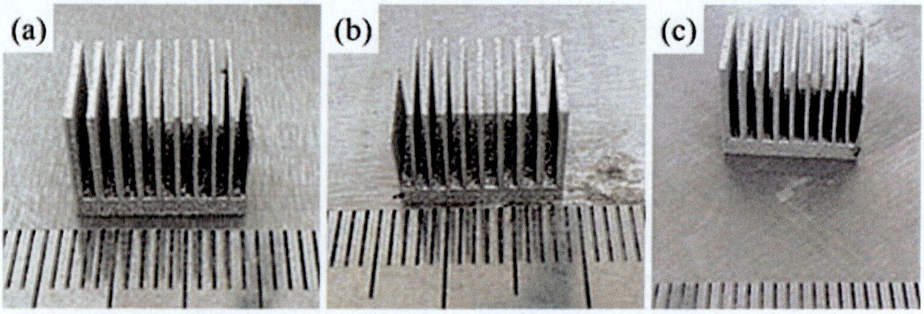

Fig. 8.4 Thin plates with different thickness: **a** along the X axis; **b** along the Y axis; **c** 45° to the X axis

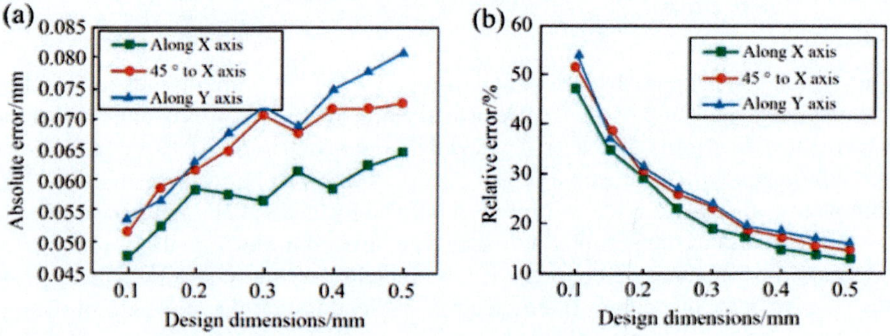

Fig. 8.5 Variation curves of thin plates thickness error: **a** absolute error curve; **b** relative error curve

The actual manufacturing size is constrained by the laser spot diameter, which means that the size of the spot diameter determines the size accuracy of the manufacturing. As shown in Fig. 8.6, when the wall thickness of thin-walled parts is smaller than the spot diameter, the actual fusion area will be larger than the cross-sectional area of the parts, and the final formed part size will also be larger than the design dimension.

8.2.3 *Sharp Corner*

During the experiment, the sharp corners with angles of 2°, 5°, 10°, 15°, 20°, 30° and height of 20 mm were formed by L-PBF with horizontal and vertical building directions, respectively. As shown in Fig. 8.7, all sharp corners were formed successfully. Figure 8.8 indicates that the absolute error of horizontal sharp corners is −0.15° to 0.2°, and the relative error is −2.2 to 3.8%. The absolute error of vertical sharp corners ranging from 0.3° to 2.6°, which increases with the increase of sharp angle; the relative error ranging from 8.64 to 14.3%, which decreases with the increase of sharp angle. The deviation of the error curve of horizontal sharp corners from the zero

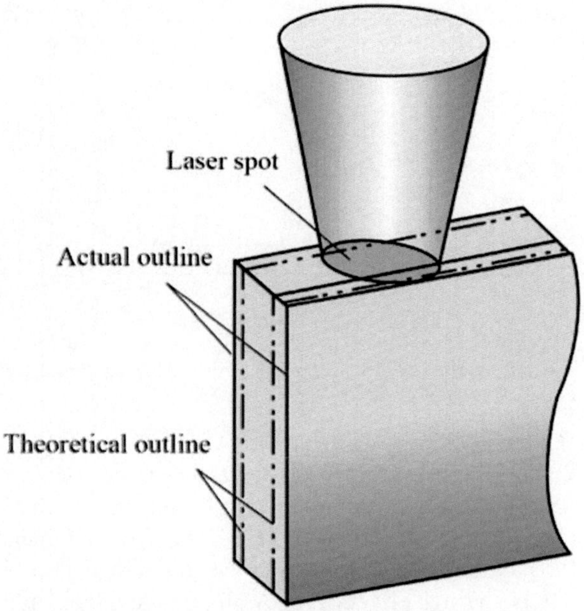

Fig. 8.6 The printable thin plate size limited by laser spot size

Fig. 8.7 Effect of LPBF build direction on sharp corners manufacturing

horizontal line is smaller than that of vertical sharp corners. Overall, the machining precision of sharp corners placed horizontally is higher than that of vertical ones.

This phenomenon can be ascribed that LPBF is based on the principle of discrete/stacking, which uses a layer-by-layer stacking method to shape the part. In this process, the model of the part needs to be layered along the stacking direction. After layering, the part model is divided into a finite number of slice layers with a certain

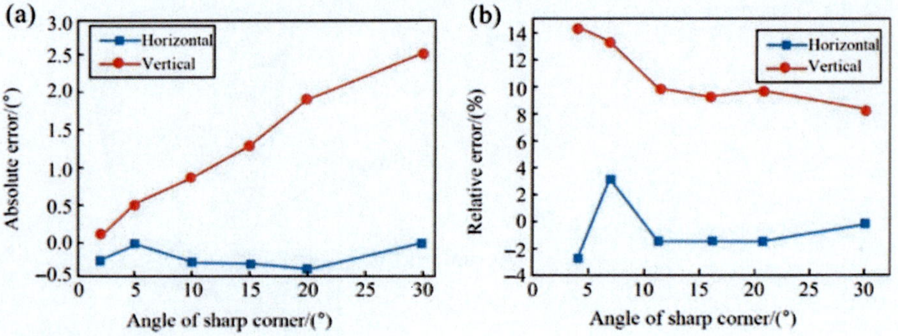

Fig. 8.8 Variation curves of angle error of sharp corners: **a** absolute error curve; **b** relative error curve

thickness. The information contained in the slice layer is only the contour of each layer and the corresponding entity, and the outer contour information between the slice layers is not counted. Therefore, in actual manufacturing, the outer surface of the part model is composed of the contour envelope surfaces of several slice layers, that is, the step effect. As shown in Fig. 8.9, the dotted line is the original contour of the vertical sharp corners, while the solid line is the outer contour obtained after the actual layering is jagged, and the horizontal sharp corners are not affected by this phenomenon. As a result, the error can be reduced by reducing the thickness of the layer, but this intrinsic error cannot be fundamentally eliminated.

Fig. 8.9 Influence of the layer thickness on the accuracy of vertical placement of sharp corners

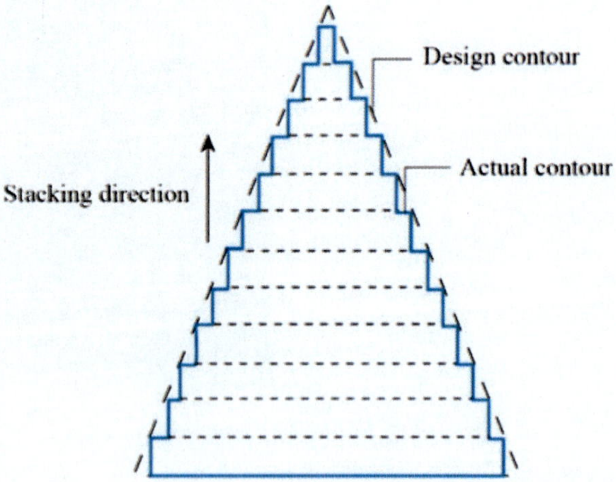

8.2.4 Cylinder

During the experiment, cylinders with diameters of 5, 3.5, 2, 1, 0.5, 0.3, 0.15, 0.1 and 0.05 mm were formed vertically by L-PBF. It can be seen from Fig. 8.10 that a cylinder with a diameter of larger than 0.1 mm can be successfully formed, but a cylinder with a diameter of 0.05 mm cannot be formed. This is because its feature is too small to generate scanning path during data processing, so its manufacturing fails. It can be seen from Fig. 8.11 that the absolute error of the cylinder diameter is 0.043–0.182 mm, and the relative error is 3.6–43%. The absolute error increases with the increase of the diameter, and the relative error decreases with the increase of the diameter and tends to be flat. Therefore, the diameter is positively related to the cylinder manufacturing accuracy.

Fig. 8.10 The photo of the LPBF manufactured cylinder sample

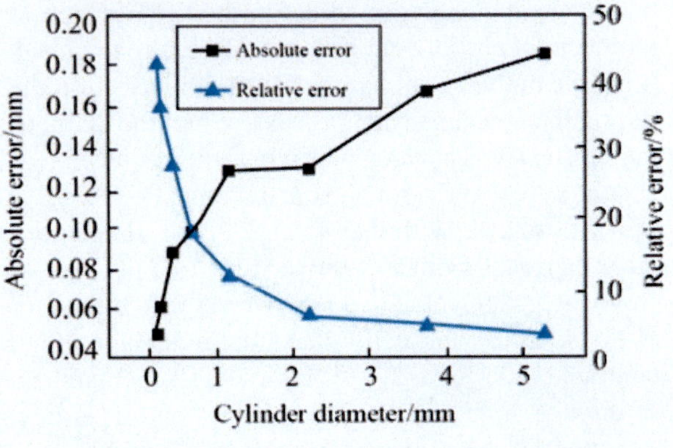

Fig. 8.11 The dimensional errors in LPBF-processed cylinder samples

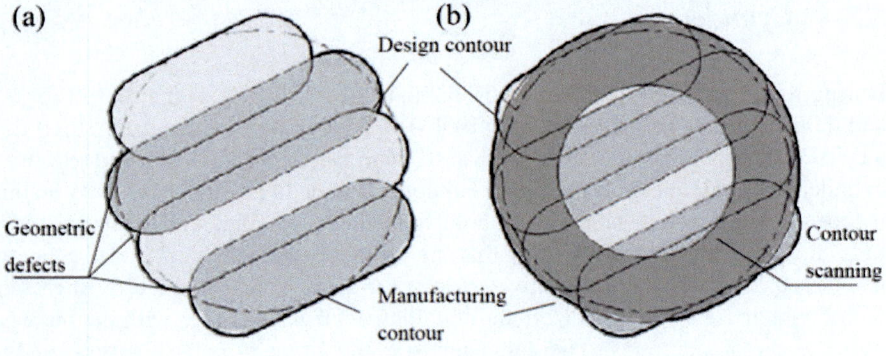

Fig. 8.12 The principle of cylinder contour error. **a** Cylinder contour error; **b** contour scanning improves accuracy

The origin of dimension error of the cylinder is mainly from the contour error. Figure 8.12a shows the filling diagram of the scanning line of the cylinder section. The design contour is round and indicated by dotted lines. Constrained by the width of the scanning melting track and the diameter of the laser spot, the actual formed contour will exceed the range of the design contour, as shown by the black solid line. Nevertheless, this kind of error can be reduced by adding a contour scanning. As shown in Fig. 8.12b, the contour scanning can make the contour surface remelted and refilled, making the contour surface smooth.

8.2.5 Round Hole Parallel to Z Axis

During the experiment, circular holes with diameters of 0.2, 0.3, 0.4, 0.5, 0.6, 0.7, 0.8, 0.9, 1, 2 and 3 mm parallel to the Z axis were formed by L-PBF. It can be seen from Fig. 8.13 that the L-PBF can form a round hole with diameters larger than 0.4 mm, but the round hole with a diameter of less than 0.3 mm fails to form. The reason is that the heat conduction effect in the L-PBF process makes the powder around the molten pool partially melted and adhered to the inside of the small hole. At the same time, the molten pool has an adsorption effect on the powder near the scanning line during the powder fusion process. When the diameter of the round hole is too small, the half molten powder will block the round hole. As shown in Fig. 8.14 that the absolute error of the round hole diameter is −0.085 to 0.06 mm, and the relative error is −21.3 to 2%. With the increase of the diameter of the circular hole, the absolute error increases from a negative value to a positive value. The relative error changes from a negative value to a positive value and then tends to be stable.

As with the vertically placed cylinder, the dimension error is mainly from the contour error. Figure 8.15 shows the schematic diagram of scanning line filling of circular hole section. The design contour is circular (dotted line in Fig. 8.15). Affected by the scanning melting track width and laser spot diameter, the actual

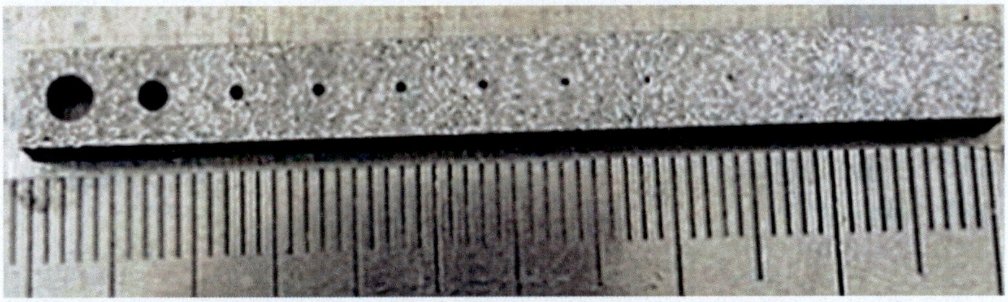

Fig. 8.13 Photo of LPBF-processed round hole parallel to Z axis

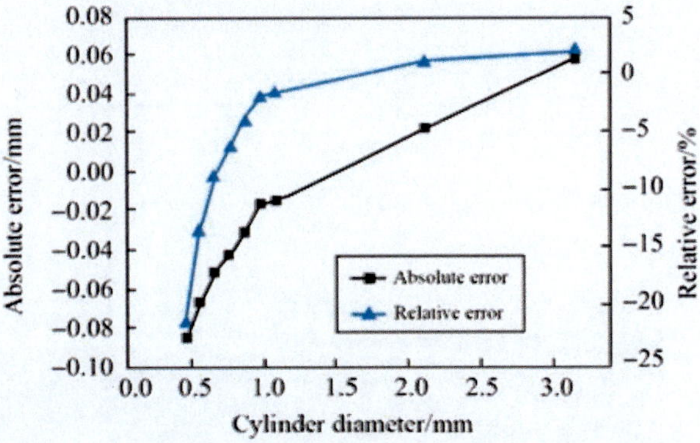

Fig. 8.14 The dimensional error of holes parallel to Z axis with different diameters

manufacturing contour is shown as the bold black solid line in Fig. 8.15. Meanwhile, powder adhesion will also affect the size accuracy of round holes. Due to the fact that the maximum particle size of the powder used is 35 μm, powder adhesion on the inner wall of a round hole will cause the hole diameter to become smaller, especially under a small hole diameter. Besides, the absolute error caused by powder adhesion should not be ignored.

8.2.6 Round Hole Perpendicular to Z Axis

During the experiment, round holes with diameters of 0.2, 0.5, 1, 2, 3, 4, 5, 6, 7 and 8 mm, which are perpendicular to the Z axis, are placed in turn and formed by L-PBF. As shown in Fig. 8.16 that L-PBFcan successfully form a round hole with a diameter of larger than 0.5 mm, while the round hole with a diameter of 0.2 mm fails to form. There is "slag hanging" at the top of the round hole, and the larger the hole diameter, the more serious the "slag hanging". As shown in Fig. 8.17a that the absolute error of the round hole diameter is −0.08 to 0.22 mm, and the relative error is −16 to 3%. With the increase of the diameter of the round hole, the absolute error changes from −0.08 mm to the positive error and gradually increases; the relative error changes

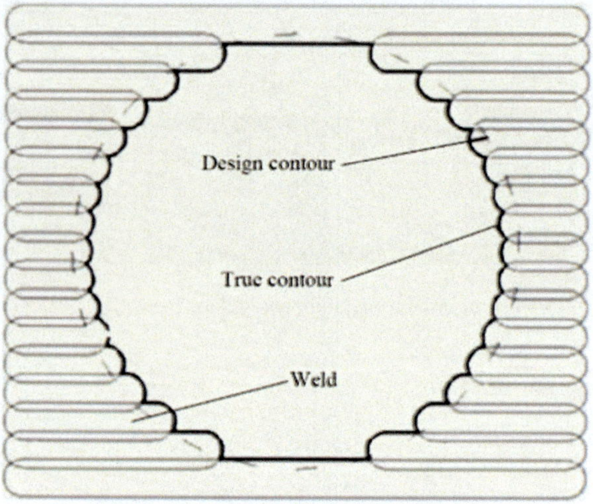

Fig. 8.15 Contour error of circular hole parallel to Z axis

Fig. 8.16 Photo of LPBF-manufactured circular holes perpendicular to Z axis

from −16% to positive error and tends to be stable. As shown in Fig. 8.17b, the round increases from 0.157 to 0.39 mm with the increasing diameter. In summary, the diameter of the round hole is negatively related to the shape accuracy.

In principle, the "slag hanging" on the top of the round hole is caused by the deep penetration of the laser. During L-PBF, the layers of the part are lapped by the laser penetrating the current formed layer and melting part of the volume of the previous formed layer, thus creating a metallurgical bond between the two layers. This part structure where the laser penetrates deeply and melts the powder layer is the overhang structure, as shown in Fig. 8.18. With the powder as the support of the overhang structure of the part, the molten pool sinks into the powder due to gravity and capillary force, resulting in "slag hanging" phenomenon in the overhang structure. The round hole is positively related to the overhang structure, so the phenomenon of "slag hanging" is more serious. "Slag hanging" will affect the dimensional accuracy

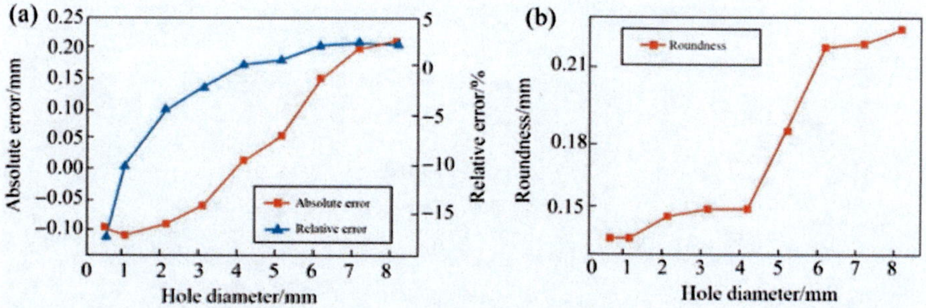

Fig. 8.17 Variation curves of diameter error of circular hole perpendicular to Z axis: **a** diameter error variation curve; **b** roundness variation curve

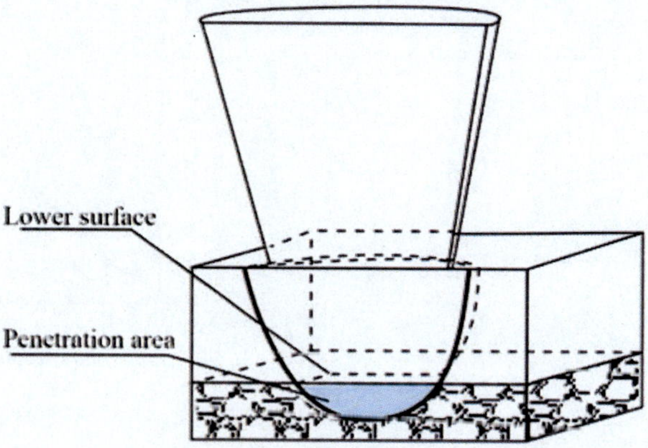

Fig. 8.18 Schematic diagram of laser deep penetration

and shape accuracy of round holes. In addition, the step effect is also one of the main reason for the diameter error of the overhanging circular hole.

8.2.7 Square Hole Perpendicular to Z Axis

During the experiment, square holes with side lengths of 0.5, 1, 2, 3, 4, 5, 6, 7 and 8 mm that perpendicular to the Z axis were formed by L-PBF. As shown in Fig. 8.19, square holes with side lengths of 1, 2 and 3 mm can be successfully formed without obvious defects. The square hole with a side length of larger than 4 mm can be formed, but due to the deep penetration of the laser, there are many "slag hanging" on the top of the hole, and the phenomenon of warping is serious. The square hole with a side length of 0.5 mm can only be barely formed due to powder adhesion and "slag hanging". Because of the serious "slag hanging" on the top of the square hole, the upper and lower heights of the square hole cannot be accurately measured, so only the span of the square hole is measured. Figure 8.20 shows the error curve of the

Fig. 8.19 Photo of LPBF-manufactured square holes perpendicular to the Z axis

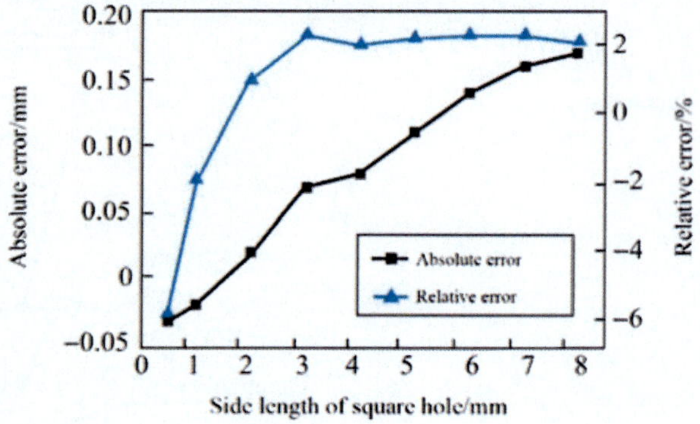

Fig. 8.20 Variation curve of span error of square hole perpendicular to Z axis

span of the square hole. As shown in Fig. 8.20 that the absolute error of square hole span is −0.03 to 0.17 mm, and the relative error is −6 to 2.33%. With the increase of square hole span, the absolute error approaches zero from −0.03 mm and becomes positive error, and then gradually increases to 0.17 mm. The relative error approaches 0 from −6% and becomes positive error, then gradually increases and tends to be stable.

8.2.8 Sphere

During the experiment, spheres with diameters of 0.5, 0.8, 1, 1.5, 2, 2.5, 3, 4, 5 and 8 mm were designed and formed by L-PBF. As shown in Fig. 8.21 that when the diameter is less than 1 mm, the shape of the sphere is not obvious and agglomerated. When the diameter is 1–1.5 mm, the shape of the upper half of the sphere is obvious, but the lower surface of the sphere is connected with the substrate, and the shape of the lower half of the sphere is not clear. When the diameter of the sphere reaches 2 mm, the surface of the sphere is uneven, and the influence of the adhered powder on the size and shape of the sphere cannot be ignored.

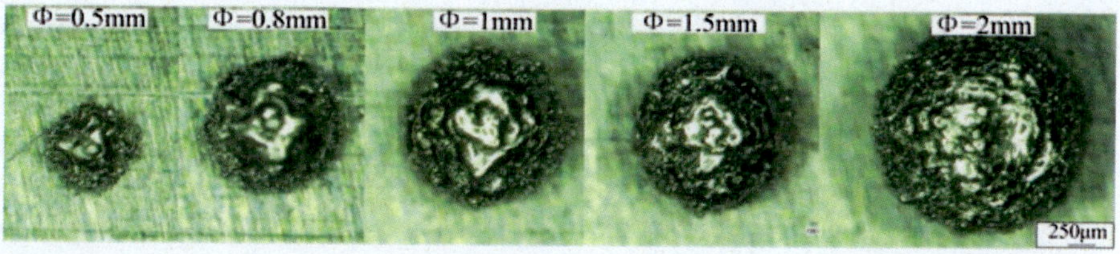

Fig. 8.21 Photo of LPBF-manufactured spheres

8.2.9 Gap Feature

During the experiment, vertical gaps, inclined gaps and curved gaps with gap dimensions of 0.02, 0.04, 0.06, 0.08, 0.1, 0.12, 0.14, 0.16, 0.18, 0.2, 0.22, 0.24, 0.26, 0.28, 0.3, 0.35 and 0.4 mm were formed by L-PBF. In Fig. 8.22, the gap dimensions are arranged from left to right in the above order. The gap features are completely blocked when the gaps are too narrow. For the vertical gap with size 0.1–0.18 mm, inclined gap with size 0.08–0.2 mm and curved gap with size 0.12–0.16 mm, the gap characteristics are blurred and a lot of powder is trapped inside the gap. For the vertical gaps with dimensions greater than 0.2 mm, inclined gaps with dimensions greater than 0.2 mm and curved gaps with dimensions greater than 0.18 mm have obvious gap characteristics and the gaps are clearly visible.

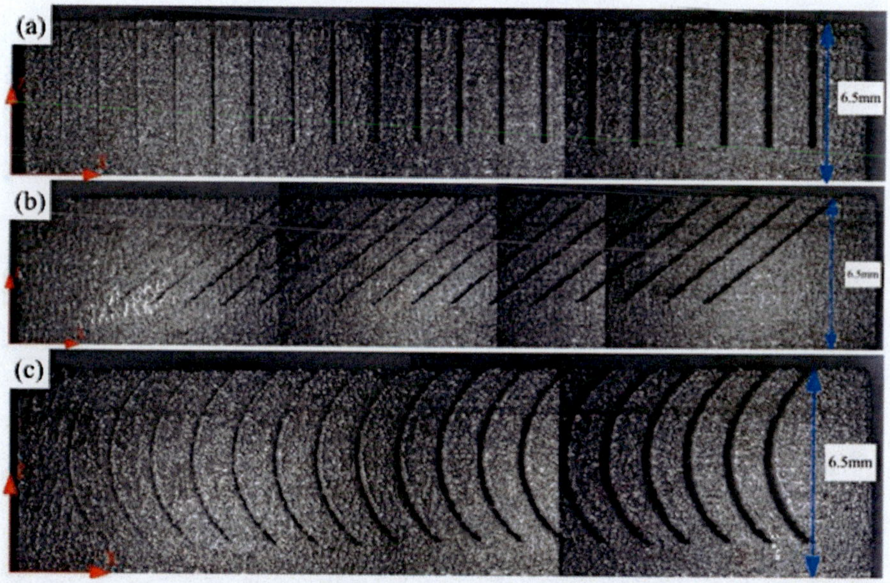

Fig. 8.22 Different gap features manufactured by LPBF: **a** vertical gaps; **b** inclined gaps; **c** curved gaps

Overall, gap manufacturing quality is influenced by many factors. Although the powders in the gap features are not melted during the manufacturing process, they are heated and form large clumps that are difficult to be removed from the gap features due to the thermal effects of spot scanning, resulting in the gap features being blocked. The delamination discretization causes dimensional errors in the manufacture of inclined surfaces and curved surfaces, and the "slag hanging" phenomenon on the overhanging surface caused by the deep penetration effect of the laser will aggravate the clogging of the inclined and curved surface gaps [1–3].

8.3 Critical Manufacturing Angle

L-PBF manufactures parts by overlapping between layers. When manufacturing geometry with inclined feature, the thickness of processing layer and the angle of inclination determine the relative area of overlapping surfaces. When the processing layer thickness is certain, the smaller the angle of inclination of the overhanging surface, the smaller the area of the overlapping lap surface, the more overhanging parts, the more overhanging objects on the overhanging surface, thus causing poor quality of the manufacturing surface and warpage defects and other problems. Therefore, there is a critical manufacturing angle when L-PBF is used to shape the geometry with inclined characteristics. On the basis of the optimized process, overhang structures with different inclination angles were designed and fabricated. Figure 8.23 shows the manufacturing effect of the overhang structure with the inclination angle reduced from 45° to 25°. As shown in Fig. 8.23 that the overhang structure with an inclination angle greater than or equal to 40° is well formed, and the overhang structure with an inclination angle of less than or equal to 35° has warpage defects. In addition, the inclination angle is negatively related to the powder adheres to the

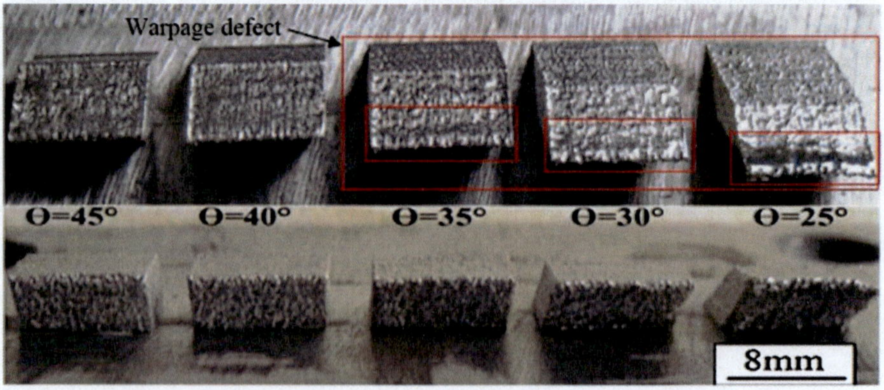

Fig. 8.23 LPBF manufactured overhang structures with different inclination angles

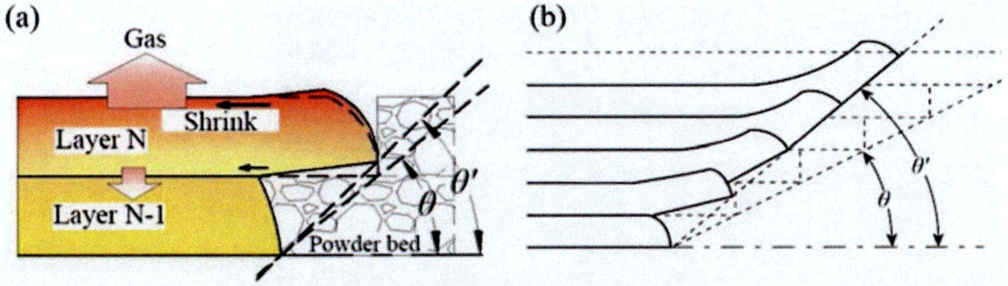

Fig. 8.24 Warpage principle in LPBF manufacturing of overhanging surfaces: **a** warpage principle; **b** warpage accumulation

lower surface of the overhang structure, which makes the quality of the lower surface worse than that of the upper surface.

Warpage defect is caused by the thermal stress produced by rapid solidification of molten pool during L-PBF. Many researchers have investigated the effect of thermal stress on selective laser sintering (SLS) and powder bed fusion (PBF) processes, and suggested that temperature regimes can explain the warpage phenomenon. When the thermal stress exceeds the yield strength of the material, plastic deformation will occur. Warpage defects in the overhanging surface could also be caused by the lack of support that does not form a strong bond between the manufacturing structure and the previous layer. The accumulation of residual stress is the main reason for the manufacture of defects in the overhang surface. Figure 8.24a shows the warping principle of L-PBF when manufacturing overhanging curved surfaces. After the scanning of a single layer with overhanging parts during manufacturing, the volume shrinkage of the melted powder during the liquid–solid change process results in the overhanging portion is warped upwards. Because of the temperature difference between the top and bottom of the scanning layer and the non-uniform thermal conductivity, the upper part of the manufacturing layer shrinks faster than the bottom, resulting in the upwarping of the overhanging layer. In this state, the angle of inclination between the overhanging part and the overhanging surface of the part formed by the previous manufacturing layer θ' greater than design angle θ. As shown in Fig. 8.24b, when one layer of the overhanging part begins to warp, the actual manufacturing layer thickness of the next layer will be affected, and greater warping will be caused, thus making the inclination angle of the part overhanging surface θ' constantly exceed the design angle θ. When the accumulated warping height is higher than the preset height of the next layer, there will be no powder coating in some manufacturing areas, and the whole workpiece will become more and more fragile. The overhanging surface may be subject to repeated laser scanning, and even the overhanging part may be separated from the whole part due to repeated collision with the powder scraper, resulting in manufacturing failure. In some case, the whole manufacturing process must be stopped, and the parts need to be redesigned or the process needs to be optimized if the warping defect is serious.

Overhanging surfaces cause the powder to overhang when formed. During laser scanning of the metal powder, there is a heat affected zone around the molten pool,

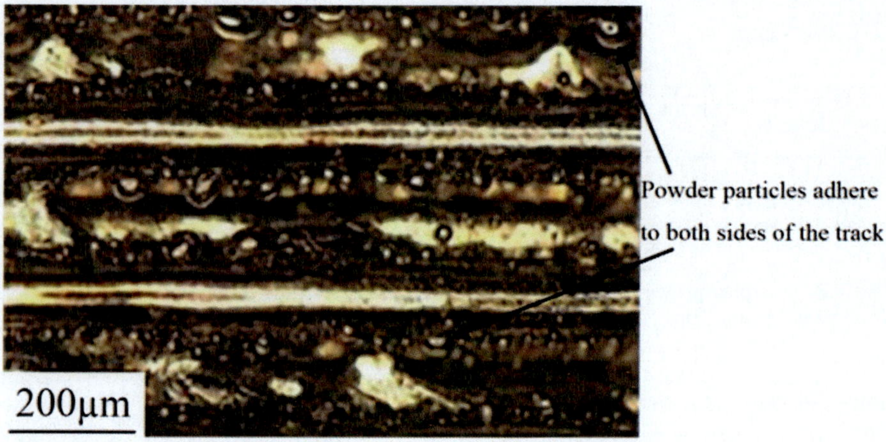

Fig. 8.25 Powder particles stuck on both sides of the track

which causes the metal powder around the pool to melt completely or to be in a more brittle state. As shown in Fig. 8.25, many metal powder particles are adhered to the sides of the track. As shown in Fig. 8.26 of manufacturing overhanging structures, when the laser irradiates the solid support area (point a), the thermal conductivity is high and prone to severe powder adhesion; when the laser irradiates the powder support area (point b), the thermal conductivity is only 1/100 of the thermal conductivity of the solid support area, which often occurs in the process of manufacturing overhanging surfaces by the L-PBF. Therefore, in the case of similar process parameters, the energy input absorbed by laser irradiation in the powder support area is much larger than that in the solid support area, resulting in a larger melt pool and sinking into the powder under the action of gravity and capillary force. Due to these reasons, slag will be formed when the overhanging structure is formed by L-PBF, and under this circumstance the dimensional accuracy will become very low [4].

8.4 Design Rules for Laser Powder Bed Fusion Manufacturing

8.4.1 Design Constraints

L-PBF can avoid most of the limitations of traditional manufacturing, and is able to directly convert complex designs into final products. Nevertheless, although L-PBF has a very high degree of manufacturing freedom, it does not mean that any part can be formed, and improper design will also cause processing failure. As a result, the constraints of manufacturing conditions should be considered carefully in the design stage to avoid the possible manufacturing failure and improve the manufacturing

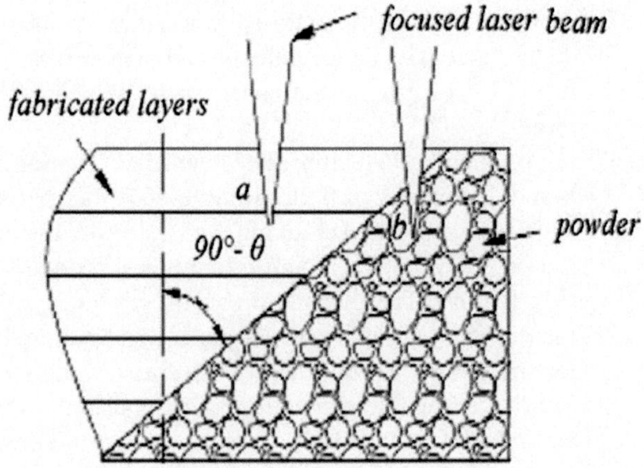

Fig. 8.26 Principle of LPBF manufacturing overhang surface

quality. Besides, designers should also consider the limitations of the process in the product design stage, and avoid the limitations of the process from the perspective of product design, so that the design freedom and the manufacturing freedom are more compatible. The shape of parts formed by L-PBF may be very complex, but it is also composed of basic geometric elements. Through the study of manufacturing geometry structure, its geometric manufacturing constraints can be roughly classified into the following categories:

(1) Thin wall feature. Due to the limitation of the focus size of the laser spot used in the L-PBF, thin-walled parts with a wall thickness smaller than the spot diameter cannot be formed. Moreover, the mechanical properties of thin-walled parts with too small wall thickness are difficult to guarantee, and have no practical value. The theoretical minimum limit size for thin walls is the width of a single melting track.
(2) Sharp corner feature. Since the shape of the laser spot can be considered as a circle, although the size of the spot is only tens to more than one hundred microns, for fine structures such as sharp corner feature, it will lead to large errors in the shape and size of the sharp corners. Excessive sharp corner feature should be avoided as much as possible in product design. There is a minimal limit to the sharp corners of fine structures.
(3) Hole feature. Because the laser spot size has a limit value and is affected by the diffusion of the heat affected zone during laser processing, the width of the melting track is larger than the spot size. If the pore size is too small, the melting track will block the hole, and the size of the hole feature perpendicular to the manufacturing direction has a minimum limit. The manufacturing effect is influenced not only by the laser spot, but also by the deep penetration of the laser for hole feature parallel to the manufacturing direction. The diameter of the hole is positively related to the overhanging area, "slag hanging" amount is negatively related to the shape accuracy and dimensional accuracy. Therefore, the hole size should be controlled at a reasonable level. The heat accumulation

effect of high aspect ratio parts is more obvious, and the accumulated thermal stress can easily cause defects such as warping and cracking, and the parts are difficult to form, so the aspect ratio of geometric features should not be too large.

(4) High aspect ratio feature. The heat accumulation effect of high aspect ratio parts is more obvious, and the accumulated thermal stress is easy to cause defects such as warping and cracking, and therefore the parts are difficult to form. Thus, the aspect ratio of geometric features should not be too large.

(5) Gap feature. The thermal diffusion caused by the laser processing may cause incompletely melted powder particles to adhere to the gap, affecting the manufacturing accuracy of the gap. As a result, the size of the gap should not be too small. Reasonable gap features can be used for the integrated design and manufacturing of the assembly-free mechanism to ensure the freedom of movement after manufacturing.

8.4.2 Design Principles

8.4.2.1 Placement of Parts

During the processing of parts, there are specific "working areas" that has high requirements for surface quality. Under this circumstance, designers should prioritize these area to ensure the manufacturing quality of the "working areas". Therefore, a reasonable position is needed to ensure the quality of the "working area" and the efficiency of the manufacturing. In the L-PBF process, the parts are stacked layer by layer through the powder bed. In addition to dissipating heat through the air, the bottom layer is also an important way to dissipate heat. When the lower layer of the molten powder is solidified metal solid, the heat spreads faster, and the molten pool solidifies quickly and is precisely combined with the lower solid. When the lower layer of the molten powder is also powder, the heat dissipation of the molten pool is relatively slow due to the lower thermal conductivity of the powder, which will cause the powder on the lower surface of the molten pool to be sintered and adhered to the lower surface of the overhanging structure, making the surface quality of overhanging structure rough, so further post treatment is needed to improve the surface quality. Surfaces with overhang angles below 30° require additional support to avoid manufacturing failure, but the support structure also affects the surface quality. In the structural optimization design, the machining position of the parts should be considered to avoid removing too much material to generate too many machining overhangs, or adding too much support structure to affect the manufacturing quality of the "working area". In the early stage of part design, when considering the influence of placement factors, priority should be given to ensuring the manufacturing quality of the "work area". Parts with complex geometric shapes may be difficult to ensure the manufacturing effect of all key surfaces during placement. As a result, making a trade-off between surface quality, structural details, processing cost and support quantity are important. Notably, the processing software Magics can be used to evaluate each placement direction at the early stage of part design to determine the most effective way, and continue to further design on this basis.

8.4.2.2 Self-supporting Structure

Subtractively manufactured parts have the most stable and rigid state at the beginning of processing, and the stiffness gradually decreases with the removal of material. In contrast, additive manufacturing (AM) is quite different. During AM, the part changes from an initial unstable state to a final stable state. Each machining cycle must ensure that the rigidity of the part can resist the force exerted on it, including the gravity of the part itself, the external force exerted by the machining equipment, and the thermal stress. In order to ensure smooth manufacturing, the parts are always in a more rigid state during processing, and designers usually choose to add support structures. However, considering the impact of the removal of the support structure on the quality of the final parts, how to reduce the support should also be considered in the product design stage. Besides, in the product design stage, designer should consider the placement of the parts during the LPBF manufacturing, adjust the structure of the manufacturing danger zone to achieve self-supporting function to reduce the support structure, or add self-supporting structure without affecting the performance of the parts. For example, it is difficult to ensure the shape accuracy of round holes, especially when the round holes are parallel to the manufacturing plane. Under this circumstance, the shape of the holes should be optimized into a water drop shape or a diamond shape. Both the water drop hole and the diamond shaped hole are the edges of the holes converging and shrinking more and more upward. There is no suspended surface with too low inclined angle. The lower part of the hole provides support for the upper part of the hole; the self-supporting structure (the part structure itself as the support carrier of the part) is used to improve the stability of the part during the manufacturing process. The self-supporting structure can not only strengthen the stiffness of the original part to avoid warping and fracture during the manufacturing process, but act as a heat transfer path to effectively reduce thermal stress as well.

In some cases, support structure is essential, such as when the overhang angle is less than 30° with overhang surface. The support is needed to fix the part, promote heat dissipation, should consider the location of the support to add, remove the support and other post-processing means of convenience, etc., to avoid damage to the parts, especially the removal of thin-walled parts of the support is easy to make the thin-walled parts of the body deformation. The support inside the part is difficult to reach with the tool, which also affects the efficiency of the support removal. Improper support design will increase the risk of destruction of the prototype, regardless of the means to remove the support, will increase the cost and prolong the manufacturing cycle, so the design should be considered by changing the design, changing the way the parts are placed, etc. to avoid adding support structure for the parts as much as possible.

8.4.2.3 Machining Allowance

In some cases where high surface quality is required, the metal parts directly formed by L-PBF may can not achieve the required manufacturing quality, as the surface may have ripples, powder adhesion, residue after removal of support, etc. Under this circumstance, post treatment is needed to improve the surface quality of the part. Therefore, designer should retain a certain amount of margin for the post treatment.

8.4.2.4 Rational Selection of Processes, Materials and Equipment

The type of process, material type and manufacturing equipment of L-PBF will have different degree of constraints on the manufacturing of the parts and affect the quality of the manufacturing of the parts. The combined effect of these factors affects the shrinkage, surface accuracy, dimensional accuracy and shape accuracy of the part. Considering these limitations, designers must choose the appropriate L-PBF process to meet the functions that specific parts and materials should have, or take into account the constraints of L-PBF in the product design process and compensate by modifying the design scheme.

8.4.2.5 Post-processing and Testing Technology

Unlike conventional manufacturing methods, which manufacture parts with relatively regular shapes, L-PBF can form parts with complex curved configurations or internal cavities, which poses new challenges for post-processing and quality inspection. Designers should consider the convenience of post-processing and quality inspection control in the early product design, especially to ensure the optimal treatment and inspection of the surface quality of the "working area" [5].

8.5 Design Rules and Process Characteristics of Porous Structure

8.5.1 Dimensional Accuracy

The smallest geometric features that can be produced by L-PBF are mainly determined by the spot size of the focused laser. When the thickness of the part profile is smaller than the size of the laser focus spot, the actual manufactured profile is larger than the design value even without considering the effect of heat conduction, resulting in a large error in dimensional accuracy. The laser source is Gaussian distributed and the size of the focused laser spot varies with the amount of defocus. And when the laser power and scanning speed is constant, the energy density near

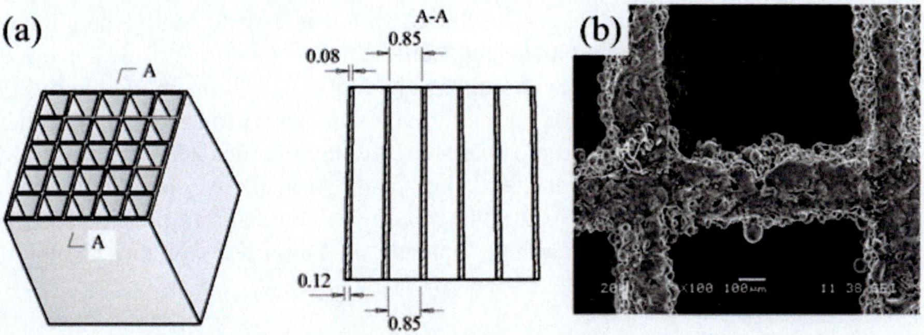

Fig. 8.27 Dimensional accuracy test. **a** Design model; **b** SEM image after manufacturing

the center of the beam of the small spot is higher than that of the center of the beam of the large spot. Therefore, even when the defocus amount changes in a small range, the heat transfer of laser energy will have an impact on the manufacturing accuracy, and the actual formed minimum contour size will also be larger than the focused laser spot diameter. Figure 8.27 shows the dimensional accuracy test, from which it can be seen that both thin wall and square holes can be made to almost identical dimensions. As shown in Fig. 8.27b, there are some incompletely melted powders adhering to the surface of the thin wall, which need to be removed by other methods, such as electrochemical methods. Powder adhesion has an effect on the dimensional accuracy. The measured values of the top thickness and bottom thickness of the thin wall are 101.3 μm and 142 μm respectively, and the absolute errors are 21.3 μm and 22 μm respectively. The powder adhesion due to heat transfer causes a certain dimensional deviation between the actual part and the model.

8.5.2 Geometric Feature Resolution

The introduction of geometric characteristic resolution can provide a basis for the design of pillar diameter and void size of porous structures. The resolution of the manufacturing geometric features of L-PBF has an important influence on the design and processing of the porous structure, and is an important parameter basis for the manufacturing of the porous structure, that is, the diameter of the pillar cannot be smaller than the minimum thickness of the thin plate and the diameter of the cylinder in the design of the porous structure. According to the thin plate experiment in Sect. 8.2.2, thin plates with thickness greater than or equal to 150 μm can be formed smoothly. As shown in the cylinder experiment in Sect. 8.2.4, the cylinder with a diameter greater than or equal to 100 μm can be formed smoothly. Thin plates and cylinders with geometric features larger than 150 μm can be formed. Considering that smaller geometrical features will lead to lower mechanical properties of the porous structure and limited by the focused spot size, etc., the minimum size should be greater than the experimental result value of 150 μm when designing the pillar of the porous structure. According to the gap manufacturing experiments in Sect. 8.2.9, it can be seen that vertical gaps with dimensions larger than 200 μm, inclined gaps with dimensions larger than 200 μm and curved gaps with dimensions larger than 180 μm are formed well and the gap characteristics are obvious and clearly visible. Therefore, when the design gap is 200 μm, it can be basically formed. Considering that the melt path has a certain width, the gap after manufacturing must be smaller than the design gap, so the design hole diameter of the porous structure cannot be less than 200 μm.

8.5.3 Inclined Angle

When the tilt angle of the cell bracket is small, it is difficult to control the fabrication quality of the porous structure. In order to avoid overhanging surfaces with small inclination, it is desirable to optimize the porous cell to meet the requirements of the L-PBF process. In this book, based on the discussion of the critical tilt angle, the porous unitary shown in Fig. 8.28 is proposed, which can also be called octahedral unitary. By adjusting the inclination angle of the pore pillar, the octahedral structure can theoretically avoid the overhanging surface with small inclination angle, thus avoiding the defects of overhanging floating slag and blocking pore space.

As shown in Fig. 8.28, the following equation is obtained by analyzing the intrinsic relationship between the unit radius *R*, pillar length *L*, pillar diameter *d* and pillar inclination angle θ:

$$R = \frac{1}{2} \times L \times \sin 2\theta - d \tag{8.1}$$

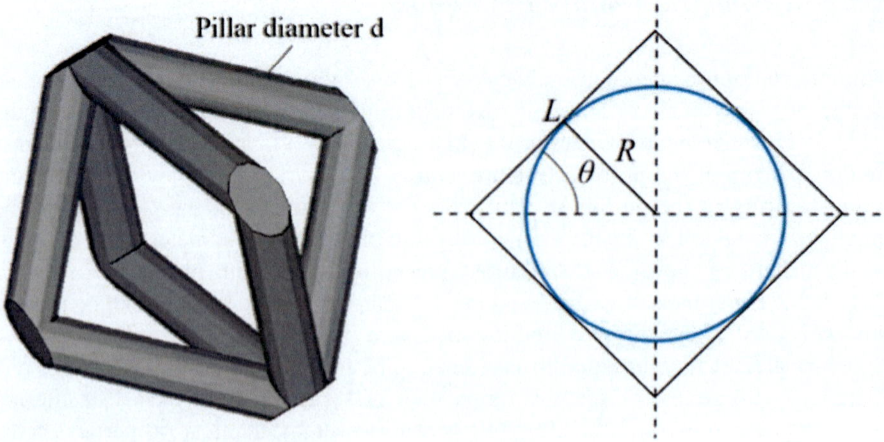

Fig. 8.28 Octahedral structural unit

Fig. 8.29 Porous structure unit body suitable for L-PBF process

where, R is unit radius, L is pillar length, θ is inclination angle of pillar, d is pillar diameter.

In Formula (8.1), the pillar diameter d mainly depends on laser processing parameters and powder particle size. Due to the thermal effect around the molten pool during manufacturing, the pillar diameter d is usually larger than the focus diameter. When $\theta = 45°$, the radius of the porous cell is the largest, that is, $R = 1/2L - d$. At the same time, it can be deduced that when $R > 0$, $1/2L\sin 2\theta > d$ is required. Combined with the experimental results in Sect. 8.5.2, the design principle of octahedral unit formed by L-PBF is obtained as follows:

$$\begin{cases} 30° < \theta < 90° \\ d \geq 0.15\,\text{mm} \\ \frac{1}{2} \times L \times \sin 2\theta > d \end{cases} \tag{8.2}$$

According to the design principles summarized above, the design of porous structure unit should consider the mutual constraints among critical inclination angle, processing resolution and geometric parameters of the unit. Although the octahedral porous structure is used as the unit body in this book, it is not the only unit body suitable for L-PBF manufacturing porous structure. Here, as long as the constraint of critical inclination angle and the limit of geometric resolution are considered, a satisfactory unit body can be designed. Figure 8.29 shows three other porous structure units which are also suitable for L-PBF process.

8.5.4 Contour Accuracy

According to the design principle of octahedral unit, the octahedral porous structure with the pillar diameter $d = 0.3$ mm, pillar length $L = 1$ mm and pillar inclination angle $\theta = 45°$ was designed. The octahedral porous structure is processed and shaped by the optimized process parameters, and the formed porous structure is shown in Fig. 8.30. As shown in Fig. 8.30a that the overall manufacturing effect of porous

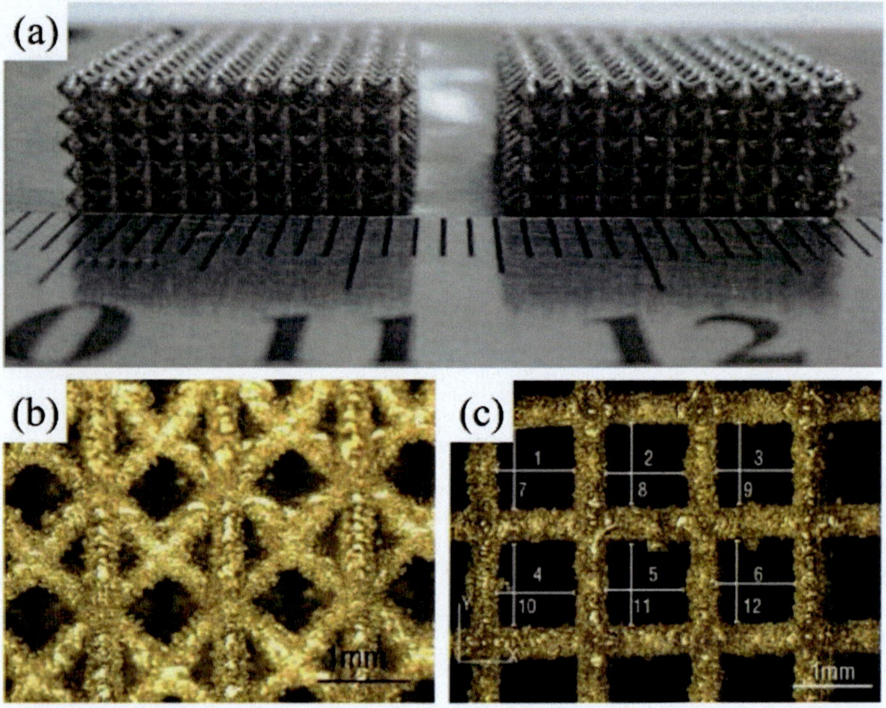

Fig. 8.30 The LPBF-processed octahedral porous structure: **a** overall view; **b** enlarged view of side surface; **c** enlarged view of upper surface

structure is good, but As shown in the enlarged figures of Fig. 8.30b, c that some fine powder particles adhere to the surface of the holes, and the surface of the pillars is not as smooth as the metal parts produced by traditional methods. There are three main reasons: (1) Due to the melting and solidification of the powder, there is always a heat affected zone on both sides of the melting track, which is easily for the powder to adhere to the surface of the melting track; (2) The cross-sectional scanning area of each layer is small, so the scanning time of micro-area laser is very short. Sometimes the powder is not completely melted, and some of the powder sticks to the periphery in the form of semi-melting. (3) When the laser scans the powder outside the preset hole, the liquid metal will inevitably penetrate the inner hole, which makes the powder easily adhere to the melting track, resulting in low dimensional accuracy and poor surface quality of the final porous structure. The most difficult thing to deal with is how to remove the powder in the hole after manufacturing. In order to ensure the surface quality and mechanical strength of the porous structure, the diameter of the hole should not be too small.

Table 8.3 shows the measurement results of pore size in Fig. 8.30c. As shown in Table 8.3 that the actual size of the gap is slightly smaller than the designed value, which may be caused by powder adhesion. The optimized process parameters are the

Table 8.3 Measurement results of pore size

X direction		Y direction	
Measuring point	Width/μm	Measuring point	Width/μm
1	965	7	976
2	999	8	979
3	961	9	984
4	930	10	976
5	946	11	965
6	987	12	972
Average	965	Average	975
Design	1000	Design	1000
Error	35	Error	25

basis for ensuring the strength of the porous structure. Only by obtaining a compact structure can the strength of the porous structure be guaranteed. The inclination angle of the pillar of the unit should not be less than the critical angle, otherwise the porous structure inside will be blocked by powder, and the diameter of pores must be large enough.

Figure 8.31 shows a porous structure with vertical and 45° inclined struts. As shown in Fig. 8.31c that the cross section of the vertical strut is approximately elliptical. This is because when the scanning path is generated, the circular profile is filled by the scanning line. During the machining process, one scanning line in the scanning data, i.e., the single melting track, is obtained by moving the focused laser spot, and the end section at the end of the melting track may not completely conform to the designed sample profile. The actual processed contours are irregular, and the densities can be improved by using interlaminar overlapping scanning, but the outer contours will still have certain geometric defects. Scanning the interface contour can be used to improve the contour accuracy. After adding contour scanning, the processed contour can be guaranteed to be circular in principle, and the contour accuracy can be improved.

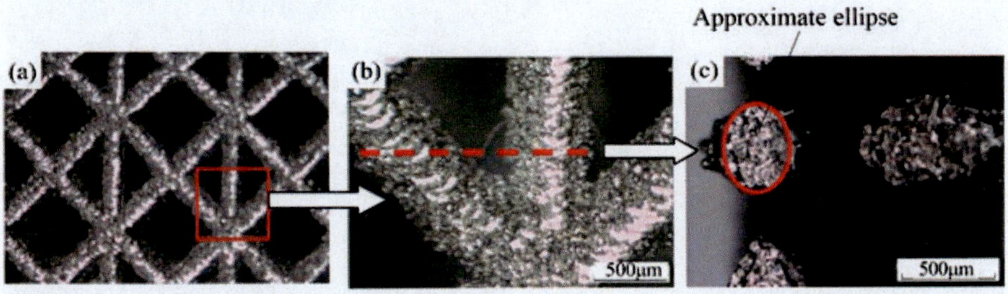

Fig. 8.31 Contour defects observation: **a** low-magnification view; **b** high-magnification view; **c** cross-section of the struts

In the process of scanning the contour, the actual contour scanning track makes the pillar surface remelted, the melted material is refilled, and the remelted pillar surface becomes smooth. In fact, one of the advantages of L-PBF is that it can process the final parts, which can be directly used after simple post-processing. The contour defects generated by the scanning of each layer also lead to the increase of the roughness of the side surface, and the surface roughness is an important factor affecting the application of PBF technology. Although the surface quality can be improved by post-processing such as polishing or shot peening, for porous structural parts with fine structural features, polishing or shot peening may destroy the fine features of the porous structure, such as pillar fractures. Therefore, the selection of appropriate process parameters, scanning strategy and metal powder particle size is crucial for the manufacture of porous structures.

8.5.5 Slag Hanging and Powder Adhesion

Slag hanging and powder adhesion are common phenomena in the process of L-PBF manufacturing porous structures, as shown in Fig. 8.32. Slag hanging and powder adhesion can lead to the reduction of the pore shape accuracy and dimensional accuracy of the porous structure, and it is difficult to completely remove the metal powder adhering to the pillar by post-processing technology. One of the main reasons for slag hanging and powder adhesion is that the laser interacts with the powder to form a melting zone and a heat affected zone. After the melting zone solidifies, a melting track is formed. The powder in the heat affected zone is not fully melted due to insufficient energy absorption, and adheres around the melting track. The overhanging structure is also an important cause of powder adhesion, because the existence of the overhanging structure will make the laser directly incident on the powder, which will cause the molten pool in this area to be too large, and the molten pool will sink into the powder due to gravity, resulting in the phenomenon of "slag

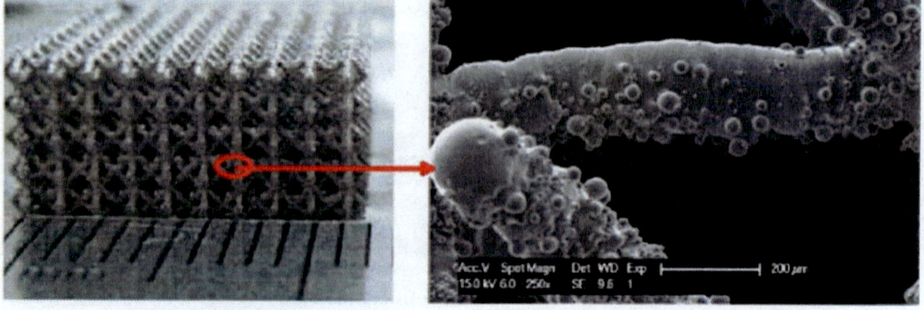

Fig. 8.32 Powder adhesion on the struts of porous structures

hanging" on the overhang surface, that is, powder adhesion. Although slag hanging and powder adhesion is an unavoidable phenomenon in the process of manufacturing porous structures by L-PBF, the powder adhesion can be minimized by controlling the structural parameters and processing parameters of the porous structure.

8.5.6 Surface Roughness of Porous Structure

The surface roughness of the parts formed by LPBF is generally high, and the surface roughness of the manufacturing pillar is affected by factors such as inclination angle and process, which is generally between 10–60 μm. Although the surface roughness of the parts formed by LPBF can be reduced by subsequent sandblasting, shot peening or simple manual grinding. However, if the parts have complex structures and fine features, the above treatment methods are no longer applicable. Due to the complexity of the porous structure, the mechanical polishing method of sandblasting (shot peening) are not suitable. Under this circumstance, the porous structure can be treated by chemical polishing and electrochemical (electrolytic) polishing. Figure 8.33 shows the SEM image of the porous structure formed by LPBF through chemical treatment and electrolytic treatment. It can be seen that in the initial stage the surface of the porous structure pillar is rough due to the adhesion of fine powder particles or burrs caused by processing. After chemical polishing, the unmelted powder particles adhering to the pillar surface can be removed, but the surface quality of chemical polishing is generally slightly lower than that of electrolytic polishing. Electropolishing can remove unmelted powder particles and burrs and reduce the surface roughness of the pillars.

Figure 8.34 shows a comparison of the surface (planar) roughness of a formed part before and after electropolishing as measured by a roughness meter. After electropolishing, the surface roughness Ra of parts can be reduced to 5–15 μm. However, compared with traditional machined parts, there is still a certain gap in surface quality, which is mainly determined by the intrinsic limitations of the L-PBF. At present, there is still some controversy regarding the use of chemical polishing or

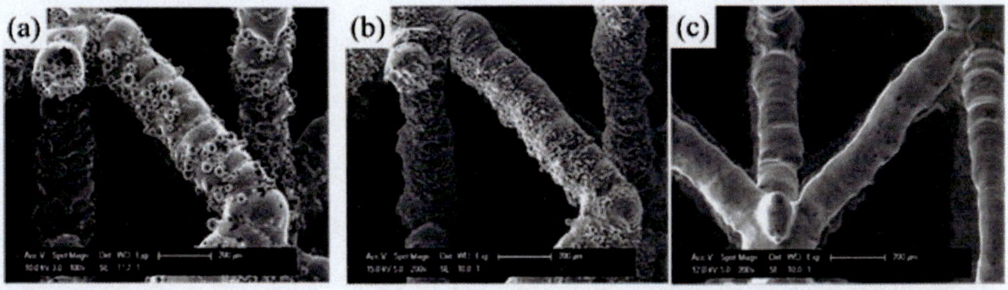

Fig. 8.33 Surface modification method of porous structure: **a** as-built; **b** after chemical polishing; **c** after electrolytic polishing

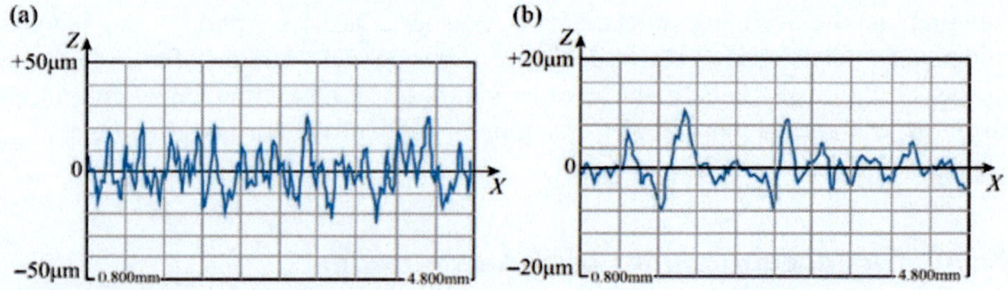

Fig. 8.34 Roughness comparison before and after electropolishing: **a** before electropolishing; **b** after electropolishing

electropolishing for biomedical implants, because this surface treatment method may deteriorate the biocompatibility. As a result, it is of certain significance to reduce the surface roughness from the aspects of design, process and materials [4, 6].

References

1. Liu Y (2015) Research on the mechanism of selective laser melting and direct manufacturing of structural features. South China University of Technology
2. Yang X, Yang Y, Liu Y et al (2015) Study on dimensional accuracy of typical geometric features manufactured by selective laser melting. Chin J Lasers 42(03):70–79
3. Su XB (2011) Study on digital design and direct fabrication of functional parts based on selective laser melting. South China University of Technology
4. Xiao DM (2013) Modeling of porous structure of implants and direct manufacturing by selective laser melting. South China University of Technology
5. Xiao ZF (2018) Research on design for additive manufacturing of lightweight complex component manufactured by selective laser melting. South China University of Technology
6. Wang D, Yang Y, Liu R et al (2013) Study on the designing rules and processability of porous structure based on selective laser fusion (SLM). J Mater Process Technol 213(10):1734–1742

Chapter 9
Advanced and Future Development of Laser Powder Bed Fusion

As one of the mainstream metal additive manufacturing (AM) technologies, laser powder bed fusion (LPBF) technique, there are four main problems at present: (1) the size of the formed parts is small, unable to meet the market demands for large-size parts, and the future development trend requires more and more size, such as aerospace, shipbuilding industry. (2) The dimensional precision of the formed products can not be compared with the products produced by CNC machine tools. (3) The forming efficiency is generally slow at present, for example, some large parts need one or two months to form, and if the situation occurs during the forming, the whole part will be scrapped, causing huge losses, therefore, the efficiency is currently restricting the development of AM. (4) A machine can only use one type of material to form parts, but in fact, there may be many different functional requirements in a part and that different positions may have different material requirements.

According to the industry demand, to solve the above 4 problems is the development trend of metal AM field: the parts size is getting bigger and bigger to meet the requirements of industrial development and the market demand for large parts; Further improve the forming efficiency, manufacturing precision and metal AM technology of various materials to meet the high requirements of suppliers and the market.

9.1 LPBF Forming Heterogeneous Materials

So far, most commercial LPBF machines are designed and manufactured based on single-material, which can easily produce complex single-material parts [1]. However, with the rapid iteration of science and technology, the traditional single-material parts are difficult to meet the increasing complexity of industrial products and the requirements of flexibility and high efficiency. Therefore, heterogeneous parts with huge application potential have attracted extensive attentions and become one of the important research directions in AM field.

© National Defense Industry Press 2024

D. Wang et al., *Laser Powder Bed Fusion of Additive Manufacturing Technology*, Additive Manufacturing Technology, https://doi.org/10.1007/978-981-99-5513-8_9

9.1.1 Introduction to Heterogeneous Material Parts

Heterogeneous material parts, also known as multi-material parts, are designed to meet desired performance requirements by placing the appropriate materials at desired locations. It should be noted that those composites formed with pre-mixed composite powders and uniform material layouts are not heterogeneous materials as referred to in this book.

Because different materials have different properties or functions, heterogeneous material parts can have outstanding mechanical, electricity, thermal, acoustic, optical and other properties that single-material parts do not have, and have great development prospects in many fields. For example, in the field of aerospace, the combustion chamber wall of the spacecraft engine at work: one side needs to withstand high temperature and thermal erosion above 2000 K, requiring the material to have excellent heat insulation and heat resistance. The other side needs to be cooled by low temperature liquid hydrogen, which requires the material to have low temperature resistance and high thermal conductivity. It is difficult or impossible for single-material parts to meet such demanding performance requirements, while heterogeneous material parts have great potential to meet the demanding requirements in the aerospace field. Heterogeneous material parts have higher strength and toughness, not only can withstand large external stress and thermal stress, but also can withstand huge temperature difference, and maintain a long service life, these are the excellent performance that single-material parts do not have [2]. In addition, in the field of national defense and military industries, modern equipments not only has complex and fine structure requirements, but also has multi-functional requirements, such as light or sound stealth, electromagnetic shielding, high temperature resistance and other special functions. Single-material parts cannot realize multi-function integration, but heterogeneous material parts have such potential.

9.1.2 LPBF of Heterogeneous Material Parts

Although heterogeneous material parts have excellent performances, it is very difficult to produce them to meet the application requirements, which limits its application in actual production and life. For example, pressure machining, powder metallurgy, casting, welding, spraying, chemical vapor deposition, self-propagating synthesis and other traditional manufacturing methods have difficulties in manufacturing of heterogeneous material parts, not only difficult to obtain complex shape and flexible material layout, but also complicated manufacturing process. However, the LPBF, which has attracted much attention in recent years, provides a new manufacturing method for heterogeneous materials parts, and also provides a function-oriented design idea. Based on the manufacturing characteristics of material superposition forming, the heterogeneous material parts with complex shape and size error less than ± 0.1 mm can be formed by LPBF, and the materials can be freely arranged with the shape to achieve desired material layout on the parts to obtain excellent performance. The LPBF can realize the gradient transition of composition when forming the heterogeneous material interface, so the forming heterogeneous material parts can reach or even exceed the bonding strength of traditional welding.

At present, some universities at home and abroad have carried out more in-depth research on the LPBF forming of heterogeneous materials parts, and successfully produced some unique heterogeneous materials parts. For example, the Metal 3D Printing AM Laboratory at South China University of Technology has fabricated several heterogeneous material parts using CuSn10 copper alloy and 4340 steel, as shown in Fig. 9.1 [3]. The Laser Machining Research Centre at the University of Manchester has fabricated heterogeneous material parts using 316L stainless steel and Cu10Sn copper alloy, as shown in Fig. 9.2.

The heterogeneous material parts fabricated by LPBF have not only complex structures, but also unique material layouts, which proves that LPBF has unique advantages in the forming of heterogeneous material parts. Therefore, the laser melt forming of heterogeneous materials has great application prospects in biomedical, aerospace and other high-tech fields. In the biomedical field, bone implants have a huge market demand, but have many requirements. However, these requirements

Fig. 9.1 Heterogenous material parts manufactured by LPBF in South China University of Technology: **a** square heat dissipation gradient gasket; **b** gears of heterogeneous materials; **c** bulk heterogeneous material parts. Reprinted with permission from a work by Wu et al. [3]

Fig. 9.2 Heterogeneous material parts manufactured by LPBF at the University of Chester: **a** The Great Sphinx; **b** miniature houses. Reprinted with permission from a work by Wei et al. [4]

can be basically met by laser melting of heterogeneous material powder bed: (1) To meet the personalized requirements of bone implants, in order to achieve a high fit with the human body. (2) Bone implants with a porous structural design to mimic natural bone and eliminate the stress shielding effect, so that patients can avoid painful revision surgery; (3) A subtle multi-material arrangement is achieved in bone implants to achieve properties close to those of natural bone. Many of the core components in the aerospace industry require complex and elaborate structures and work in extremely harsh environments, making them complex and costly to manufacture. On the one hand, forming these heterogeneous material parts through LPBF will simplify the manufacturing process and reduce costs. On the other hand, excellent environmental adaptability can be achieved through flexible material layout. For example, the National Aeronautics and Space Administration (NASA) is conducting a project called RAMPT (Rapid Analysis and Manufacturing Propulsion Technology). One of the key directions of this project is to promote bimetallic and polymetallic AM technology. In this project, LPBF has been successfully applied to the formation of rocket combustors in combination with other AM techniques to produce lightweight polymetallic thrust chamber components (Fig. 9.3). The results of this project show that the heterogeneous material AM technology can form a continuous cooling channel between the chamber and the nozzle, and can be lightweight through proper material layout.

However, some problems still need to be solved in order to obtain the wide application of LPBF for heterogeneous material parts in these high-tech fields. One of the key issues is the compatibility of different materials: because different materials have different chemical, metallurgical and thermal properties, the interface bonding strength is weak as these properties are incompatible, thus easy to produce cracking, delamination and other defects. The heterogeneous material parts manufactured by the author's team through LPBF are shown in Fig. 9.4b, and the material composition is shown in Fig. 9.4a. Figure 9.5a shows that 18Ni(300) layer and CoCr layer have a good bonding interface, indicating that the material of 18Ni(300) and CoCr is well matched. Figures 9.5b, c show that the heterogeneous material parts have solidification defects (cracks, holes, etc.) at the interface, especially the 316L/CuSn10 interface not only has many holes, but also a large number of micro cracks, which indicates that the material compatibility of 316L and CuSn10 is poor [6]. Because the thermal expansion coefficient of CuSn10 is larger than that of 316L, when both 316L and CuSn10 are formed in full compact state, CuSn10 tears 316L and forms a large number of microcracks at the interface. In addition, due to the differences in the thermal physical properties of different materials (such as laser absorption rate, melting temperature, heat capacity, linear thermal expansion coefficient and thermal conductivity, etc.), the thermal behavior and solidification behavior at the interface are extremely complicated, so the interface defects have various forms and complex causes. During the LPBF process, brittle intermetallic compounds may be formed at the interface, which will lead to the brittleness and fracture of the interface. Based on the above problems, the interface of heterogeneous material parts is often the weak point of its mechanical properties, which will adversely affect the overall performance and limit their application in many fields.

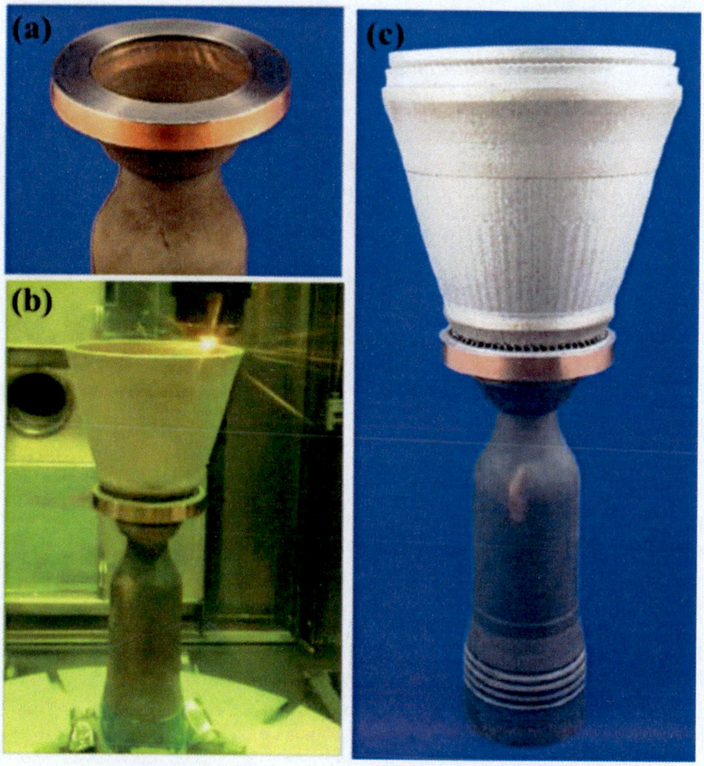

Fig. 9.3 Coupling process of LPBFed combustor and DED nozzles: **a** LPBFed copper alloy combusts with bimetallic joints prepared for DED; **b** the coupled process of DED; **c** complete coupling of heterogeneous material parts. Reprinted with permission from a work by Gradl et al. [5]

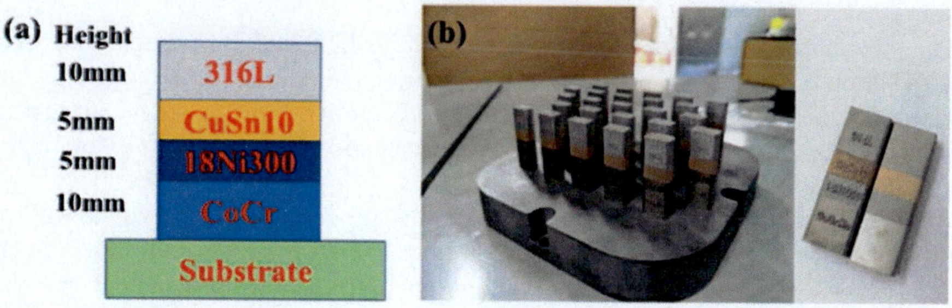

Fig. 9.4 Heterogeneous material parts manufactured by LPBF: **a** material composition; **b** a printed copy

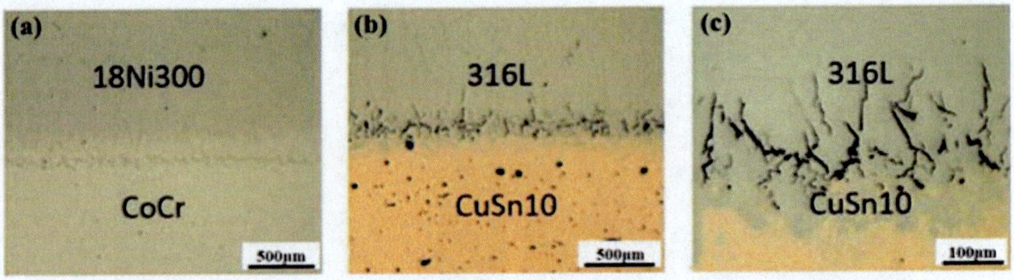

Fig. 9.5 Surface features and defects of heterogeneous material parts: **a** good bonding interface; **b** pores; **c** crack

Therefore, interface performance control and process matching of heterogeneous material parts are major challenges for LPBF of heterogeneous material. To study the interface bonding process of different material combinations, to understand the physical and chemical properties of the interface, to explore the matching principle of heterogeneous materials and the control method of interface properties, are the core and key problems to be solved in the LPBF of heterogeneous materials. In addition, in the intelligent manufacturing of heterogeneous materials, the precise presetting of materials, powder separation and corresponding control software are the key problems to be solved [7, 8]. In terms of interface forming and interface properties, it is necessary to explore the mechanism of interface defects of heterogeneous materials, summarize the interface forming methods and process optimization flow, so as to effectively solve the problems caused by material incompatibility and improve the interface properties [9–13].

9.2 LPBF of Precious Metal Materials

The precious metals mainly refer to eight metallic elements, including Gold, Silver and Platinum group metals (Ruthenium, Rhodium, Palladium, Osmium, Iridium and Platinum). Most of these metals have a beautiful color with strong chemical stability, and are not easy to react with other chemical substances. Precious metals are mainly used in the jewelry field.

Jewelry has also become a pioneer in the LPBF process, the reasons are as follows: First, there is no fixed specification in the jewelry industry. Second, most jewelry designers use CAD software, and they often subcontract. Third, jewelry designers are also good at finishing and polishing jewelry. Fourth, jewelry designers are also used to making some custom projects, and they are eager to do some free and unusual design works. In view of this, the continuous growth of the jewelry industry also continuously drives the development of AM to the direction of rare metal materials.

The traditional jewelry manufacturing process has to go through a variety of processes, such as daguerreotype, mould pressing, mould opening, wax injection and mould repairing. Material, site, equipment, manpower and time costs is larger.

On the contrary, AM of precious metals becomes more and more popular because it has a strong advantage of personalized customization as well as the characteristics of saving production time and cost. In terms of precision and freedom of design, the advantages of direct metal printing over traditional lost wax casting are becoming more and more obvious.

LPBF is often used to fabricate Titanium, Aluminum or stainless steel. Gold or Silver alloys are difficult to make directly by laser melting because of their high reflectivity and thermal conductivity. Due to its high technical threshold, the LPBF of precious metals still has the following problems:

(1) Surface accuracy and smoothness are not ideal. The surface precision and smoothness of LPBF of jewelry products are much worse than that of traditional lost wax casting jewelry, and a lot of post-treatment processes are needed.
(2) The cost of precious metal powder is high. The precious metal powder that can be used for AM requires very high purity and particle size, so the cost is high.
(3) Precious metals have high reflectivity and thermal conductivity. Mixing precious metals with other metals or adhesives may solve the problem.
(4) Not suitable for mass customization.
(5) Practical jewelry design cases that can be used for reference are limited. Jewelry design based on AM is still relatively scarce.

Despite all the difficulties, human exploration of AM of precious metals has made considerable progress. In 2014, EOS introduced a device that can directly LPBF of precious metals jewelry and high-end watches.

In March 2016, the British company Cooksongold developed Platinum powder, and fabricated the world's first Platinum jewelry by LPBF in collaborated with the Platinum Council (PGI). The former works on bracelets based on ancient Chinese vases, and the latter is inspired by bone structures.

In March 2017, Germany's FEM Institute for Precious Metals and Metal Chemistry announced that it had developed a new method to significantly improve the quality of LPBF of Gold. In their study, the gold is mixed with Iron and Germanium, because the Iron and Germanium can be oxidized during forming, thus reducing the reflectivity of the powder.

Using LPBF to form precious metal jewelry can not only shorten the time, but also improve the design freedom of designers. Because it allows the designer to not have to consider the process. Once precious metal AM equipment is widely used, the whole jewelry process will omit the wax mold and casting.

9.3 LPBF of Oversized Parts

At present, due to the limitation of optical devices, the size of metal AM equipments at home and abroad is generally small, but the size of most complex components such as turbine blades and engines which are difficult to be processed exceeds the forming capacity of metal AM equipments, which makes it difficult for LPBF to exploit the advantages in the aerospace field. Therefore, in order to improve the forming efficiency and increase the forming size on the basis of the traditional LPBF, the multi-beam LPBF technology comes into being. Multi-beam LPBF technology has good forming process, high material utilization rate, wide applicability and direct manufacturing of metal parts with complex shapes. Besides, it can effectively reduce the internal stress, warping, spheroidization and cracks and other defects in parts. Thus, it can be used in the fields of aerospace, automotive and other fields with complex shapes, has a broad application With prospects and a wide range of applications [14].

At present, commercial large LPBF device, almost all use multiple lasers, multi-mirror partition simultaneously scanning forming scheme, faced with many technical problems such as laser focusing spot positioning accuracy, laser power consistency, multi-beam seamless splicing, multi-quadrant processing coincidence area manufacturing quality control and optical system stability control under high power long time working conditions. In a multi-laser AM device, when multiple lasers work in relatively close distances, the emission of laser from one laser will affect another depends on their relative positions in the inert gas flow. When a laser is downwind of another, its laser beam will be affected by the upwind laser melting. In view of the influence mechanism of different splicing methods on the surface dimensional accuracy and internal defects of the formed parts. Zhang et al. [15] found that in the staggered splicing method, the splicing interface between the two adjacent layers is staggered with each other, which can ensure the effective fusion of the adjacent laser partitions and effectively limit the influence of remelting in the process of laser repeat scanning, restrain the bad fusion defects at the splicing interface, and final surface of the formed parts is flat. The dimensional accuracy and metallurgical quality of the formed parts can be improved effectively by using staggered splicing method to achieve splicing between different laser scanning areas under the condition of multi-beam laser selective melting [15].

The research and development of new multi-laser AM device is the focus of equipment manufacturers. The configuration of the current mainstream large-size LPBF equipments in the market is listed in Table 9.1.

For the first time, Shanghai TanZhen Company proposed a new AM method of four laser beam scanning. Its TS series four-laser AM device TZ SLM500A can reach a forming size of 500 mm × 500 mm × 1000 mm with forming accuracy of ±0.05 mm. Compared with the single laser system, the multi-laser system greatly reduces the multiple preparation time at the early forming stage. Although the powder laying stroke is long and the forming parts are many, the single-layer time caused by the four-laser additive manufacturing has not been extended.

The Aviation Industry Manufacturing Institute of Aviation Industry Corporation of China (AVIC) has successfully developed the largest multi-laser LPBF device in China, which provides a good platform for solving the problem of lightweight structure with large complex shape. The device has broken through a number of key technology such as dual-beam real-time collaborative control, laser galvanometer dynamic focusing control, path data file parsing technology and laser selective melting control. Equipped with dual laser head and push–pull two-way processing station, the forming size of parts can reach 810 mm × 450 mm × 700 mm.

Table 9.1 Configurations of mainstream large-size LPBF devices

Models	Maximum molding size	Configuration
Tan Zhen TZ-SLM500A	500 mm × 500 mm × 1000 mm	Four laser and four vibration mirror
Laseradd DiMetal-500	500 mm × 250 mm × 300 mm	Double laser and double vibrator
Farsoon FS421M	425 mm × 425 mm × 420 mm	Double laser and double vibrator
E Plus 3D	455 mm × 455 mm × 500 mm	Double laser and double vibrator
Bright Laser Technologies BLT-S600	600 mm × 600 mm × 600 mm	Four laser and four vibration mirror
SLM Solution SLM 500	500 mm × 280 mm × 365 mm	Four laser and four vibration mirror
EOS M400	400 mm × 400 mm × 400 mm	Four laser and four vibration mirror

In 2020, Bright Laser Technologies launched large size LPBF device BLT S600, the maximum forming size reaches 600 mm × 600 mm × 600 mm, which can meet the size requirements of various application scenarios, and solve the integrated forming problems of complex structures such as large-scale special-shaped space surface features, multi-feature cross-scale structures, hollow mesh and spatial continuous topological envelope in high-end applications. The device adopts multi-beam splicing technology, and four fiber lasers with 500 W can co-print, which can improve the printing efficiency by more than 60% compared with a single laser device. In addition, the device adopts bidirectional powder laying technology to eliminate the ineffective time of unidirectional powder laying and effectively improve the forming efficiency of parts.

9.4 High Resolution Laser Powder Bed Fusion

A developing trend of metal AM technology is to develop high-precision desktop level micro-scale AM devices to meet the urgent needs of some small and medium-sized enterprises and related scientific research institutions.

The mainstream metal AM equipments increasing focus spot between 100–200 μm, processing of 20–100 μm, with a thick layer of powder particle size is 20–40 μm on average. These several key technical parameters limits the size accuracy of forming parts ≥100 μm and surface roughness Ra of ≥15 μm. That is, the mainstream metal AM devices can fabricate large size parts but with poor resolution.

Some high-end applications, such as automobile fuel nozzles, electronic components, and medical applications, etc., need more precision of metal AM. The development of high resolution laser micro-melting metal AM equipment is an important step for metal AM technology to step forward to the micro-scale. Many tiny components are difficult to produce using traditional processes, but demand for such products has increased dramatically in recent years. The development trend of this demand is reflected in three aspects: individuation, functional integration and miniaturization.

The laser micro-melting AM system integrates the advantages of AM and micromechanical systems to produce parts with high resolution and precision, and can directly produce functional structural parts without assembly. These advantages come from the system's micro-light spots, fine powder and ultra-low machining layer thickness. In Austria, Vienna University of Technology and other institutions have developed nanostructured AM equipment, which can manufacture integrated micro mobile mechanism, providing solutions for personalized, functional integration and micro machinery, but it is only limited to resin materials.

Scientists and Engineers from the Hanover, Germany (LZH) laser center successfully developed selective laser micro melting (μSLM) technology, which has been used to fabricate materials such as Platinum, Nickel titanium alloy, stainless steel to make medical implants.

9.5 Quality Control and Feedback from Molten Pool Monitoring

Quality assurance and process monitoring are necessary means to upgrade AM technology from model manufacturing level to first class workshop manufacturing level. During the LPBF process, the laser is used as the energy source to melt the powder to form a molten pool, and the metals in the molten pool will flow. With the removal of the laser, the molten pool solidifies to form a single cladding layer. The characteristics of the molten pool and the single cladding layer affect the quality of the final parts, so it is necessary to further understand the molten pool.

At present, the monitoring system of laser AM process mainly focuses on the online detection of molten pool physical parameters and the detection of component defects and the reduction of these defects through feedback control [16]. The laser monitoring process is mainly divided into two parts: One is data acquisition, the other is data processing. Data acquisition mainly consists of two parts: molten pool morphology and molten pool temperature. The morphology of molten pool is usually obtained by CCD camera or infrared camera, and the temperature of molten pool is usually measured by photodiode or pyrometer. Data processing means that the measured data will be processed and transmitted to the controller, and the controller will configure and update the operating parameters of the system, and effectively control the operation process of the system, so as to improve the quality of the products. It is worth noting that the controller uses a variety of control methods: traditional PID control, fuzzy control, artificial intelligence control such as neural network control and so on. At present the most mature use is the traditional PID control. The current research focus is various artificial intelligence control methods.

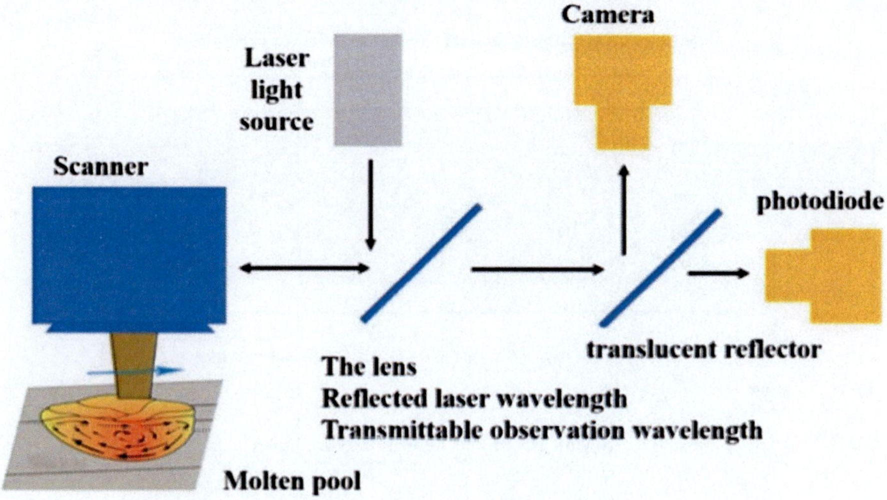

Fig. 9.6 Schematic diagram of coaxial melt pool monitoring system. Reprinted with permission from a work by Berumen et al. [16]

Berumen et al. [16] used high-speed cameras to detect the molten pool size and photodiodes to detect the average radiation of molten pools, and established a set of coaxial monitoring system for the LPBF process. The schematic diagram of the system is shown in Fig. 9.6. The results show that this system can effectively monitor the deviation information of molten pool in the manufacturing process, and the measurement results can be used as feedback to control the manufacturing process.

Current research shows that many systems have not been applied in the actual industrial production process due to the complexity of the integration of control systems and production processes, the limitations of measurement tools and sensors, and the difficulty of real time control. It is believed that in the near future, with the continuously deepening of research, laser AM monitoring technology will be more mature development and practical applications.

9.6 Hybrid of Additive and Subtraction + Intelligent Additive Manufacturing

The process of additive and subtraction hybrid manufacturing is a process of combining additive and subtraction on a single machine. Because of this dual natures, hybrid manufacturing machines can use either of these two processes to produce parts. Starting production using AM can be more efficient than milling alone, often providing wider design freedom.

The hybrid processing system of additive and subtraction materials is mainly applied in the field of metal parts processing. Usually, the metal AM technology is

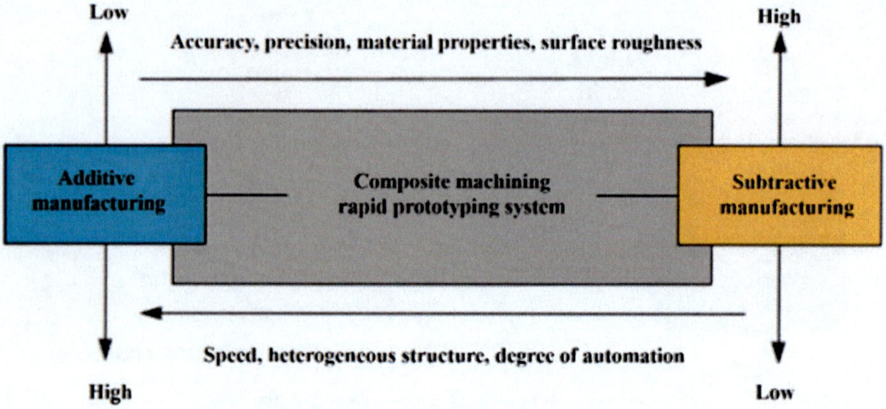

Fig. 9.7 Advantages and disadvantages of adding or removing materials

powder or wire, which is mainly processed by laser and plasma. Powder materials are suitable for fine parts and small parts, while wire is suitable for large-sized structural parts. CNC finishing was achieved to ensure the desired accuracy by eliminating the stellar effect generated by the deposition of certain layers. As shown in Fig. 9.7, the organic integration of subtraction manufacturing and additive manufacturing can effectively complement each other's advantages and disadvantages, improve production efficiency and reduce cost.

Additive and subtraction hybrid manufacturing technology not only has the advantages of AM of rapid forming speed, high material utilization rate and easy forming of complex structures, but also has the advantages of CNC of high quality and high precision. It is one of the important means of integrally forming complex metal components. Matsuura's Lumex Avance-25 composite machining machine integrates laser melting AM with milling, as shown in Fig. 9.8 [17]. The Lumex Avance-25 is a machine tool that performs selective laser melting and then finishes the entire part or part of its surface with high-speed milling to achieve high precision and surface quality. As shown in Fig. 9.9, Lumex Avance-25 hybrid machine is to 3D print every 10 layers (about 0.5–2 mm) to form a metal sheet, and then with high speed milling (spindle speed of 45,000 r/min) on its contour finish machining, then repeats this procedure and finally superimposed into high-precision and complex structure parts.

The greatest advantage of AM and milling is that complex die can be made without assembly. The traditional manufacturing method is to decompose the complex die into several components, and then assemble them together, which is not only time-consuming and laborious, but also inevitably has some errors, which reduces the accuracy of the die. On the machine tool integrated with AM and milling, the complex die with deep groove and thin wall can be processed at one time, which completely changes the design and manufacturing process of complex die [18].

In recent years, with the integration of additive and subtraction manufacturing technologies, as well as the development of automation technology and artificial intelligence technology, AM equipment has been upgraded to intelligent direction.

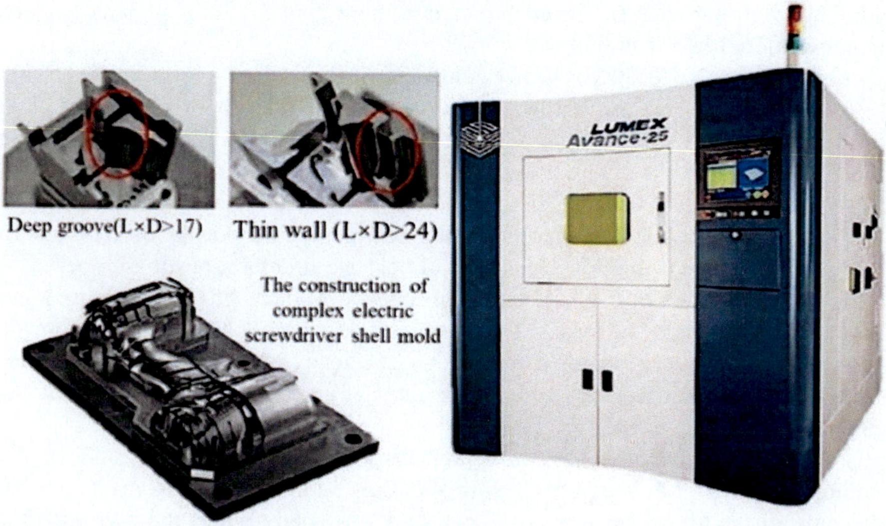

Fig. 9.8 Lumex Avance-25 composite machining machine tool and its processed parts. Reprinted with permission from a work by Zhang et al. [17]

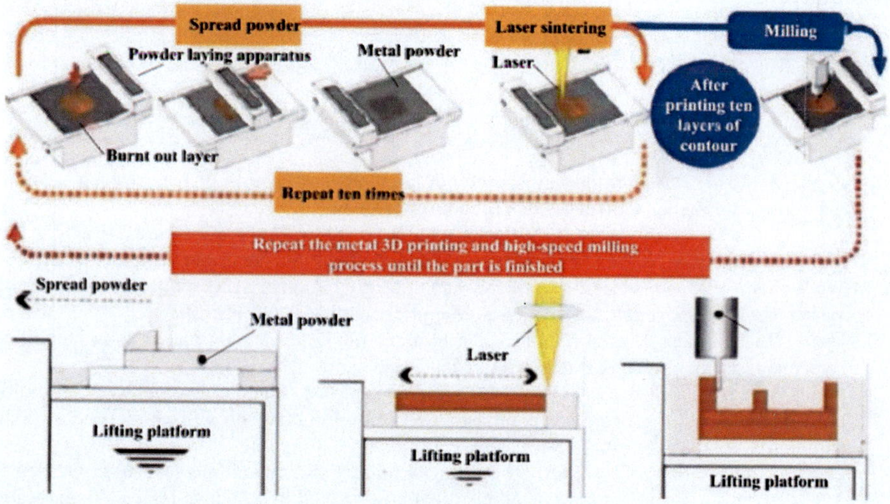

Fig. 9.9 Lumex Avance-25 composite processing machine. Reprinted with permission from a work by Zhang et al. [17]

Velo3D, based in San Francisco, USA, has developed a new smart metal AM system of powder bed fusion that breaks the "45° rule" based on the LPBF field and can fabricate structures with angles as low as 10° without the need for supports. Large sizes up to 40 mm can also be constructed without the need for supports compared to conventional AM techniques.

At Formnext 2019, EOS is showcasing a production-oriented solution. The solution consists of a variety of hardware and software modules that simplify and parallelize the workflow of the front and back parts of the forming process, especially when running multiple 3D printing systems. The shared module has high efficiency, scalability and cost advantages in the manufacture of high quality metal parts.

A key part of the LPBF technology is the treatment of the powder used to produce parts. The quality of the powder is crucial because it determines the properties of parts and is the only way to guarantee a high-quality product. Advanced Technologies handles automated powder production with the Russel Lamas Prosievestation powder screening and recovery system, increasing the efficiency and reliability of the process. The system is fully modular and can be directly integrated into the construction process of existing powder bed systems.

At present, there are a lot of research results in the hardware system of AM equipment, which can integrate the corresponding additive into CNC machine tools, but most of the transformation equipments are designed to meet the corresponding experimental requirements, and the hybrid machining tools that can be put into commercial use are relatively few. Breakthroughs in single processes can promote the development of hybrid processes. However, in order to fully realize the hybrid manufacturing process, a number of future research advances need to be addressed.

References

1. Zhang Z (2014) Research on key technology of rapid prototyping of heterogeneous multi-material parts. Wuhan University of Technology, Wuhan
2. Chen C, Gu D, Dai D et al (2019) Laser additive manufacturing of layered TiB2/Ti6Al4V multi-material parts: understanding thermal behavior evolution. Opt Laser Technol 119:105666
3. Wu W, Yang Y, Mao G et al (2019) Design and implementation of SLM additive manufacturing system for heterogeneous material parts. Manuf Technol Mach Tools 10:13
4. Wei C, Li L, Zhang X et al (2018) 3D printing of multiple metallic materials via modified selective laser melting. CIRP Ann 67(1):245–248
5. Gradl PR, Protz C, Fikes J et al (2020) Lightweight thrust chamber assemblies using multi-alloy additive manufacturing and composite overwrap. In: AIAA propulsion and energy 2020 forum, p 3787
6. Chen J, Yang Y, Song C et al (2019) Interfacial microstructure and mechanical properties of 316L/CuSn10 multi-material bimetallic structure fabricated by selective laser melting. Mater Sci Eng A 752:75–85
7. Wang D, Wang Y, Yang Y (2019) A multi-material molding device based on metal 3D printing. CN208853710U, 2019-05-14
8. Wang D, Wang Y, Yang Y (2018) A multi-material forming device and method based on metal 3D printing. CN108907189A, 2018-11-30
9. Tan C, Zhou K, Ma W et al (2018) Interfacial characteristic and mechanical performance of maraging steel-copper functional bimetal produced by selective laser melting based hybrid manufacture. Mater Des 155:77–85

10. Wang P, Lao CS, Chen ZW et al (2020) Microstructure and mechanical properties of Al-12Si and Al-3.5 Cu-1.5 Mg-1Si bimetal fabricated by selective laser melting. J Mater Sci Technol 36:18–26
11. Demir AG, Previtali B (2017) Multi-material selective laser melting of Fe/Al-12Si components. Manuf Lett 11:8–11
12. Chivel Y (2016) New approach to multi-material processing in selective laser melting. Phys Procedia 3:891–898
13. Sing SL, Lam LP, Zhang DQ et al (2015) Interfacial characterization of SLM parts in multi-material processing: intermetallic phase formation between AlSi10Mg and C18400 copper alloy. Mater Charact 107:220–227
14. She B (2019) Basic research on multibeam laser selective melting of TA15 alloy. Huazhong University of Science and Technology, Wuhan
15. Zhang S, Wang M, Wang C, Lin X, Huang W (2019) Effect of splicing method on forming characteristics of TC4 in multi-beam SLM forming. Appl Laser 39(04):544–549
16. Berumen S, Bechmann F, Lindner S et al (2010) Quality control of laser-and powder bed-based additive manufacturing (AM) technologies. Phys Procedia 5:617–622
17. Zhang S (2015) Hybrid machining with additive and cutting technologies. Mach Build Autom 44(06):1–7
18. Zhu Z, Dhokia VG, Nassehi A et al (2013) A review of hybrid manufacturing processes–state of the art and future perspectives. Int J Comput Integr Manuf 26(7):596–615